Networking

Networking

12th Edition

by Doug Lowe

Networking For Dummies®, 12th Edition

Published by: **John Wiley & Sons, Inc.**, 111 River Street, Hoboken, NJ 07030-5774, www.wiley.com

For general information on our other products and services, please contact our Customer Care Department within the U.S. at 877-762-2974, outside the U.S. at 317-572-3993, or fax 317-572-4002. For technical support, please visit https://hub.wiley.com/community/support/dummies.

Wiley publishes in a variety of print and electronic formats and by print-on-demand. Some material included with standard print versions of this book may not be included in e-books or in print-on-demand. If this book refers to media such as a CD or DVD that is not included in the version you purchased, you may download this material at http://booksupport.wiley.com. For more information about Wiley products, visit www.wiley.com.

Library of Congress Control Number: 2020933431

ISBN 978-1-119-64850-5 (pbk); ISBN 978-1-119-64851-2 (ebk); ISBN 978-1-119-64849-9 (ebk)

Manufactured in the United States of America

10 9 8 7 6 5 4 3 2 1

Contents at a Glance

Table of Contents

Introduction

Welcome to the 12th edition of *Networking For Dummies*, the book that's written especially for people who have this nagging feeling in the back of their minds that they should network their computers but haven't a clue about how to start or where to begin.

Do you often copy a spreadsheet to a flash drive just so you can give it to someone else in your office? Are you frustrated because you can't use the fancy color laser printer that's on the financial secretary's computer? Do you wait in line to use the computer that has the customer database? You need a network!

Or maybe you already have a network, but you have just one problem: Someone promised that a network would make your life easier, but it's instead turned your computing life upside down. Just when you had this computer thing figured out, someone popped into your office, hooked up a cable, and said, "Happy networking!" Makes you want to scream.

Regardless, you've found the right book. Help is here, within these humble pages.

This book talks about networks in everyday (and often irreverent) terms. The language is friendly; you don't need a graduate education to get through it. And the occasional potshot helps unseat the hallowed and sacred traditions of networkdom, bringing just a bit of fun to an otherwise dry subject. The goal is to bring the lofty precepts of networking down to earth, where you can touch them and squeeze them and say, "What's the big deal? I can do this!"

About This Book

This isn't the kind of book you pick up and read from start to finish, as if it were a cheap novel. If I ever see you reading it at the beach, I'll kick sand in your face. This book is more like a reference, the kind of book you can pick up, turn to just about any page, and start reading. Each chapter covers a specific aspect of networking, such as printing from the network, hooking up network cables, or setting up security so that bad guys can't break in. Just turn to the chapter you're interested in and start reading.

Each chapter is divided into self-contained chunks, all related to the major theme of the chapter. For example, the chapter on hooking up the network cable contains nuggets like these:

- What is Ethernet?
- All about cables
- To shield or not to shield
- Wall jacks and patch panels
- Switches

You don't have to memorize anything in this book. It's a need-to-know book: You pick it up when you need to know something. Need to know what 100BaseT is? Pick up the book. Need to know how to create good passwords? Pick up the book. Otherwise, put it down and get on with your life.

Feel free to skip the sidebars that appear throughout the book; these shaded gray boxes contain interesting info that isn't essential to your understanding of the subject at hand. The same goes for any text I mark with the Technical Stuff icon.

If you need to type something, you see the text you need to type like this: **Type this stuff**. In this example, you type **Type this stuff** at the keyboard and then press Enter. An explanation usually follows, just in case you're scratching your head and grunting, "Huh?"

Within this book, you may note that some web addresses break across two lines of text. If you're reading this book in print and want to visit one of these web pages, simply key in the web address exactly as it's noted in the text, pretending as though the line break doesn't exist. If you're reading this as an e-book, you've got it easy — just click the web address to be taken directly to the web page.

Foolish Assumptions

I'm making only two assumptions about who you are: You're someone who works with a computer, and you either have a network or you're thinking about getting one. I hope that you know (and are on speaking terms with) someone who knows more about computers than you do. My goal is to decrease your reliance on that person, but don't throw away his phone number yet.

Is this book useful for Macintosh users? Absolutely. Although the bulk of this book is devoted to showing you how to link Windows-based computers to form a network, you can find information about how to network Macintosh computers as well.

Windows 10? Gotcha covered. You'll find plenty of information about how to network with the latest and greatest Microsoft desktop operating system.

Windows Server 2019? No worries. You'll find plenty of information about the newest version of Microsoft's server operating system.

Icons Used in This Book

Those nifty little pictures in the margin aren't there just to pretty up the place. They also have practical functions.

TECHNICAL STUFF

Hold it — technical details lurk just around the corner. Read on only if you have a pocket protector.

TIP

Pay special attention to this icon; it lets you know that some particularly useful tidbit is at hand — perhaps a shortcut or a little-used command that pays off big.

REMEMBER

Did I tell you about the memory course I took?

WARNING

Danger, Will Robinson! This icon highlights information that may help you avoid disaster.

Beyond the Book

In addition to the material in the print or e-book you're reading right now, this product also comes with some access-anywhere goodies on the web. Check out the free Cheat Sheet for links to useful websites for networking information, private IP address ranges for networks, and more. To get this Cheat Sheet, simply go to `www.dummies.com` and type **Networking For Dummies Cheat Sheet** in the Search box.

Where to Go from Here

Yes, you can get there from here. With this book in hand, you're ready to plow right through the rugged networking terrain. Browse through the Table of Contents and decide where you want to start. Be bold! Be courageous! Be adventurous! Above all, have fun!

1 Getting Started with Networking

IN THIS PART . . .

Find out what a network is and what you can do with one.

Set up Windows and Mac computers to work on a network.

Access network resources such as shared storage and network printers.

Use Microsoft Office and other software on a network.

IN THIS CHAPTER

» Considering why networking is useful (and is everywhere)

» Telling the difference between servers and clients

» Looking under the hood at the network operating system

» Asking "How does it work when a network works if a network works for me?" (Say what?)

» Assessing how networks change computing life

» Identifying (and offering sympathy to) the network administrator

» Comparing servers to clients: What have they got that you don't got?

Chapter 1
Let's Network!

Computer networks get a bad rap in the movies. Beginning in the 1980s, the *Terminator* movies featured Skynet, a computer network that becomes self-aware, takes over the planet, builds deadly terminator robots, and sends them back through time to kill everyone unfortunate enough to have the name Sarah Connor. In the *Matrix* movies, a vast and powerful computer network enslaves humans and keeps them trapped in a simulation of the real world. And in the 2015 blockbuster *Spectre*, James Bond goes rogue (again) to prevent the Evil Genius Ernst Blofeld from taking over the world (again) by linking the computer systems of all the world's intelligence agencies together to form a single all-powerful evil network that spies on everybody.

Fear not. These bad networks exist only in the dreams of science fiction writers. Real-world networks are much more calm and predictable. Although sophisticated networks do seem to know a lot about you, they don't think for themselves and they don't evolve into self-awareness. And although they can gather a sometimes disturbing amount of information about you, they aren't trying to kill you, even if your name is Sarah Connor.

Now that you're over your fear of networks, you're ready to breeze through this chapter. It's a gentle, even superficial, introduction to computer networks, with a slant toward the concepts that can help you use a computer that's attached to a network. This chapter goes easy on the details; the detailed and boring stuff comes later.

Defining a Network

A *network* is nothing more than two or more computers connected by a cable or by a wireless radio connection so that they can exchange information.

Of course, computers can exchange information in ways other than networks. Most of us have used what computer nerds call the *sneakernet.* That's where you copy a file to a flash drive or other portable storage device and then walk the data over to someone else's computer. (The term *sneakernet* is typical of computer nerds' feeble attempts at humor.)

The whole problem with the sneakernet is that it's slow, and it wears a trail in your carpet. One day, some penny-pinching computer geeks discovered that connecting computers with cables was cheaper than replacing the carpet every six months. Thus, the modern computer network was born.

You can create a simple computer network by hooking together all the computers in your office with cables and using the computer's *network interface* (an electronic circuit that resides inside your computer and has a special jack on the computer's backside). Then you tweak a few simple settings in the computer's operating system (OS) software, and *voilà!* You have a working network. That's all there is to it.

If you don't want to mess with cables, you can create a wireless network instead. In a wireless network, the computers use wireless network adapters that communicate via radio signals. All modern laptop computers have built-in wireless network adapters, as do most desktop computers. (If yours doesn't, you can purchase a separate wireless network adapter that plugs into one of the computer's USB ports.)

Figure 1-1 shows a typical network with four computers. You can see that all four computers are connected by a network cable to a central network device (in this case, a home router). This component, common in small networks, actually consists of three distinct but related network devices:

- **Router:** Connects your computers to the Internet
- **Switch:** Allows you to connect two or more computers together with cables
- **Wireless access point:** Lets you connect computers and other devices to your network without using cables

In the figure, you can see that two computers — Bart's gaming computer and Homer's old 1989 computer — are connected via cables to the switch component of the home router. You can also see that Lisa connects her laptop to the network wirelessly. Marge also connects her iPad to the network wirelessly.

You can also see in the figure that Homer's computer has a printer attached to it. Because of the network, Bart, Lisa, and Marge can also use this printer.

Finally, you can see that the entire network is connected to the Internet via the router.

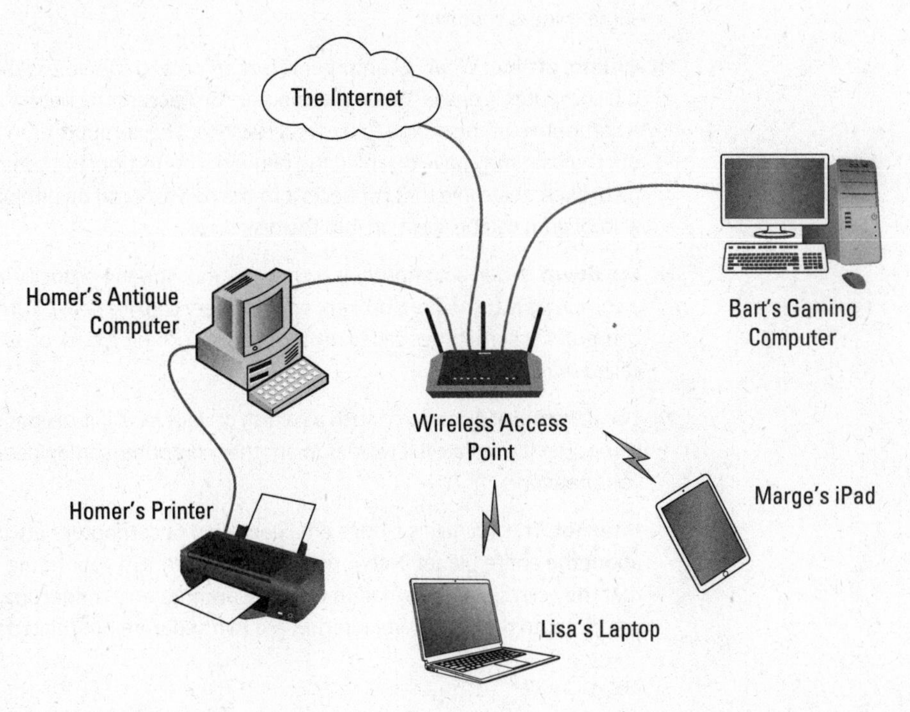

FIGURE 1-1: A typical network.

Computer networking has its own strange vocabulary. Although you don't have to know every esoteric networking term, it helps to be acquainted with a few of the basic buzzwords:

- **LAN:** Networks are often called LANs, short for *local area network.* In Figure 1-1, the LAN consists of the home router and the computers and iPad that are connected to it directly via cable or wirelessly.

TECHNICAL STUFF

  LAN is the first *TLA* — or *three-letter acronym* — of this book. You don't really need to remember it or any of the many TLAs that follow. In fact, the only three-letter acronym you need to remember is TLA. You might guess that the acronym for *four-letter acronym* is *FLA.* Wrong! A four-letter acronym is an *ETLA,* which stands for *extended three-letter acronym.* After all, it just wouldn't be right if the acronym for *four-letter acronym* had only three letters.

- **WAN:** The second TLA in this book is WAN. The WAN is part of the network that connects to the Internet. WAN stands for *wide area network*.

TECHNICAL STUFF

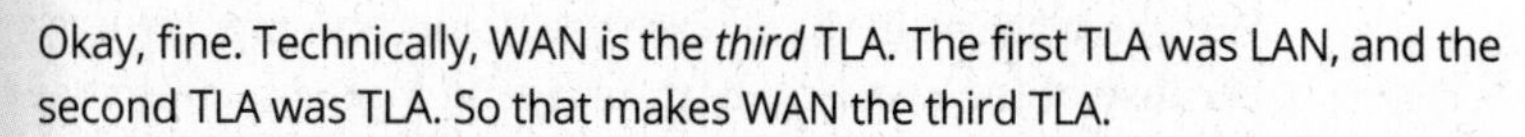

  Okay, fine. Technically, WAN is the *third* TLA. The first TLA was LAN, and the second TLA was TLA. So that makes WAN the third TLA.

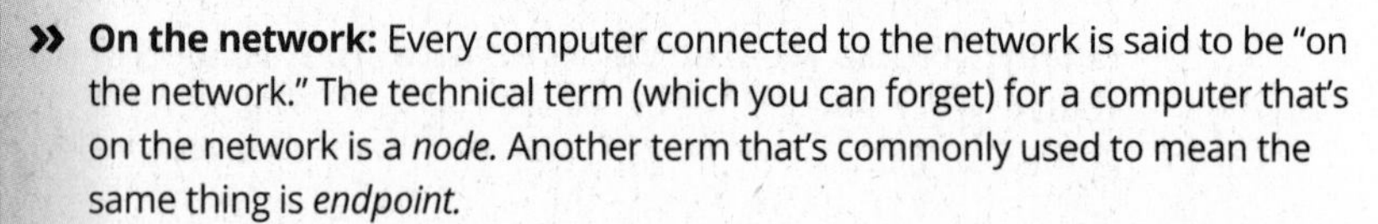

- **On the network:** Every computer connected to the network is said to be "on the network." The technical term (which you can forget) for a computer that's on the network is a *node.* Another term that's commonly used to mean the same thing is *endpoint.*
- **Online, offline:** When a computer is turned on and can access the network, the computer is *online.* When a computer can't access the network, it's *offline.* A computer can be offline for several reasons. The computer can be turned off, the user may have disabled the network connection, the computer may be broken, the cable that connects it to the network can be unplugged, or a wad of gum can be jammed into the disk drive.
- **Up, down:** When a computer is turned on and working properly, it's *up.* When a computer is turned off, broken, or being serviced, it's *down.* Turning off a computer is sometimes called *taking it down.* Turning it back on is sometimes called *bringing it up.*
- **Local, remote:** A resource such as a disk drive is *local* if it resides in your computer. It's *remote* if it resides in another computer somewhere else on your network.
- **Internet:** The *Internet* is a huge amalgamation of computer networks strewn about the entire planet. Networking the computers in your home or office so that they can share information with one another and connecting your computer to the worldwide Internet are two separate but related tasks.

Why Bother with a Network?

Frankly, computer networks are a bit of a pain to set up. So why bother? Because the benefits of having a network outweigh the difficulties of setting up one.

You don't have to be a PhD to understand the benefits of networking. In fact, you learned everything you need to know in kindergarten: Networks are all about sharing. Specifically, networks are about sharing four: files, resources, programs, and messages.

Sharing files

Networks enable you to share information with other computers on the network. Depending on how you set up your network, you can share files with your network friends in several different ways. You can send a file from your computer directly to a friend's computer by attaching the file to an email message and then mailing it. Or you can let your friend access your computer over the network so that your friend can retrieve the file directly from your hard drive. Yet another method is to copy the file to a disk on another computer and then tell your friend where you put the file so that your friend can retrieve it later. One way or the other, the data travels to your friend's computer over the network cable and not on a CD or DVD or flash drive, as it would in a sneakernet.

Sharing resources

You can set up certain computer resources — such as hard drives or printers — so that all computers on the network can access them. For example, the printer attached to Homer's computer in Figure 1-1 is a *shared resource,* which means that anyone on the network can use it. Without the network, Bart, Lisa, and Marge would have to buy their own printers.

Hard drives can be shared resources, too. In fact, you must set up a hard drive as a shared resource to share files with other users. Suppose that Bart wants to share a file with Lisa, and a shared hard drive has been set up on Homer's computer. All Bart has to do is copy his file to the shared hard drive in Homer's computer and tell Lisa where he put it. Then, when Lisa gets around to it, she can copy the file from Homer's computer to her own.

You can share other resources, too, such as an Internet connection. In fact, sharing an Internet connection is one of the main reasons why many networks are created.

Sharing programs

Rather than keep separate copies of programs on each person's computer, putting programs on a drive that everyone shares is sometimes best. For example, if ten computer users all use a particular program, you can purchase and install ten copies of the program, one for each computer. Or you can purchase a ten-user license for the program and then install just one copy of the program on a shared drive. Each of the ten users can then access the program from the shared hard drive.

In most cases, however, running a shared copy of a program over the network is unacceptably slow. A more common way of using a network to share programs is to copy the program's installation files to a shared network location. Then you can use that copy to install a separate copy of the program on each user's local hard drive.

The advantage of installing a program from a shared network drive is that you don't have to download the software separately for each computer on which you want to install the software. And the system administrator can customize the network installation so that the software is installed the same way on each user's computer. (However, these benefits are significant only for larger networks. If your network has fewer than about ten computers, you're probably better off downloading and installing the program separately on each computer.)

WARNING

Remember that purchasing a single-user copy of a program and then putting it on a shared network location so that everyone on the network can access it is illegal. If five people use the program, you need to either purchase five copies of the program or purchase a network license that specifically allows five or more users.

TIP

Many software manufacturers sell their software with a concurrent usage license, which means that you can install the software on as many computers as you want, but only a certain number of people can use the software at any given time. Usually, special licensing software that runs on one of the network's server computers keeps track of how many people are currently using the software. This type of license is frequently used with more specialized (and expensive) software, such as accounting systems or computer drafting systems.

Sharing messages

Another benefit of networking is that networks enable computer users to communicate with one another over the network by sharing messages. Those messages can come in many forms. Email and instant-messaging programs are the most common. But you can also exchange audio or video messages. For example, you can hold online meetings over the network. Network users who have inexpensive video cameras (webcams) attached to their computers can have videoconferences. You can even play a friendly game of Hearts over a network — during your lunch break, of course.

Servers and Clients

The network computer that contains the hard drives, printers, and other resources that are shared with other network computers is a *server.* This term comes up repeatedly, so you have to remember it. Write it on the back of your left hand.

Any computer that's not a server is a *client.* You have to remember this term, too. Write it on the back of your right hand.

Only two kinds of computers are on a network: servers and clients. Look at your left hand and then look at your right hand. Don't wash your hands until you memorize these terms.

The distinction between servers and clients in a network has parallels in sociology — in effect, a sort of class distinction between the "haves" and "have-nots" of computer resources:

- Usually, the most powerful and expensive computers in a network are the servers. There's a good technical reason: All users on the network share the server's resources.
- The cheaper and less-powerful computers in a network are the clients. *Clients* are the computers used by individual users for everyday work. Because clients' resources don't have to be shared, they don't have to be as fancy.
- Most networks have more clients than servers. For example, a network with ten clients can probably get by with one server.
- In many networks, a clean line of demarcation exists between servers and clients. In other words, a computer functions as either a server or a client, not both. For the sake of an efficient network, a server can't become a client, nor can a client become a server.
- Other (usually smaller) networks can be more evenhanded by allowing any computer in the network to be a server and allowing any computer to be both server and client at the same time.

Dedicated Servers and Peers

In some networks, a server computer is a server computer and nothing else. It's dedicated to the sole task of providing shared resources, such as hard drives and printers, to be accessed by the network client computers. This type of server is a *dedicated server* because it can perform no other task than network services.

Some smaller networks take an alternative approach by enabling any computer on the network to function as both a client and a server. Thus, any computer can share its printers and hard drives with other computers on the network. And while a computer is working as a server, you can still use that same computer for other functions, such as word processing. This type of network is a *peer-to-peer network* because all the computers are thought of as *peers*, or equals.

Here are some points to ponder concerning the differences between dedicated server networks and peer-to-peer networks while you're walking the dog tomorrow morning:

- Peer-to-peer networking features are built into Windows. Thus, if your computer runs Windows, you don't have to buy any additional software to turn your computer into a server. All you have to do is enable the Windows server features.
- The network server features that are built into desktop versions of Windows (such as Windows 10) aren't particularly efficient because these versions of Windows weren't designed primarily to be network servers.

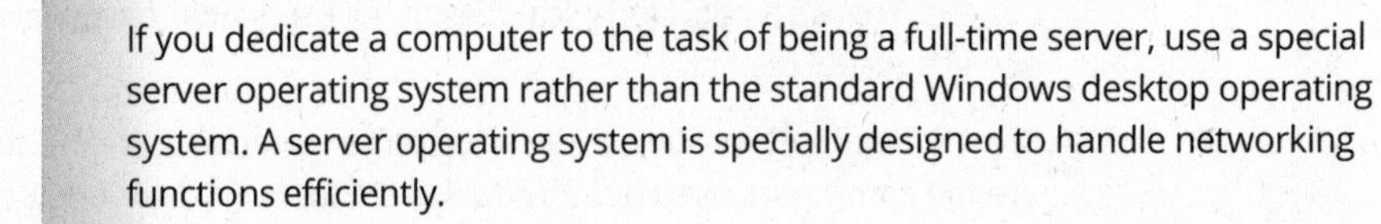

  If you dedicate a computer to the task of being a full-time server, use a special server operating system rather than the standard Windows desktop operating system. A server operating system is specially designed to handle networking functions efficiently.

  - The most commonly used server operating systems are the server versions of Windows.

    As of this writing, the current server version of Windows is Windows Server 2019. However, many companies still use the previous version (Windows Server 2016), and a few even use its predecessors, Windows Server 2012 and Windows Server 2008.
  - Another popular server operating system is *Linux*. Linux is popular because it is free. However, it requires more expertise to set up than Windows Server.
- Many networks are both peer-to-peer *and* dedicated-server networks at the same time. These networks have
  - At least one server computer that runs a server operating system such as Windows Server 2019
  - *Client* computers that use the server features of Windows to share their resources with the network

- Besides being dedicated, your servers should also be sincere.

What Makes a Network Tick?

To use a network, you don't really have to know much about how it works. Still, you may feel a little bit better about using the network if you realize that it doesn't work by voodoo. A network may seem like magic, but it isn't. The following list describes the inner workings of a typical network:

- **Network interface:** Inside any computer attached to a network is a special electronic circuit called the *network interface.* The network interface has either an external jack into which you can plug a network cable — or, in the case of a wireless network interface, an antenna.
- **Network cable:** The network cable physically connects the computers. It plugs into the network interface card (NIC) on the back of your computer.

  The type of network cable most commonly used is twisted-pair cable, so named because it consists of several pairs of wires twisted together in a certain way. Twisted-pair cable superficially resembles telephone cable. However, appearances can be deceiving. Most phone systems are wired using a lower grade of cable that doesn't work for networks.

  For the complete lowdown on networking cables, see Chapter 7.

  TIP

  Network cable isn't necessary when wireless networking is used. For more information about wireless networking, see Chapter 8.
- **Network switch:** Networks built with twisted-pair cabling require one or more switches. A *switch* is a box with a bunch of cable connectors. Each computer on the network is connected by cable to the switch. The switch, in turn, connects all the computers to each other. For more information about network switches, see Chapter 7.

  TECHNICAL STUFF

  In the early days of twisted-pair networking, devices known as hubs were used rather than switches. The term hub is sometimes used to refer to switches, but true hubs went out of style sometime around the turn of the century.

  TECHNICAL STUFF

  In networks with just a few computers, the network switch is often combined with another networking device called a router. A router is used to connect two networks. Typically, a router is used to connect your network to the Internet. By combining a router and a switch in a single box, you can easily connect several computers to the Internet and to each other.
- **Network software:** Of course, the software makes the network work. To make any network work, a whole bunch of software has to be set up just right. For peer-to-peer networking with Windows, you have to play with the Control Panel to get networking to work. And a network operating system such as Windows Server 2019 requires a substantial amount of tweaking to get it to work just right.

It's Not a Personal Computer Anymore!

If I had to choose one point that I want you to remember from this chapter more than anything else, it's this: After you hook up your personal computer (PC) to a network, it's not a "personal" computer anymore. You're now part of a network of computers, and in a way, you've given up one of the key concepts that made PCs so successful in the first place: independence.

I got my start in computers back in the days when mainframe computers ruled the roost. *Mainframe computers* are big, complex machines that used to fill entire rooms and had to be cooled with chilled water. I worked with an IBM System 370 Model 168. It had a whopping 8MB of memory. (The computer on which I'm writing this book has 3,000 times as much memory.)

Mainframe computers required staffs of programmers and operators in white lab coats just to keep them going. The mainframes had to be carefully managed. A whole bureaucracy grew up around managing them.

Mainframe computers used to be the dominant computers in the workplace. Personal computers changed all that: They took the computing power out of the big computer room and put it on the user's desktop, where it belongs. PCs severed the tie to the centralized control of the mainframe computer. With a PC, a user could look at the computer and say, "This is mine — all mine!" Mainframes still exist, but they're not nearly as popular as they once were.

But networks have changed everything all over again. In a way, it's a change back to the mainframe-computer way of thinking: central location, distributed resources. True, the network isn't housed in a separate building. But you can no longer think of "your" PC as your own. You're part of a network — and like the mainframe, the network has to be carefully managed.

Here are several ways in which a network robs you of your independence:

- **You can't just indiscriminately delete files from the network.** They may not be yours.
- **You're forced to be concerned about network security.** For example, a server computer has to know who you are before it allows you to access its files. So you have to know your user ID and password to access the network. This precaution prevents some 15-year-old kid from hacking his way into your office network by using its Internet connection and stealing all your computer games.

- **You may have to wait for shared resources.** Just because Bart sends something to Homer's printer doesn't mean that it immediately starts to print. Lisa may have sent a two-hour print job before that. Bart will just have to wait.
- **You may have to wait for access to documents.** You may try to retrieve an Excel spreadsheet file from a network drive, only to discover that someone else is using it. Like Bart, you just have to wait.
- **You don't have unlimited storage space.** If you copy a 100GB video file to a server's drive, you may get calls later from angry co-workers complaining that no room is left on the server's drive for their important files.
- **Your files can become infected from viruses given to you by someone over the network.** You may then accidentally infect other network users.
- **You have to be careful about saving sensitive files on the server.** If you write an angry note about your boss and save it on the server's hard drive, your boss may find the memo and read it.
- **The server computer must be up and running at all times.** For example, if you turn Homer's computer into a server computer, Homer can't turn his computer off when he's out of the office. If he does, you can't access the files stored on his computer.
- **If your computer is a server, you can't just turn it off when you're finished using it.** Someone else may be accessing a file on your hard drive or printing on your printer.

The Network Administrator

Because so much can go wrong — even with a simple network — designating one person as network administrator is important. This way, someone is responsible for making sure that the network doesn't fall apart or get out of control.

The network administrator doesn't have to be a technical genius. In fact, some of the best network administrators are complete idiots when it comes to technical stuff. What's important is that the administrator is organized. That person's job is to make sure that plenty of space is available on the file server, that the file server is backed up regularly, and that new employees can access the network, among other tasks.

The network administrator's job also includes solving basic problems that the users themselves can't solve — and knowing when to call in an expert when something really bad happens. It's a tough job, but somebody's got to do it. Here are a few tips that might help:

- Part 4 of this book is devoted entirely to the hapless network administrator. So if you're nominated, read the chapters in that part. If you're lucky enough that someone *else* is nominated, celebrate by buying her a copy of this book.
- In small companies, picking the network administrator by drawing straws is common. The person who draws the shortest straw loses and becomes administrator.
- Of course, the network administrator can't be a *complete* technical idiot. I was lying about that. (For those of you in Congress, the word is *testifying.*) I exaggerated to make the point that organizational skills are more important than technical skills. The network administrator needs to know how to do various maintenance tasks. Although this knowledge requires at least a little technical know-how, the organizational skills are more important.

What Have They Got That You Don't Got?

With all this technical stuff to worry about, you may begin to wonder whether you're smart enough to use your computer after it's attached to the network. Let me assure you that you are. If you're smart enough to buy this book because you know that you need a network, you're more than smart enough to use the network after it's put in. You're also smart enough to install and manage a network yourself. It isn't rocket science.

I know people who use networks all the time. They're no smarter than you are, but they do have one thing that you don't have: a certificate. And so, by the powers vested in me by the International Society for the Computer Impaired, I present you with the certificate in Figure 1-2, confirming that you've earned the coveted title Certified Network Dummy, better known as CND. This title is considered much more prestigious in certain circles than the more stodgy CNE or MCSE badges worn by real network experts.

Congratulations, and go in peace.

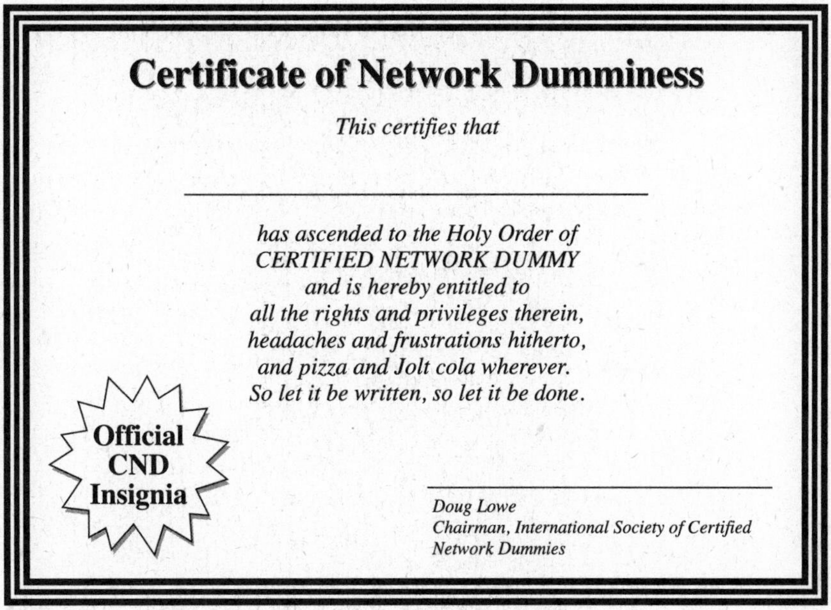

Certificate of Network Dumminess

This certifies that

____

has ascended to the Holy Order of
CERTIFIED NETWORK DUMMY
and is hereby entitled to
all the rights and privileges therein,
headaches and frustrations hitherto,
and pizza and Jolt cola wherever.
So let it be written, so let it be done.

Official CND Insignia

____

Doug Lowe
Chairman, International Society of Certified Network Dummies

FIGURE 1-2: Your official CND certificate.

» Changing the computer name

» Joining a domain

Chapter 2

Configuring Windows and Mac Clients

Among the most basic aspects of using a network is configuring your computer to connect to the network. In particular, you have to configure each client computer's network interface so that it works properly, and you have to install the right protocols so that the clients can communicate with other computers on the network.

Fortunately, the task of configuring client computers for the network is child's play in Windows. For starters, Windows automatically recognizes your network interface card when you start up your computer. All that remains is to make sure that Windows properly installed the network protocols and client software.

With each version of Windows, Microsoft has simplified the process of configuring client network support. In the first half of this chapter, I describe the steps for configuring networking for Windows 10. The procedures for previous versions of Windows are similar.

Configuring a Mac computer for networking is just as easy. The second half of this chapter shows you how to dial in networking using Apple's latest incarnation of the macOS, known as Catalina. The procedures are similar for previous macOS versions.

Configuring Windows Network Connections

Windows usually detects the presence of a network adapter automatically; typically, you don't have to install device drivers manually for the adapter. When Windows detects a network adapter, Windows automatically creates a network connection and configures it to support basic networking protocols. You may need to change the configuration of a network connection manually, however.

The following steps show you how to configure your network adapter on a Windows 10 system:

1. **Click the Start icon (or press the Start button on the keyboard), and then tap or click Settings.**

   The Settings page appears, as shown in Figure 2-1.

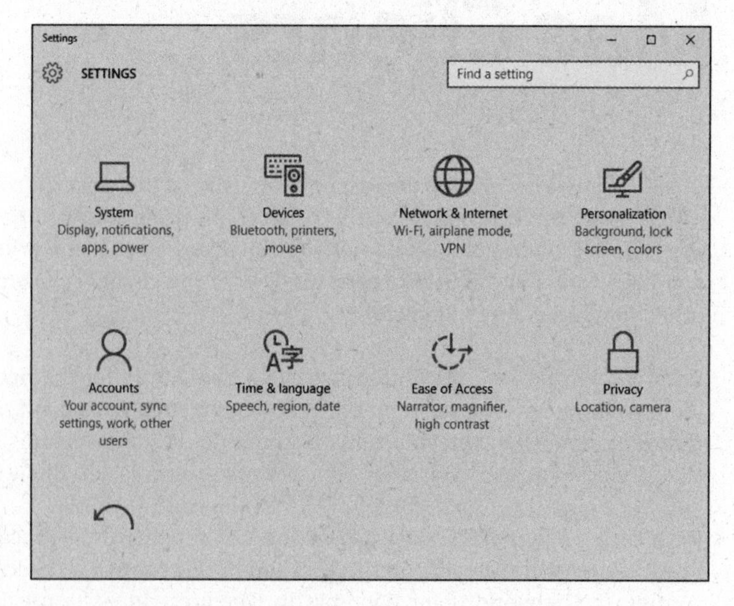

FIGURE 2-1: The Settings page.

2. **Click Network & Internet.**

   The Network & Internet page appears, as shown in Figure 2-2.

3. **Click Ethernet.**

   The Ethernet settings page appears, as shown in Figure 2-3.

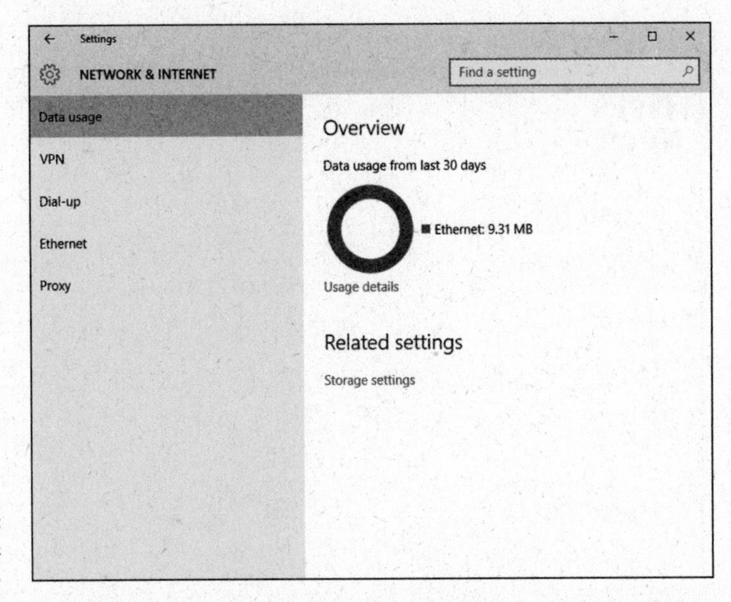

FIGURE 2-2: The Network & Internet page.

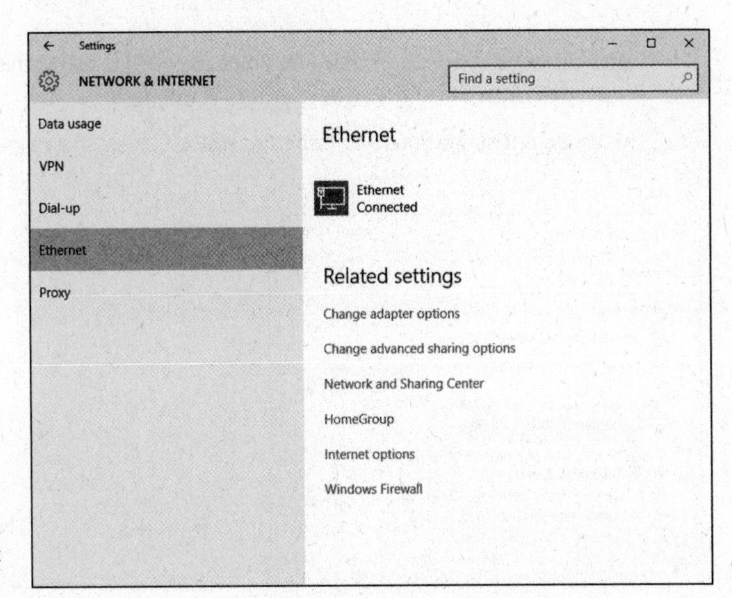

FIGURE 2-3: The Ethernet settings page.

4. **Click Change Adapter Options.**

   The Network Connections page appears, as shown in Figure 2-4. This page lists each of your network adapters. In this case, only a single wired Ethernet adapter is shown. If the device has more than one adapter, additional adapters will appear on this page.

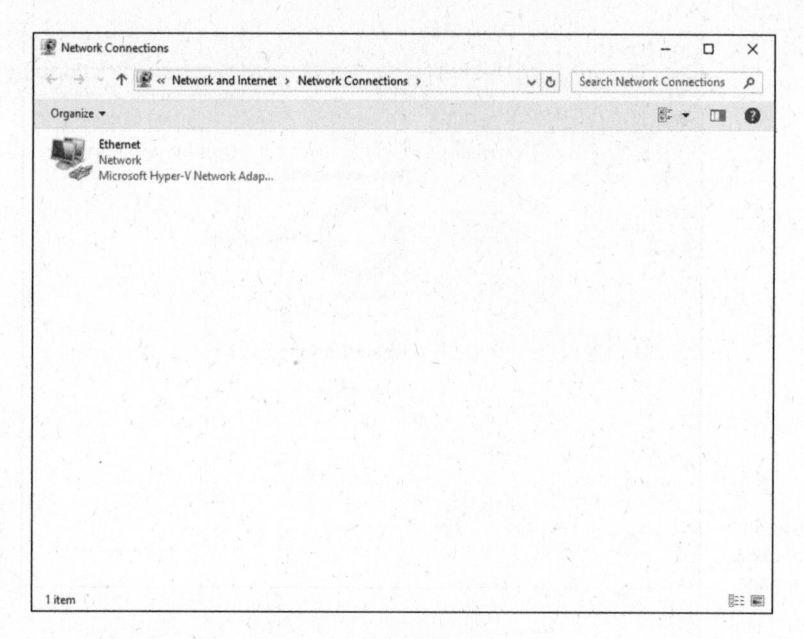

FIGURE 2-4: The Network Connections page.

5. **Right-click the connection that you want to configure and then choose Properties from the contextual menu that appears.**

   This action opens the Ethernet Properties dialog box, as shown in Figure 2-5.

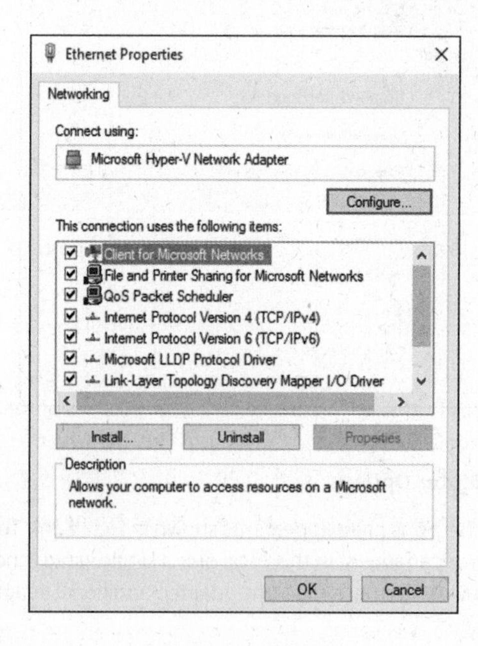

FIGURE 2-5: The Ethernet Properties dialog box.

6. **To configure the network adapter card settings, click Configure.**

The Properties dialog box for your network adapter appears, as shown in Figure 2-6. This dialog box has seven tabs that let you configure the adapter:

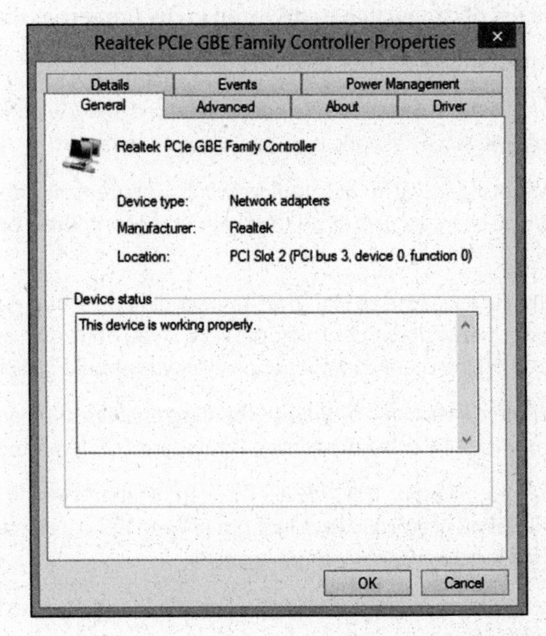

FIGURE 2-6: The Properties dialog box for a network adapter.

- *General:* Shows basic information about the adapter, such as the device type and status.
- *Advanced:* Lets you set a variety of device-specific parameters that affect the operation of the adapter.
- *About:* Displays information about the device's patent protection.
- *Driver:* Displays information about the device driver that's bound to the NIC and lets you update the driver to a newer version, roll back the driver to a previously working version, or uninstall the driver.
- *Details:* With this tab, you can inspect various properties of the adapter such as the date and version of the device driver. To view the setting of a particular property, select the property name from the drop-down list.
- *Events:* Lists recent events that have been logged for the device.
- *Power Management:* Lets you configure power management options for the device.

TIP

When you click OK to dismiss the dialog box, the network connection's Properties dialog box closes and you're returned to the Network Connections page (refer to Figure 2-4). Right-click the network adapter and choose Properties again to continue the procedure.

7. **Review the list of connection items listed in the Properties dialog box.**

   The most important items you commonly see are:

   - *Client for Microsoft Networks:* This item is required if you want to access a Microsoft Windows network. It should always be present.
   - *File and Printer Sharing for Microsoft Networks:* This item allows your computer to share its files or printers with other computers on the network.

TIP

     This option is usually used with peer-to-peer networks, but you can use it even if your network has dedicated servers. If you don't plan to share files or printers on the client computer, however, you should disable this item.

   - *Internet Protocol Version 4 (TCP/IPv4):* This item enables the client computer to communicate by using the version 4 standard TCP/IP protocol.
   - *Internet Protocol Version 6 (TCP/IPv6):* This item enables version 6 of the standard TCP/IP protocol. Typically, both IP4 and IP6 are enabled, even though most networks rely primarily on IP4.

8. **If a protocol that you need isn't listed, click the Install button to add the needed protocol.**

   A dialog box appears, asking whether you want to add a network client, protocol, or service. Click Protocol and then click Add. A list of available protocols appears. Select the one you want to add; then click OK.

9. **To remove a network item that you don't need (such as File and Printer Sharing for Microsoft Networks), select the item, and click the Uninstall button.**

   For security reasons, you should make it a point to remove any clients, protocols, or services that you don't need.

10. **To configure TCP/IP settings, click Internet Protocol Version 4 (TCP/IPv4); click Properties to display the TCP/IP Properties dialog box (shown in Figure 2-7); adjust the settings; and then click OK.**

    The TCP/IP Properties dialog box lets you choose among these options:

    - *Obtain an IP Address Automatically:* Choose this option if your network has a DHCP server that assigns IP addresses automatically. Choosing this option dramatically simplifies administering TCP/IP on your network. (See Chapter 6 for more information about DHCP.)

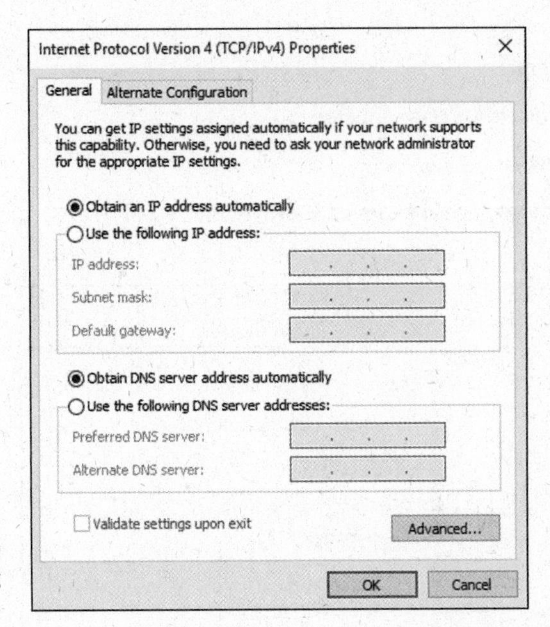

FIGURE 2-7: Configuring TCP/IP.

- *Use the Following IP Address:* If your computer must have a specific IP address, choose this option and then type the computer's IP address, subnet mask, and default gateway address. (For more information about these settings, see Chapter 6.)
- *Obtain DNS Server Address Automatically:* The DHCP server can also provide the address of the Domain Name System (DNS) server that the computer should use. Choose this option if your network has a DHCP server. (See Chapter 6 for more information about DNS.)
- *Use the Following DNS Server Addresses:* Choose this option if a DNS server isn't available. Then type the IP addresses of the primary and secondary DNS servers.

Joining a Windows Computer to a Domain

When Windows first installs, it isn't joined to a domain network. Instead, it's available as part of a workgroup, which is an unmanaged network suitable only for the smallest of networks with just a few computers and without dedicated servers. To use a computer in a domain network, you must join the computer to the domain. Here are the steps for Windows 10:

1. **Click the Start icon (or press the Start button on the keyboard), and then tap or click Settings.**

   The Settings page appears (refer to Figure 2-1).

2. **Click System.**

   The System settings page appears.

3. **Click About.**

   The PC settings page appears, as shown in Figure 2-8.

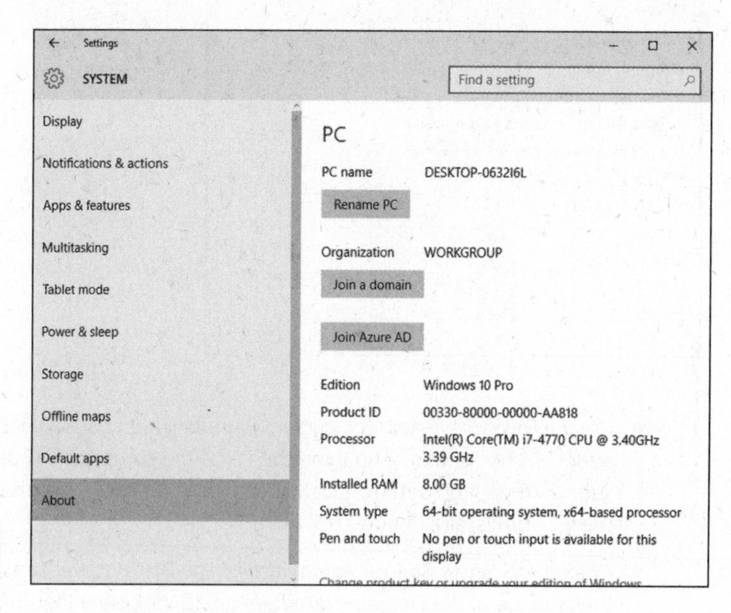

FIGURE 2-8: The PC settings page.

4. **(Optional) To change the name of the computer, click Rename PC.**

   You're prompted to enter a new name and then reboot the computer.

   Before you join a domain, you should ensure that the computer's name won't be the same as the name of a computer that's already a member of the domain. If it is, you should first change the name.

TIP

   If you do change the computer's name, repeat the procedure from Step 1 after the reboot.

5. **Click Join a Domain.**

   The Join a Domain dialog box appears, as shown in Figure 2-9.

6. **Enter the domain name and click Next.**

   You're prompted for the user name and password of a user who has administration privileges on the domain, as shown in Figure 2-10.

7. **Click OK.**

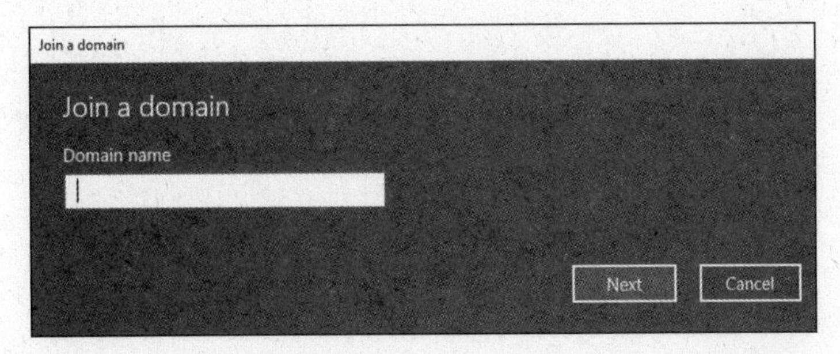

FIGURE 2-9: Joining a domain.

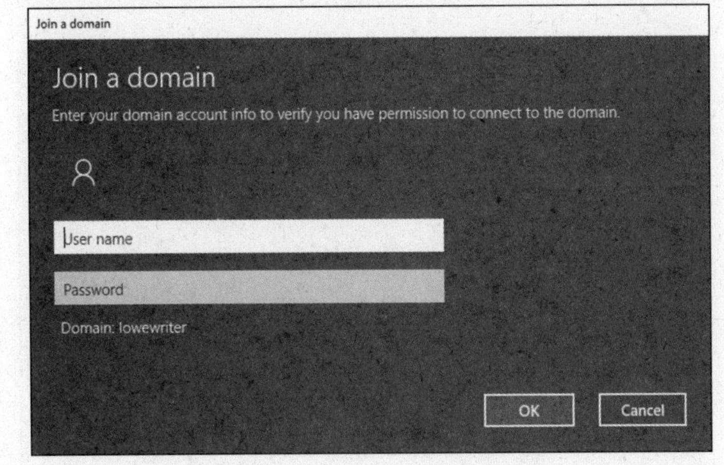

FIGURE 2-10: You must provide domain administrator credentials to join a domain.

8. **Enter the username and password for an Administrator account when prompted.**

   You're asked to provide this information only if a computer account hasn't already been created for the client computer.

9. **When informed that you need to restart the computer, click Restart Now.**

   The computer is restarted and added to the domain.

Configuring Mac Network Settings

Every Macintosh ever built, even an original 1984 model, includes networking support. Newer Macintosh computers have better built-in networking features than older Macintosh computers, of course. The newest Macs include built-in Gigabit Ethernet connections or 802.11ac wireless connections, or both. Support for these network connections is pretty much automatic, so all you have to do is plug your Mac into a network or connect to a wireless network, and you're ready to go.

Most network settings on macOS are automatic. If you want, you can look at and change the default network settings by following these steps:

1. **Choose ⇨ System Preferences and then click Network.**

   The Network preferences page appears, as shown in Figure 2-11.

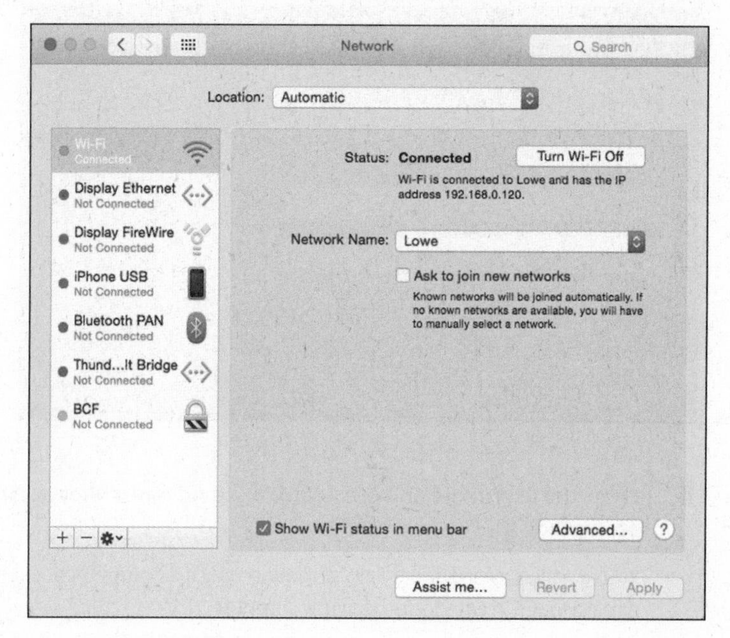

FIGURE 2-11: Network preferences.

2. **Click Advanced.**

   The advanced network settings are displayed, as shown in Figure 2-12.

3. **Click the TCP/IP tab.**

   This brings up the TCP/IP settings, as shown in Figure 2-13. From this page, you can view the currently assigned IP address for the computer. And, if you want, you can assign a static IP address by changing the Configure IPv4 drop-down from Using DHCP to Manually. Then, you can enter your own IP address, subnet mask, and router address. (For more information about IP addresses, refer to Chapter 6.)

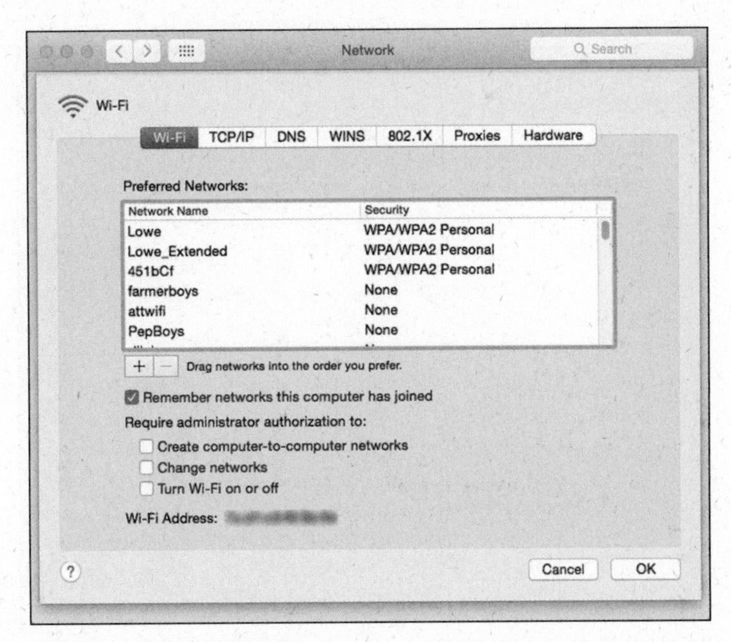

FIGURE 2-12: Advanced network settings.

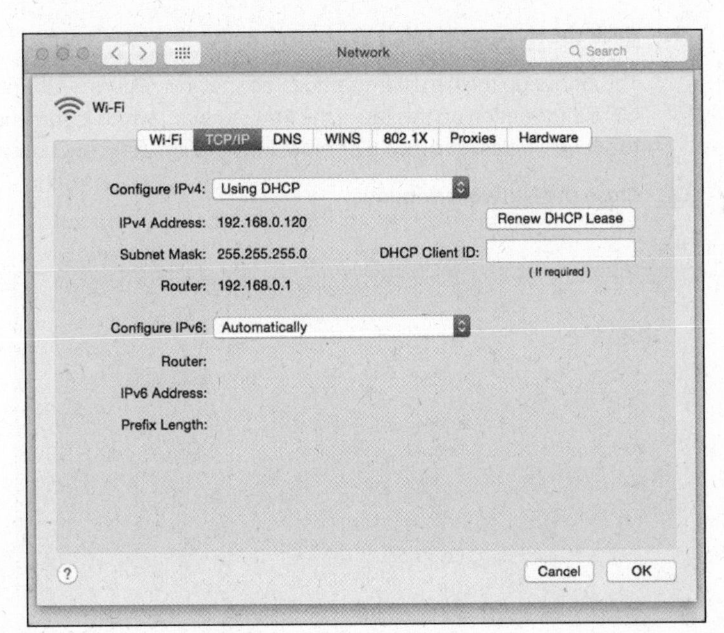

FIGURE 2-13: Mac network TCP/IP settings.

4. **Click the DNS tab.**

   This brings up the DNS settings, as shown in Figure 2-14. Here, you can see the DNS servers being used, and you can add additional DNS servers if you want.

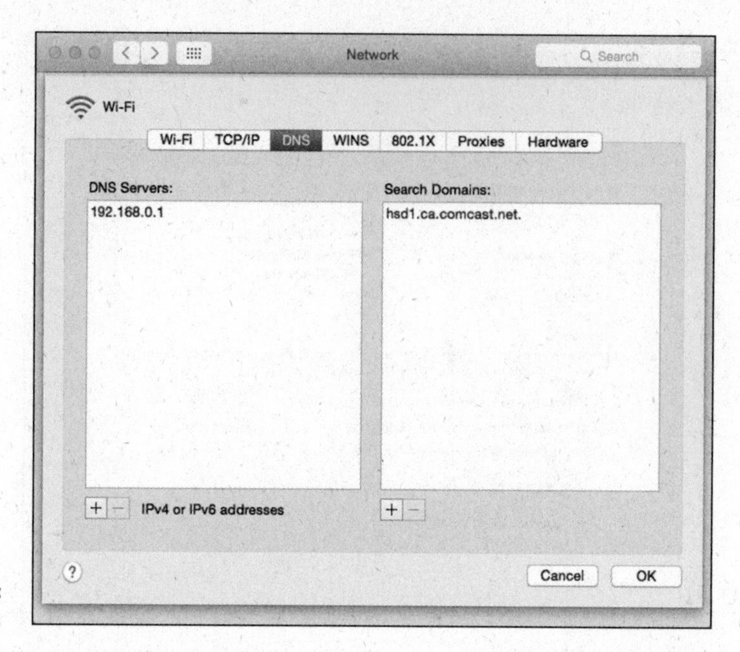

FIGURE 2-14: DNS settings.

5. **Click the Hardware tab.**

   This brings up the hardware settings, as shown in Figure 2-15. The most useful bit of information on this tab is the MAC address, which is sometimes needed to set up wireless security. (For more information, refer to Chapter 8.)

6. **Close the Network window.**

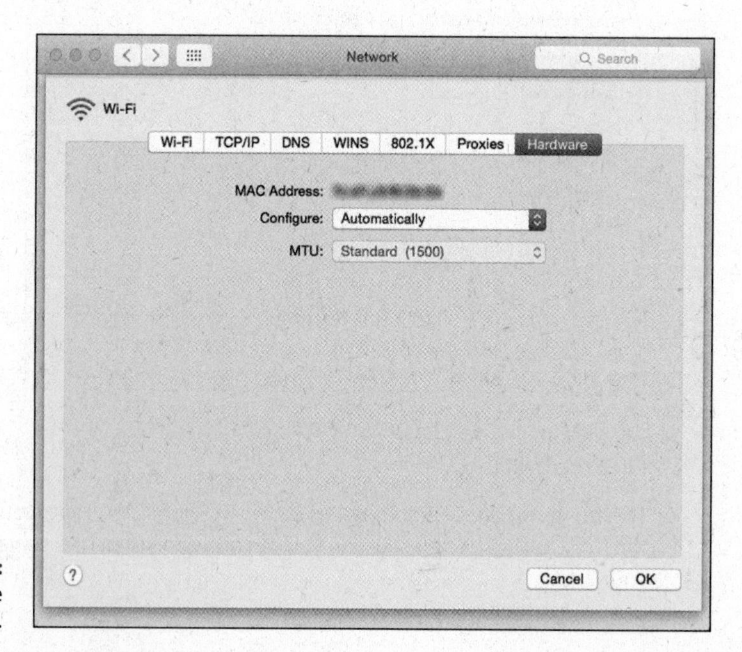

FIGURE 2-15: Hardware settings.

Joining a Mac Computer to a Domain

If you're using a Mac in a Windows domain environment, you can join the Mac to the domain by following these steps:

1. **Choose ⇨ System Preferences and then click Users & Groups.**

   This brings up the Users & Groups page, as shown in Figure 2-16.

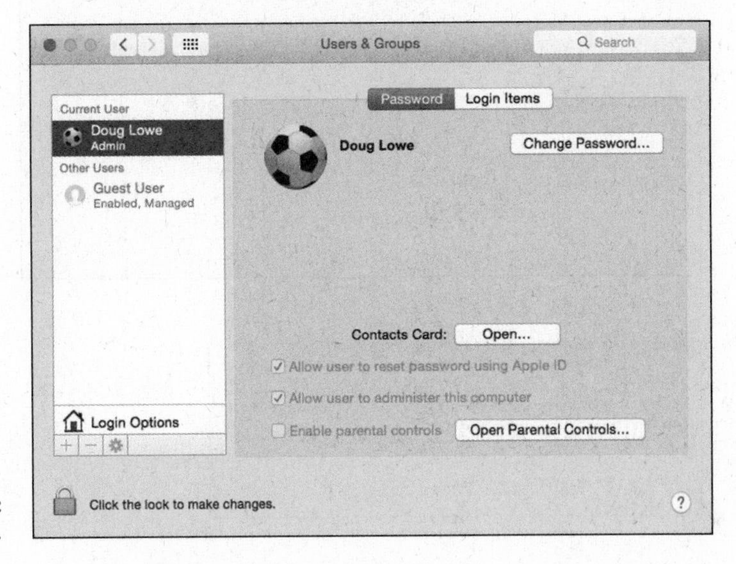

FIGURE 2-16: Users & Groups.

2. **Select the user account you want to join to the domain and then click Login Options.**

   The Login Options page appears, as shown in Figure 2-17.

3. **If the lock icon at the lower left of the page is locked, click it and enter your password when prompted.**

   By default, the user login options are locked to prevent unauthorized changes. This step unlocks the settings so that you can join the domain.

4. **Click the Join button.**

   You're prompted to enter the name of the domain you want to join, as shown in Figure 2-18.

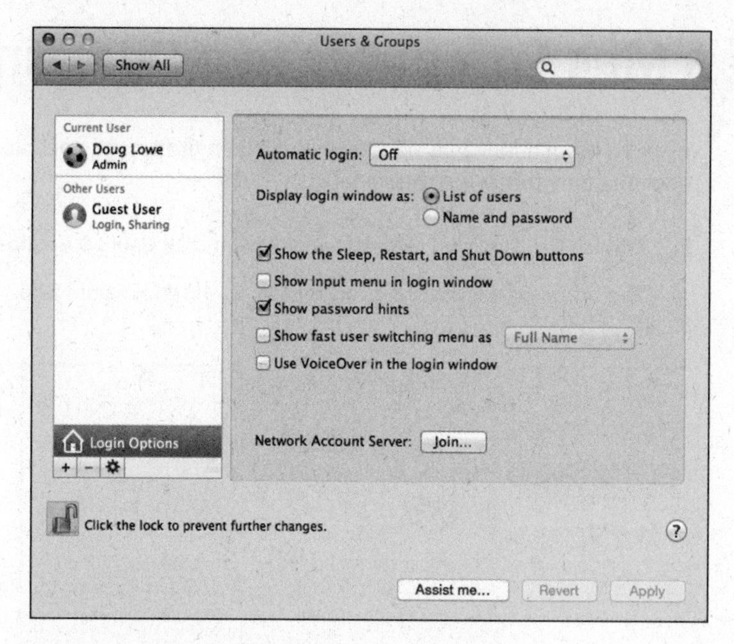

FIGURE 2-17: Login Options.

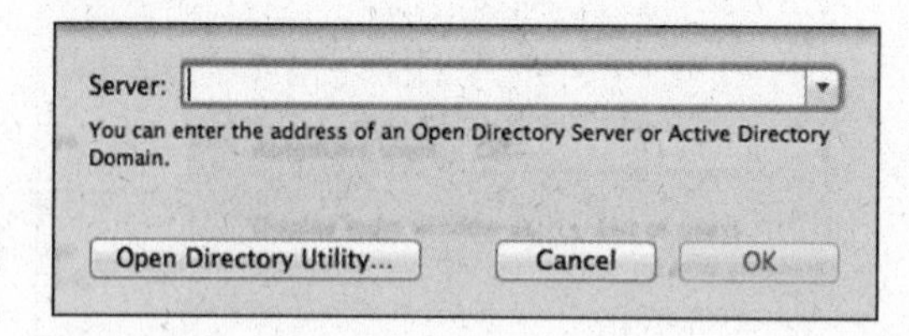

FIGURE 2-18: Joining a domain.

5. **Enter the name of the domain you want to join.**

   When you enter the domain name, the dialog box expands to allow you to enter domain credentials to allow you to join the domain, as shown in Figure 2-19.

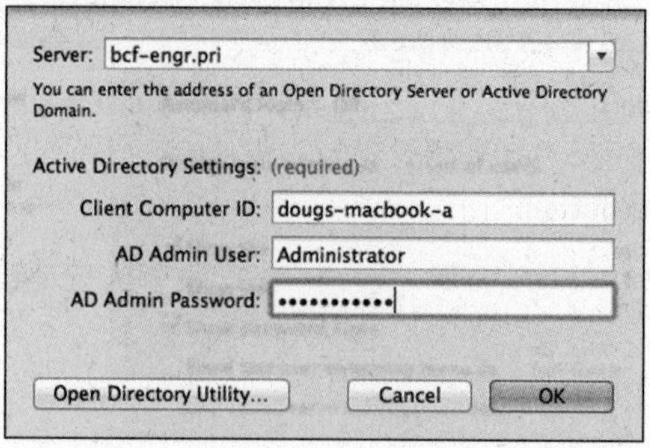

FIGURE 2-19: Authenticating with the domain.

6. **Enter the name and password of a domain administrator account, and then click OK.**

   You return to the Login Options page, which shows that you've successfully joined the domain (see Figure 2-20).

7. **Close the Users & Groups window.**

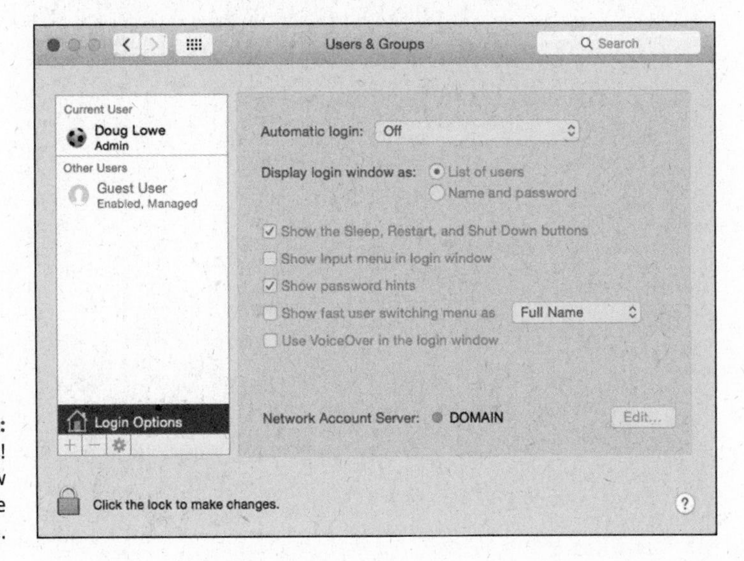

FIGURE 2-20: Congratulations! You have now joined the domain.

IN THIS CHAPTER

- » Using local resources and network resources
- » Playing the name game
- » Logging on to a network
- » Using shared folders
- » Going places with networks
- » Mapping your network drives
- » Using a network printer
- » Logging off the network

Chapter 3
Life on the Network

After you hook up your PC to a network, it's not an island anymore, separated from the rest of the world like some kind of isolationist fanatic waving a "Don't tread on me" flag. The network connection changes your PC forever. Now your computer is part of a system, connected to other computers on the network. You have to worry about annoying network details, such as using local and shared resources, logging on and accessing network drives, using network printers, logging off, and who knows what else.

Oh, bother.

This chapter brings you up to speed on what living with a computer network is like. Unfortunately, this chapter gets a little technical at times, so you may need your pocket protector.

Distinguishing between Local Resources and Network Resources

In case you don't catch this statement in Chapter 1, one of the most important differences between using an isolated computer and using a network computer lies in the distinction between local resources and network resources. *Local resources* are items — such as hard drives, printers, CD or DVD drives, and flash drives — that are connected directly to your computer. You can use local resources whether you're connected to the network or not. *Network resources*, on the other hand, are the hard drives, printers, optical drives, and other devices that are connected to the network's server computers. You can use network resources only after your computer is connected to the network.

Whenever you use a computer network, you need to know which resources are local resources (belong to you) and which are network resources (belong to the network). In most networks, your `C:` drive is a local drive, as is your Documents folder. If a printer is sitting next to your PC, it's probably a local printer. You can do anything you want with these resources without affecting the network or other users on the network (as long as the local resources aren't shared on the network). Keep these points in mind:

- You can't tell just by looking at a resource whether it's a local resource or a network resource. The printer that sits right next to your computer is probably your local printer, but then again, it may be a network printer. The same statement is true for hard drives: The hard drive in your PC is probably your own, but it (or part of it) may be shared on the network, thus enabling other users to access it.
- Because dedicated network servers are full of resources, you may say that they're not only dedicated (and sincere), but also resourceful. (Groan. Sorry. This is yet another in a tireless series of bad computer-nerd puns.)

What's in a Name?

Just about everything on a computer network has a name: The computers themselves have names, the people who use the computers have names, the hard drives and printers that can be shared on the network have names, and the network itself has a name. Knowing all the names used on your network isn't essential, but you do need to know some of them.

Here are some additional details about network names:

- **Every person who can use the network has a username (sometimes called a *user ID*).** You need to know your username to log on to the network. You also need to know the usernames of your buddies, especially if you want to steal their files or send them nasty notes.

  You can find more information about usernames and logging on in the section "Logging on to the Network," later in this chapter.

WARNING

- **Letting folks on the network use their first names as their usernames is tempting but not a good idea.** Even in a small office, you eventually run into a conflict. (And what about Mrs. McCave — made famous by Dr. Seuss — who had 23 children and named them all Dave?)

TIP

  Create a consistent way of creating usernames. For example, you may use your first name plus the first two letters of your last name. Then Lisa's username is Lisasi, and Bart's is Bartsi. Or you may use the first letter of your first name followed by your complete last name. Then Lisa's username is lsimpson, and Bart's is bsimpson. (In most networks, capitalization doesn't matter in usernames. Thus, bsimpson is the same as BSimpson.)

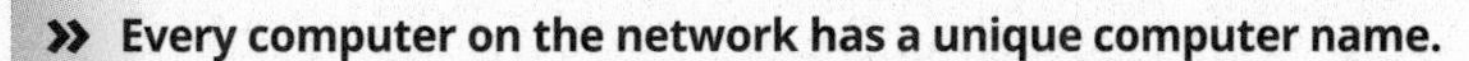

- **Every computer on the network has a unique computer name.**

TIP

  You don't have to know the names of all the computers on the network, but it helps if you know your own computer's name and the names of any server computers you need to access.

  The computer's name is sometimes the same as the username of the person who uses the computer, but that's usually a bad idea because in many companies, people come and go more often than computers. Sometimes the names indicate the physical location of the computer, such as office-12 or back-room. Server computers often have names that reflect the group that uses the server most, like acctng-server or cad-server.

  Some network nerds like to assign techie-sounding names, like BL3K5-87a. And some like to use names from science-fiction movies; HAL (from *2001: A Space Odyssey*), M5 or Data (from *Star Trek*), or Overmind (from *Teenage Mutant Ninja Turtles*) come to mind. Cute names like Herbie aren't allowed. (However, Tigger and Pooh are entirely acceptable — recommended, in fact. Networks are what Tiggers like the best.)

REMEMBER

  Usually, the sensible approach to computer naming is to use names that have numbers, such as computer001 or computer002.

- **Network resources, such as shared disk folders and printers, have names.** For example, a network server may have two printers, named `laser` and `inkjet` (to indicate the type of printer), and two shared disk folders, named `AccountingData` and `MarketingData`.

TIP

REMEMBER

» **Server-based networks have a username for the network administrator.**

If you log on using the administrator's username, you can do anything you want: add new users, define new network resources, change Lisa's password, anything. The administrator's username is usually something clever such as Administrator.

» **The network itself has a name.**

The Windows world has two basic types of networks:

- *Domain networks* are the norm for large corporate environments that have dedicated servers with IT staff to maintain them.
- *Workgroup networks* are more common in homes or in small offices that don't have dedicated servers or IT staff.

A domain network is known by — you guessed it — a *domain name.* And a workgroup network is identified by — drum roll, please — a *workgroup name.* Regardless of which type of network you use, you need to know this name to gain access to the network.

Logging on to the Network

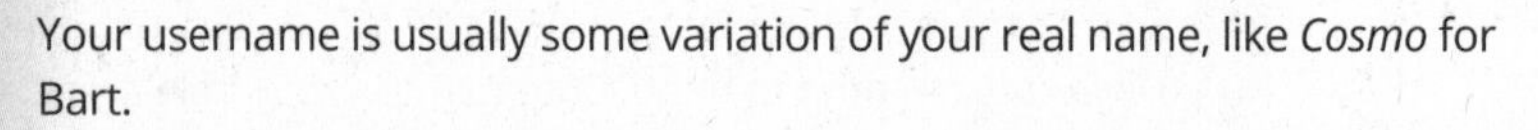

To use network resources, you must connect your computer to the network, and you must go through the supersecret process of logging on, which is how you let the network know who you are so that it can decide whether you're one of the good guys.

Logging on is a little bit like cashing a check. You must have two forms of identification:

REMEMBER

» **Your username:** The name by which the network knows you.

Your username is usually some variation of your real name, like *Cosmo* for Bart.

Everyone who uses the network must have a username.

» **Your password:** A secret word that only you and the network know. If you type the correct password, the network believes that you are who you say you are.

Every user has a different password, and the password should be a secret.

In the early days of computer networking, you had to type a logon command at a stark MS-DOS prompt and then supply your user ID and password. Nowadays, the glory of Windows is that you get to log on to the network through a special network logon screen. Figure 3-1 shows the Windows 10 version of this dialog box.

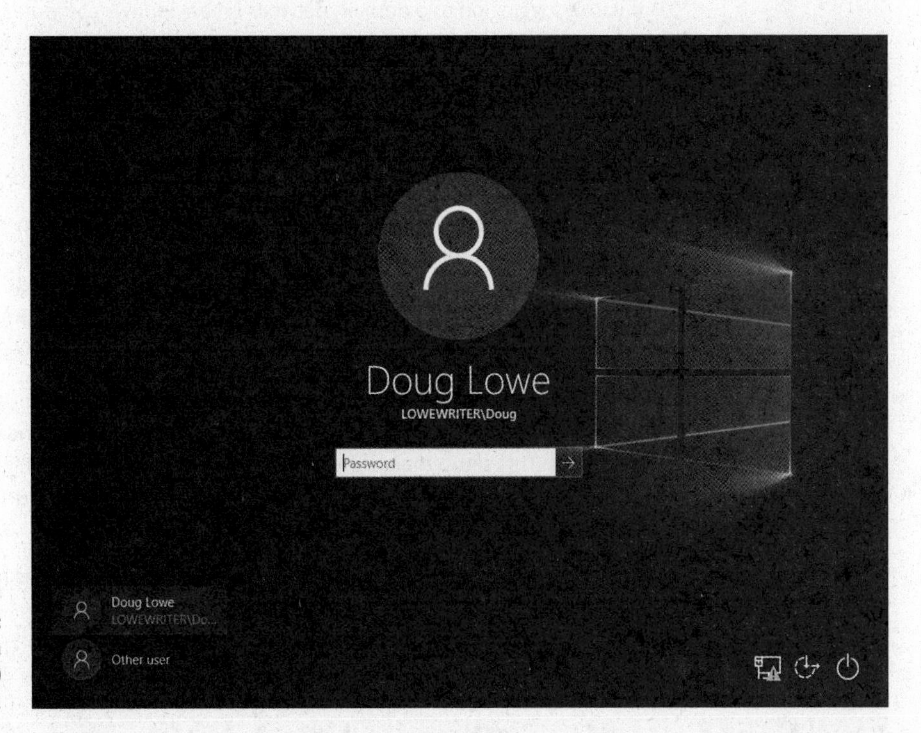

FIGURE 3-1: Logging in to a Windows 10 system.

TIP

Here are some more logon points to ponder:

- The terms *user ID* and *logon name* are sometimes used instead of *username.* They all mean the same thing.
- As long as we're talking about words that mean the same thing, *log in* and *log on* mean the same thing, as do (respectively) *log out* and *log off* as ways of saying, "I'm outta here." Although you see both out there in the world, this book uses *log on* and *log off* throughout — and if there's any exception, the book says why and grouses about it a bit.
- As far as the network's concerned, you and your computer aren't the same thing. Your username refers to you, not to your computer. That's why you have a username and your computer has a computer name. You can log on to the network by using your username from any computer that's attached to the network. Other users can log on at your computer by using their own usernames.

When others log on at your computer by using their own usernames, they can't access any of your network files that are protected by your password. However, they *can* access any local files that you haven't protected. Be careful which people you allow to use your computer.

» If you're logging on to a domain network on a Windows computer, you must type the domain name before your username, separated from it by a backslash. For example:

```
lowewriter\dlowe
```

Here, the domain name is lowewriter, and the username is dlowe.

Note that Windows remembers the domain and username from your last login, so ordinarily all you have to enter is your password. To log on to a different domain or as a different user, you must click Switch User. Then you can click the Other User icon and enter a different domain name and username, along with the password for the user you want to log on as.

» Your computer may be set up so that it logs you on automatically whenever you turn it on. In that case, you don't have to type your username and password. This setup makes the task of logging on more convenient but takes the sport out of it. And it's a terrible idea if you're the least bit worried about bad guys getting into your network or personal files.

» Guard your password with your life. I'd tell you mine, but then I'd have to shoot you.

Understanding Shared Folders

Long ago, in the days Before Network (B.N.), your computer probably had just one hard drive, known as the `C:` drive. Maybe it had two — `C:` and `D:`. The second drive might be another hard disk, or possibly a CD-ROM or DVD-ROM drive. Even to this day, the descendants of those drives are physically located inside your PC. They're your *local drives.*

Now that you're on a network, however, you may have access to drives that aren't located inside your PC but are located instead in one of the other computers on the network. These network drives can be located on a dedicated server computer or, in the case of a peer-to-peer network, on another client computer.

In some cases, you can access an entire network drive over the network. But in most cases, you can't access the entire drive. Instead, you can access only certain folders on the network drives. Either way, the shared drives or folders are known in Windows terminology as *shared folders.*

Here's where it gets confusing: The most common way to access a shared folder is to assign a drive letter to it. Suppose that a server has a shared folder named Marketing. You can assign drive letter *M* to this shared folder. Then you access the Marketing folder as drive `M:`. The `M:` drive is then called a *network drive* because it uses the network to access data in a shared folder. Assigning a drive letter to a shared folder is *mapping a drive.*

Shared folders can be set up with restrictions on how you can use them. For example, you may be granted full access to some shared folders so that you can copy files to or from them, delete files on them, or create or remove folders on them. On other shared folders, your access may be limited in certain ways. For example, you may be able to copy files to or from the shared folder but not delete files, edit files, or create new folders. You may also be asked to enter a password before you can access a protected folder. The amount of disk space you're allowed to use on a shared folder may also be limited. For more information about file-sharing restrictions, see Chapter 13.

In addition to accessing shared folders that reside on other people's computers, you can designate your computer as a server to enable other network users to access folders that you share. To find out how to share folders on your computer with other network users, see Chapter 4.

Four Good Uses for a Shared Folder

After you know which shared network folders are available, you may wonder what you're supposed to do with them. This section describes four good uses for a network folder.

Store files that everybody needs

A shared network folder is a good place to store files that more than one user needs to access. Without a network, you have to store a copy of the file on everyone's computer, and you have to worry about keeping the copies synchronized (which you can't do, no matter how hard you try). Or you can keep the file on a disk and pass it around. Or you can keep the file on one computer and play Musical Chairs; whenever someone needs to use the file, he goes to the computer that contains the file.

On a network, you can keep one copy of the file in a shared folder on the network, and everyone can access it.

Store your own files

You can also use a shared network folder as an extension of your own hard drive storage. For example, if you filled up all the free space on your hard drive with pictures, sounds, and movies that you downloaded from the Internet, but the network server has billions and billions of gigabytes of free space, you have all the drive space you need. Just store your files on the network drive!

Here are a few guidelines for storing files on network drives:

- **Using the network drive for your own files works best if the network drive is set up for private storage that other users can't access.** That way, you don't have to worry about the nosy guy down in Accounting who likes to poke around in other people's files.
- **Don't overuse the network drive.** Remember that other users have probably filled up their own hard drives, so they want to use the space on the network drive too.
- **Before you store personal files on a network drive, make sure that you have permission.** A note from your mom will do.
- **On domain networks, a drive (typically, drive H:) is commonly mapped to a user's home folder.** The *home folder* is a network folder that's unique for each user. You can think of it as a network version of Documents. If your network is set up with a home folder, use it rather than Documents for any important work-related files. That's because the home folder is usually included in the network's daily backup schedule. By contrast, most networks do *not* back up data you store in Documents.

Make a temporary resting place for files on their way to other users

"Hey, Lisa, could you send me a copy of last month's baseball stats?"

"Sure, Bart." But how? If the baseball stats file resides on Lisa's local drive, how does Lisa send a copy of the file to Bart's computer? Lisa can do it by copying the file to a network drive. Then Bart can copy the file to his local hard drive.

Here are some tips to keep in mind when you use a network drive to exchange files with other network users:

- **Remember to delete files that you saved to the network drive after they're picked up!** Otherwise, the network drive quickly fills up with unnecessary files.

» **Create a folder on the network drive specifically intended for holding files en route to other users.** I like to name this folder PITSTOP.

TIP

In many cases, it's easier to send files to other network users by email than by using a network folder. Just send a message to the other network user and attach the file you want to share. The advantage of sending a file by email is that you don't have to worry about details like where to leave the file on the server and who's responsible for deleting the file.

Back up your local hard drive

If enough drive space is available on the file server, you can use it to store backup copies of the files on your hard drive. Just copy the files that you want to back up to a shared network folder.

Obviously, if you copy *all* your data files to the network drive — and everybody else follows suit — it can fill up quickly. Check with the network manager before you start storing backup copies of your files on the server. The manager may have already set up a special network drive that's designed just for backups. And if you're lucky, your network manager may be able to set up an automatic backup schedule for your important data so that you don't have to remember to back it up manually.

I hope that your network administrator also routinely backs up the contents of the network server's disk to tape. (Yes, *tape* — see Chapter 22 for details.) That way, if something happens to the network server, the data can be recovered from the backup tapes.

Oh, the Network Places You'll Go

Windows enables you to access network resources, such as shared folders, by browsing the network. In Windows 7, choose Network from the Start menu. In Windows 10, open Windows Explorer (click File Explorer on the taskbar) and then click Network. Figure 3-2 shows the Windows 10 version of the network browser.

The network shown in Figure 3-2 consists of just two computers: the desktop client computer running Windows 10 (WIN10-01) and a server computer running Windows Server 2016 (WIN1601). In an actual network, you would obviously see more than just two computers.

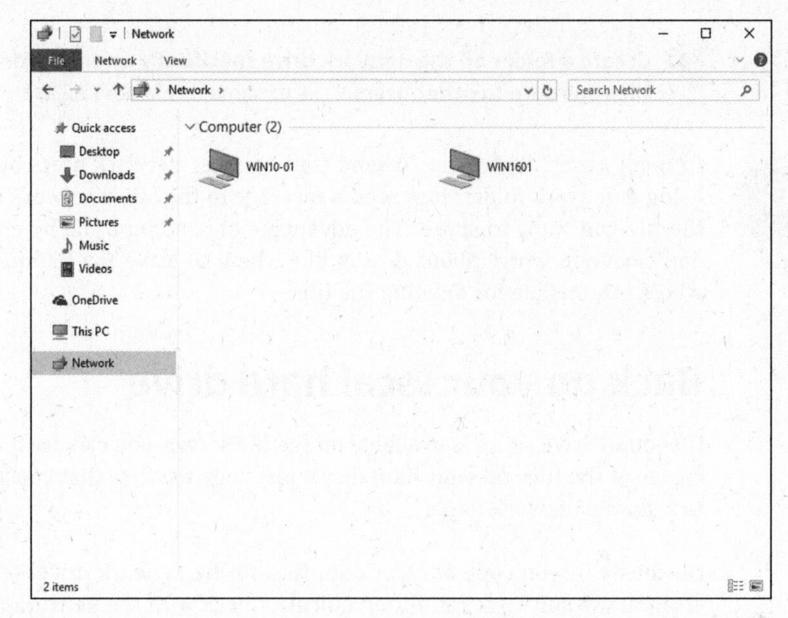

FIGURE 3-2: Browsing the network in Windows 10.

You can open a computer by double-clicking its icon to reveal a list of shared resources available on the computer. For example, Figure 3-3 shows the resources shared by the WIN1601 computer.

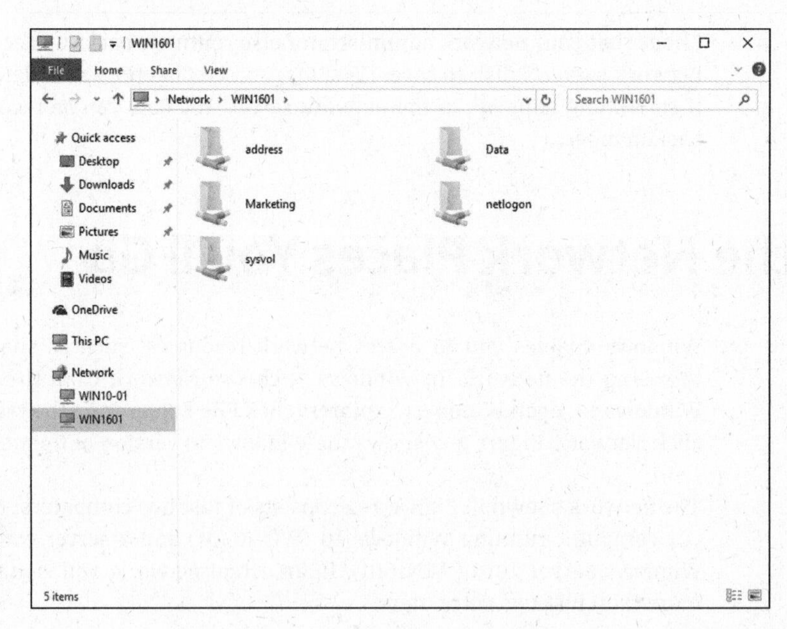

FIGURE 3-3: The resources available on a server computer.

You can also browse the network from any Windows application program. For example, you may be working with Microsoft Word and want to open a document file that's stored in a shared folder on your network. All you have to do is use the Open command to bring up the dialog box, and then choose Network in the Navigation pane to view the available network devices.

Mapping Network Drives

TIP

If you often access a particular shared folder, you may want to use the special trick known as mapping to access the shared folder more efficiently. *Mapping* assigns a drive letter to a shared folder. Then you can use the drive letter to access the shared folder as though it were a local drive. In this way, you can access the shared folder from any Windows program without having to browse the network.

For example, you can map a shared folder named Data on the server named Win1601 Files to drive `K:` on your computer. Then, to access files stored in the shared Data folder, you look on drive `K:`.

To map a shared folder to a drive letter, follow these steps:

1. **Open File Explorer.**
   - *Windows 7:* Choose Start ➪ Computer.
   - *Windows 8, 8.1, and 10:* Open the desktop and click the File Explorer icon on the taskbar, and then click Computer in the Location list on the left side of the screen.
2. **Open the Map Network Drive dialog box.**
   - *Windows 7:* Access this dialog by clicking the Map Network Drive button located on the toolbar.
   - *Windows 8 and 8.1:* Click Map Network Drive on the Ribbon.
   - *Windows 10:* Click the Computer tab, and then click Map Network Drive.

   Figure 3-4 shows the Map Network Drive dialog box for Windows 10. The dialog box for earlier versions of Windows is similar.
3. **(Optional) Change the drive letter in the Drive drop-down list.**

   You probably don't have to change the drive letter that Windows selects (in Figure 3-4, drive `Z:`). If you're picky, though, you can select the drive letter from the Drive drop-down list.

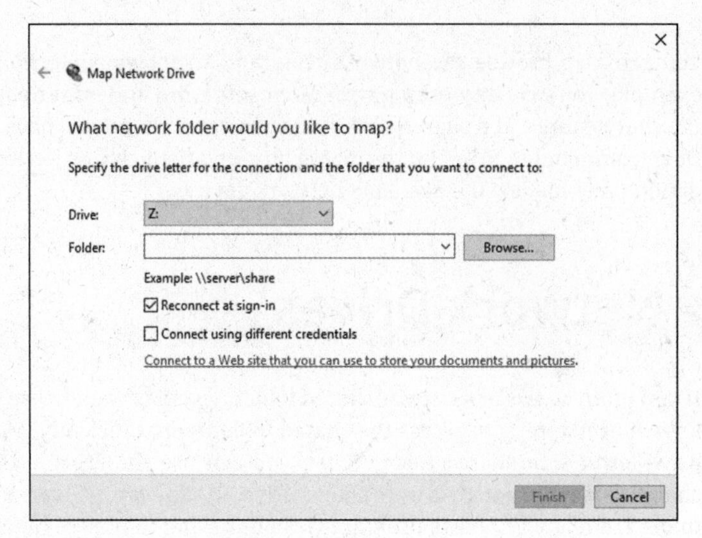

FIGURE 3-4: The Map Network Drive dialog box.

4. **Click the Browse button.**

   The dialog box shown in Figure 3-5 appears.

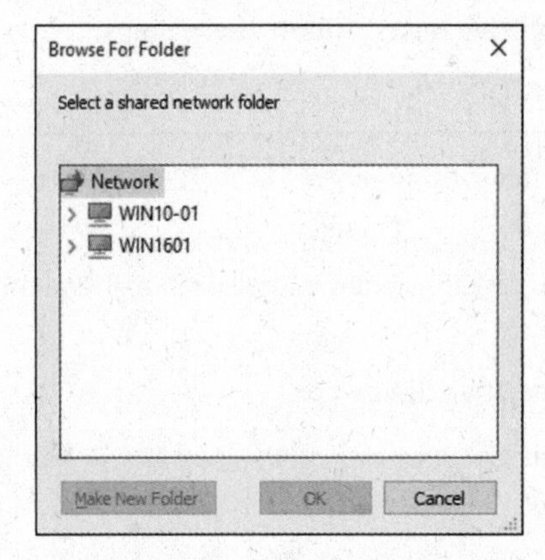

FIGURE 3-5: Browsing for the folder to map.

5. **Use the Browse for Folder dialog box to find and select the shared folder you want to use.**

   You can navigate to any shared folder on any computer in the network.

6. **Click OK.**

   The Browse for Folder dialog box is dismissed, and you return to the Map Network Drive dialog box (refer to Figure 3-4).

7. **(Optional) If you want this network drive to be automatically mapped each time you log on to the network, select the Reconnect at Sign-in check box.**

   If you leave the Reconnect at Sign-in check box deselected, the drive letter is available only until you shut down Windows or log out of the network. If you select this option, the network drive reconnects automatically each time you log on to the network.

TIP

   Be sure to select the Reconnect at Sign-in check box if you use the network drive often.

8. **Click OK.**

   You return to the This PC folder, as shown in Figure 3-6. Here, you can see the newly mapped network drive.

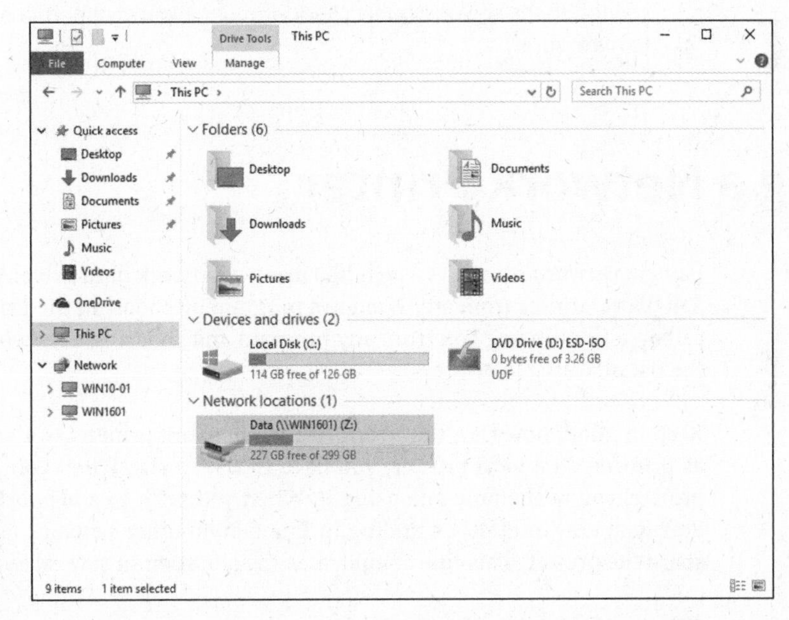

FIGURE 3-6: The This PC folder shows a mapped network drive.

Your network administrator may have already set up your computer with one or more mapped network drives. If so, you can ask her to tell you which network drives have been mapped. Or you can just open the This PC folder and have a look.

Here are a few additional tips:

» **Assigning a drive letter to a network drive is called *mapping the drive,* or *linking the drive,* by network nerds.** "Drive Q: is mapped to a network drive," they say.

- **Network drive letters don't have to be assigned the same way for every computer on the network.** For example, a network drive that's assigned drive letter *M* on your computer may be assigned drive letter *Z* on someone else's computer. In that case, your drive `M:` and the other computer's drive `Z:` refer to the same data. This arrangement can be confusing. If your network is set up this way, put pepper in your network administrator's coffee.
- **Accessing a shared network folder through a mapped network drive is much faster than accessing the same folder by browsing the network.** Windows has to browse the entire network to list all available computers whenever you browse the network. By contrast, Windows doesn't have to browse the network to access a mapped network drive.
- **If you select the Reconnect at Sign-in option for a mapped drive (refer to Figure 3-4), you receive a warning message if the drive isn't available when you log on.** In most cases, the problem is that your computer can't connect to the server. Double-check all your network connections and settings to make sure.

Using a Network Printer

Using a network printer is much like using a network hard drive: You can print to a network printer from any Windows program by choosing the Print command to call up a Print dialog box from any program and choosing a network printer from the list of available printers.

Keep in mind, however, that printing on a network printer isn't exactly the same as printing on a local printer; you have to take turns. When you print on a local printer, you're the only one using it. When you print to a network printer, however, you are (in effect) standing in line behind other network users, waiting to share the printer. This line complicates the situation in several ways:

- **If several users print to the network printer at the same time, the network has to keep the print jobs separate from one another.** If it didn't, the result would be a jumbled mess, with your 268-page report getting mixed in with the payroll checks. That would be bad. Fortunately, the network takes care of this situation by using the fancy *print spooling* feature.
- **Network printing works on a first-come, first-served basis.** Invariably, when I get in line at the hardware store, the person in front of me is trying to buy something that doesn't have a product code on it. I end up standing there for hours waiting for someone in Plumbing to pick up the phone for a price

check. Network printing can be like that. If someone sends a two-hour print job to the printer before you send your half-page memo, you have to wait.

- **You may have access to a local printer and several network printers.** Before you were forced to use the network, your computer probably had just one printer attached to it. You may want to print some documents on your cheap (oops, I mean *local*) inkjet printer but use the network laser printer for important stuff. To do that, you have to find out how to use your programs' functions for switching printers.

Adding a network printer

Before you can print to a network printer, you have to configure your computer to access the network printer that you want to use. From the Start menu, open the Control Panel and then double-click the Printers icon. If your computer is already configured to work with a network printer, an icon for the network printer appears in the Printers folder. You can tell a network printer from a local printer by the shape of the printer icon. Network printer icons have a pipe attached to the bottom of the printer.

If you don't have a network printer configured for your computer, you can add one by using the Add Printer Wizard. Just follow these steps:

1. **Open the Control Panel.**

   - *Windows 7 or earlier:* Choose Start ⇨ Control Panel.
   - *Windows 8 and later:* Press the Windows key, type **Control**, and then click the Control Panel icon.

2. **Click Devices and Printers.**

3. **Click the Add a Printer button on the toolbar.**

   This step starts the Add Printer Wizard, as shown in Figure 3-7.

4. **Click the printer you want to use.**

   If you can't find the printer you want to use, ask your network administrator for the printer's *UNC path,* which is the name used to identify the printer on the network, or its IP address. Then click The Printer That I Want Isn't Listed and enter the UNC or IP address for the printer when prompted.

5. **Click Next to add the printer.**

   The wizard copies to your computer the correct printer driver for the network printer. (You may be prompted to confirm that you want to add the driver. If so, click Install Driver to proceed.)

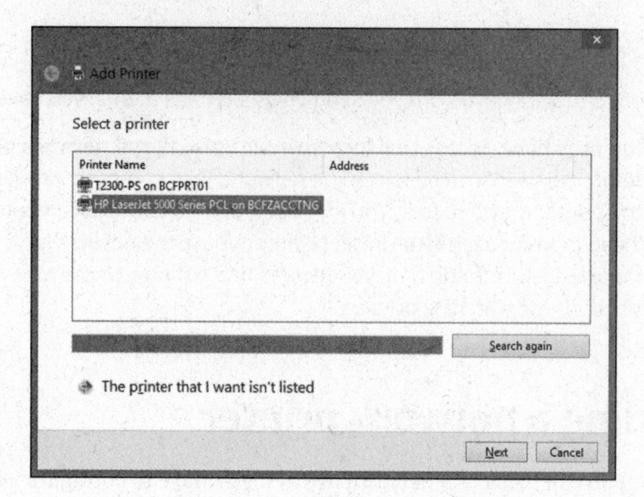

FIGURE 3-7: The Add Printer Wizard asks you to pick a printer.

The Add Printer Wizard displays a screen that shows the printer's name and asks whether you want to designate the printer as your default printer.

6. **(Optional) Designate the printer as your default printer.**
7. **Click Next to continue.**

   A final confirmation dialog box is displayed.

8. **Click Finish.**

   You're done!

TIP

Many network printers, especially newer ones, are connected directly to the network by using a built-in Ethernet card. Setting up these printers can be tricky. You may need to ask the network administrator for help in setting up this type of printer. (Some printers that are connected directly to the network have their own web addresses, such as `Printer.SimpsonFamily.com`. If that's the case, you can often set up the printer in a click or two: Use your browser to go to the printer's web page and then click a link that enables you to install the printer.)

Printing to a network printer

After you install the network printer in Windows, printing to the network printer is a snap. You can print to the network printer from any Windows program by using the Print command to summon the Print dialog box, which is usually found on the File menu. For example, Figure 3-8 shows the Print dialog box for WordPad (the free text-editing program that comes with Windows). The available printers are listed near the top of this dialog box. Choose the network printer from this list.

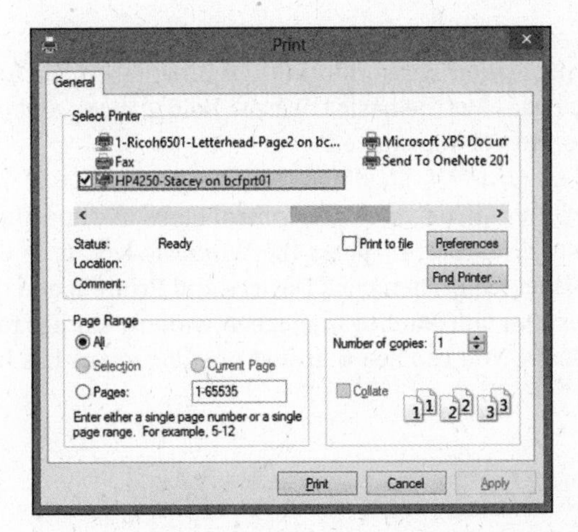

FIGURE 3-8: A typical Print dialog box.

Playing with the print queue

After you send your document to a network printer, you usually don't have to worry about it. You just go to the network printer, and voilà! Your printed document is waiting for you.

That's what happens in the ideal world. In the real world, where you and I live, all sorts of things can happen to your print job between the time you send it to the network printer and the time it prints:

- » You discover that someone else already sent a 50 trillion–page report ahead of you that isn't expected to finish printing until the national debt is paid off.
- » The price of a framis valve suddenly goes up by $2, rendering foolish the recommendations you made in your report.
- » Your boss calls and tells you that his brother-in-law will be attending the meeting, so won't you please print an extra copy of the proposal for him? Oh, and a photocopy won't do. Originals only, please.
- » You decide to take lunch, so you don't want the output to print until you get back.

Fortunately, your print job isn't totally beyond your control just because you already sent it to the network printer. You can easily change the status of jobs that you already sent. You can change the order in which jobs print, hold a job so that it doesn't print until you say so, or cancel a job.

You can probably make your network print jobs do other tricks, too: shake hands, roll over, and play dead. But the basic tricks — hold, cancel, and change the print order — are enough to get you started.

To play with the printer queue, open the Control Panel by choosing Start ⇨ Control Panel in Windows 7 or earlier; or press the Windows key, type **Control**, and the click the Control Panel icon. Then click Devices and Printers and double-click the icon for the printer that you want to manage. A window similar to the one shown in Figure 3-9 appears. You can see that just one document has been sent to the printer.

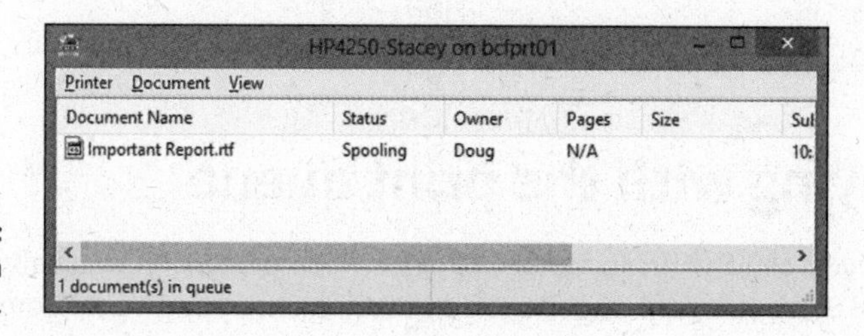

FIGURE 3-9: Managing a print queue.

To manipulate the print jobs that appear in the print queue or in the printer itself, use these tricks:

- **To temporarily stop a job from printing:** Select the job and choose Document ⇨ Pause Printing. Choose the same command again to release the job from its state of frustration and print it out, already.
- **To delete a print job:** Select the job and choose Document ⇨ Cancel Printing.
- **To stop the printer:** Choose Printer ⇨ Pause Printing. To resume, choose the command again.
- **To delete all print jobs:** Choose Printer ⇨ Purge Print Documents.
- **To cut to the front of the line:** Drag to the top of the list the print job that you want to print.

All these tips apply to your print jobs only. Unfortunately, you can't capriciously delete other people's print jobs.

Logging off the Network

After you finish using the network, log off. Logging off the network makes the network drives and printers unavailable. Your computer is still physically connected to the network (unless you cut the network cable with pruning shears; it's a bad idea — don't do it!), but the network and its resources are unavailable to you.

Here are a few other tips to keep in mind when you log off:

- After you turn off your computer, you're automatically logged off the network. After you start your computer, you have to log on again.

REMEMBER

  Logging off the network is a good idea if you're going to leave your computer unattended for a while. As long as your computer is logged in to the network, anyone can use it to access the network. And because unauthorized users can access it under your user ID, you get the blame for any damage they do.

- In Windows, you can log off the network by clicking the Start button and choosing the Log Off command. This process logs you off the network without restarting Windows:
  - *In Windows 7:* Click Start and then click the right-facing arrow that appears next to the little padlock icon.
  - *In Windows 8, 8.1, and 10:* Press Ctrl+Alt+Del and then choose Sign Out.

- » Sharing folders with network users
- » Working in the Public folder
- » Sharing your printer
- » Using Office on a network
- » Working with files offline

Chapter 4
More Ways to Use Your Network

Chapter 3 introduces you to the basics of using a network: logging on, accessing data on shared network folders, printing, and logging off. In this chapter, I go beyond these basics. You find out how to turn your computer into a server that shares its own files and printers, how to use one of the most popular network computer applications — email — and how to work with Office on a network.

Sharing Your Stuff

As you probably know, networks consist of two types of computers: client computers and server computers. In the economy of computer networks, *client computers* are the consumers — the ones that use network resources, such as shared printers and disk drives. *Servers* are the providers — the ones that offer their own printers and hard drives to the network so that the client computers can use them.

This chapter shows you how to turn your humble Windows client computer into a server computer so that other computers on your network can use your printer and any folders that you decide you want to share. In effect, your computer functions as both a client and a server at the same time. A couple of examples show how:

- It's a **client** when you send a print job to a network printer or when you access a file stored on another server's hard drive.
- It's a **server** when someone else sends a print job to your printer or accesses a file stored on your computer's hard drive.

Enabling File and Printer Sharing

Before you can share your files or your printer with other network users, you must set up a Windows File and Printer Sharing feature. Without this feature installed, your computer can be a network client but not a server.

If you're lucky, the File and Printer Sharing feature is already set up on your computer. To find out, open Windows Explorer and right-click Desktop in the Navigation pane. If the menu includes a Share With command, File and Printer Sharing is already set up, so you can skip the rest of this section. If you can't find a Share With command, follow these steps:

1. **Click the Start button, type** Network and Sharing Center, **and press Enter.**

   This step opens the Network and Sharing Center.

2. **Click Change Advanced Sharing Settings.**

   The Advanced Sharing Settings page is displayed.

3. **Click the down arrow next to the network you want to enable file and printer sharing for.**

   - *For a home computer:* Click the down arrow next to Home or Work (Windows 7) or Private (Windows 8 and later).
   - *For a computer in a public location:* Click the down arrow next to Guest or Public.
   - *For a computer connected to a domain network:* Click the down arrow next to Domain.

   Figure 4-1 shows the settings for a Domain network. The settings for a Home, Guest, or Public computer are the same.

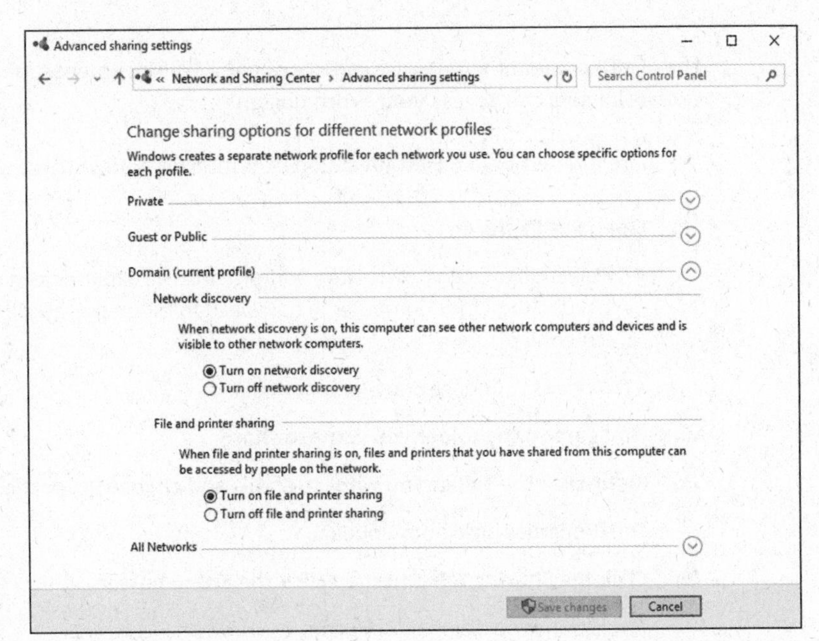

FIGURE 4-1: Enabling file and printer sharing.

WARNING

Do *not* enable file or printer sharing for the Public network. Enabling file or printer sharing on a public network exposes your computer's data to other users on the same public network.

4. **Select the Turn on File and Printer Sharing option.**
5. **Click the Save Changes button.**

   This action saves your changes and closes the Advanced Sharing Settings page.

Sharing a Folder

To enable other network users to access files that reside on your hard drive, you must designate a folder on the drive as a *shared* folder. Note that you can also share an entire drive, if you so desire. If you share an entire drive, other network users can access all the files and folders on the drive. If you share a folder, network users can access only those files that reside in the folder you share. (If the folder you share contains other folders, network users can access files in those folders, too.)

WARNING

Don't share an entire hard drive unless you want to grant *everyone on the network* the freedom to sneak a peek at every file on your hard drive. Instead, you should share just the folder or folders containing the specific documents that you want others to be able to access. For example, if you store all your Word documents in the My Documents folder, you can share your My Documents folder so that other network users can access your Word documents.

To share a folder on a desktop version of Windows, follow these steps:

1. **Open File Explorer.**
   - *Windows 10:* Open the desktop and click the File Explorer icon on the taskbar; then click Computer in the Location list on the left side of the screen.
   - *Windows 7:* Choose Start ➪ Computer.
2. **Navigate to the folder you want to share.**
3. **Right-click the folder you want to share and choose Properties.**

   The Properties dialog box appears.
4. **Click the Sharing tab and then click the Share button.**

   The File Sharing dialog box appears, as shown in Figure 4-2.

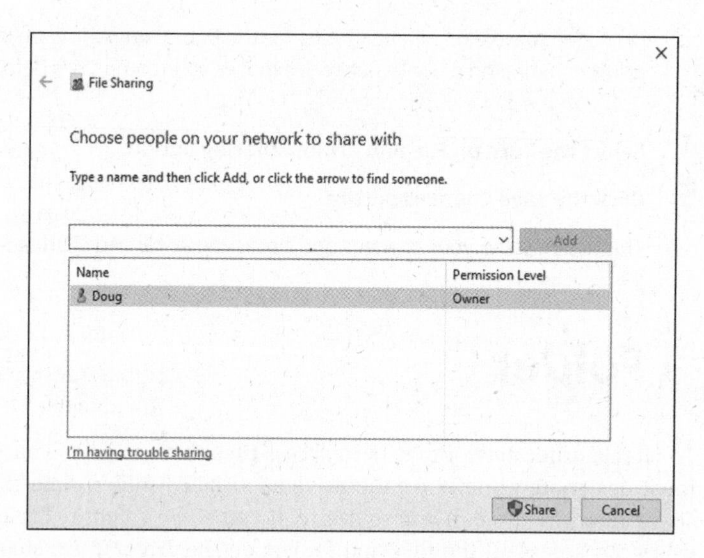

FIGURE 4-2: The File Sharing dialog box.

5. **Click the arrow in the drop-down list, choose Everyone, and then click Add.**

   This action designates that anyone on your network can access the shared folder.

   If you prefer, you can limit access to just certain users. To do so, select each person you want to grant access to and then click Add.

6. **Select the level of access you want to grant each user.**

   You can use the drop-down list in the Permission Level column to choose from three levels of access:

   - *Reader:* A reader can open files but can't modify or create new files or folders.
   - *Contributor:* A contributor can add files to the share but can change or delete only her own files.
   - *Owner:* An owner has full access to the shared folder. He or she can create, change, or delete any file in the folder.

7. **Click Share.**

   A confirmation dialog box appears to confirm that the folder has been shared.

Using the Public Folder

Windows includes an alternative method of sharing files on the network: the Public folder. The *Public folder* is a folder that's automatically designated for public access. Files you save in this folder can be accessed by other users on the network and by any user who logs on to your computer.

Before you can use the Public folder, you must enable it. Just follow the steps listed in the section "Enabling File and Printer Sharing" earlier in this chapter, but choose the Turn on Sharing option in All Networks (Windows 10) or Public Sharing Settings (earlier versions of Windows).

After you enable Public folder sharing, you can access the Public folder on your own computer in Windows 7 by choosing Start⇨Computer, expanding the Libraries item in the left pane, and then expanding the Documents, Music, Pictures, or Videos items. In Windows 8 and later, open the desktop, click the File Explorer icon on the taskbar, expand the Libraries item in the left pane, and then expand the Documents, Music, Pictures, or Videos items.

Figure 4-3 shows an example of a Public folder.

As you can see, the Public folder includes several predefined subfolders designed for sharing documents, downloaded files, music, pictures, and videos. You can use these subfolders if you want, or you can create your own subfolders to help organize the data in your Public folder.

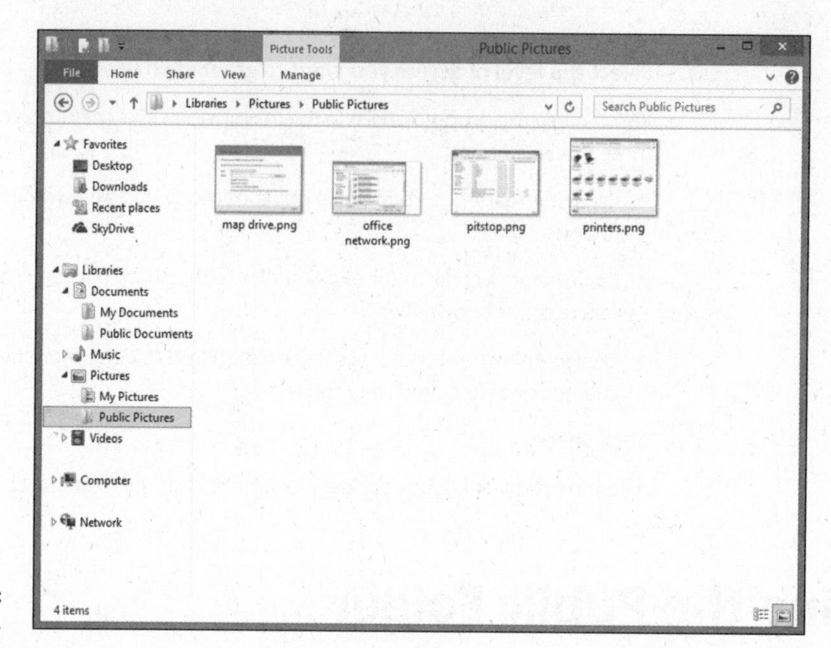

FIGURE 4-3: The Public folder.

TIP

To access the Public folder of another computer, use the techniques that I describe in Chapter 3 to either browse to the Public folder or map it to a network drive.

Sharing a Printer

Sharing a printer is much more traumatic than sharing a hard drive. When you share a hard drive, other network users access your files from time to time. When they do, you hear your drive click a few times, and your computer may hesitate for a half-second or so. The interruptions caused by other users accessing your drive are sometimes noticeable but rarely annoying.

When you share a printer, you get to see Murphy's Law in action: Your co-worker down the hall is liable to send a 140-page report to your printer just moments before you try to print a 1-page memo that has to be on the boss's desk in two minutes. The printer may run out of paper — or worse, jam — during someone else's print job — and you're expected to attend to the problem.

Although these interruptions can be annoying, sharing your printer makes a lot of sense in some situations. If you have the only decent printer in your office or workgroup, everyone will bug you to let them use it anyway. You may as well share the printer on the network. At least this way, they won't line up at your door to ask you to print their documents for them.

To share a printer, follow these steps:

1. **Open the Control Panel.**
   - *Windows 10, 8.1, and 8:* Press the Windows key, type **Control**, and then click the Control Panel icon.
   - *Windows 7:* Choose Start ➪ Control Panel.
2. **Click Devices and Printers.**
3. **Right-click the printer that you want to share and choose Printer Properties.**

   The Properties dialog box for the printer appears.
4. **Click the Sharing tab.**

   The Sharing tab appears, as shown in Figure 4-4. Notice that the options for sharing the printer are disabled.
5. **Select the Share This Printer option.**
6. **(Optional) Change the share name if you don't like the name suggested by Windows.**

   Because other computers will use the share name to identify the shared printer, pick a descriptive name.
7. **Click OK.**

   You return to the Printers folder. The icon for the printer is modified to indicate that it has been shared.

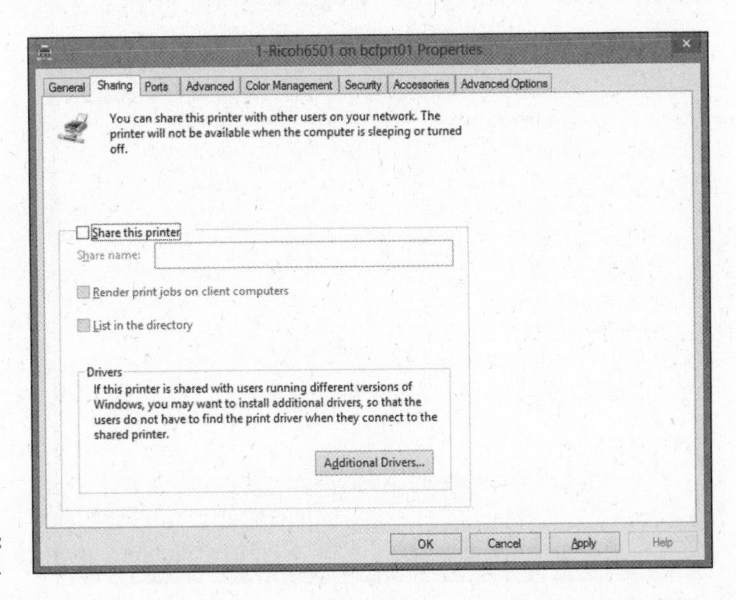

FIGURE 4-4: Sharing a printer.

To take your shared printer off the network so that other network users can't access it, follow Steps 1–4 in the preceding set of steps. Deselect the Share This Printer check box and then click OK.

Using Microsoft Office on a Network

Microsoft Office is far and away the most popular suite of application programs used on personal computers, and it includes the most common types of application programs used in an office: a word processing program (Word), a spreadsheet program (Excel), a presentation program (PowerPoint), and an excellent email program (Outlook). Depending on the version of Office you purchase, you may also get a database program (Access), a desktop publishing program (Publisher), a set of Ginsu knives (KnifePoint), and a slicer and dicer (ActiveSalsa).

TIP

To get the most from using Office on a network, you should download the Office Deployment Kit. This tool allows you to create simple click-to-run installers for Office applications. You can easily find it by searching for Office Deployment Kit in your favorite search engine.

Accessing network files

Opening a file that resides on a network drive is almost as easy as opening a file on a local drive. All Office programs use File⇨Open to summon the Open dialog box, as shown in its Excel incarnation in Figure 4-5. (The Open dialog box is nearly identical in other Office programs.)

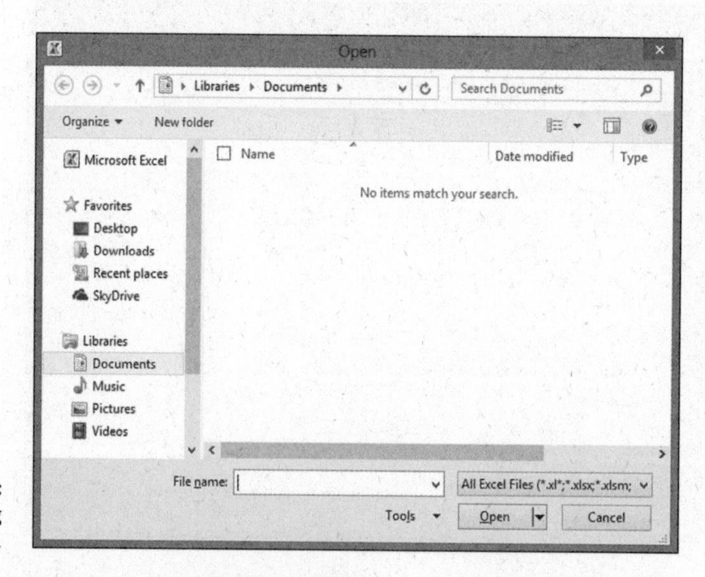

FIGURE 4-5: The Open dialog box in Excel.

To access a file that resides on a network volume that's mapped to a drive letter, all you have to do is use the drop-down list at the top of the dialog box to select the network drive.

You can map a network drive directly from the Open dialog box by navigating to the folder you want to map, right-clicking the folder, and choosing Map Network Drive.

TIP

If you try to open a file that another network user has opened already, Office tells you that the file is already in use and offers to let you open a read-only version of the file. You can read and edit the read-only version, but Office doesn't let you overwrite the existing version of the file. Instead, you have to use the Save As command to save your changes to a new file.

Using workgroup templates

Although an occasional sacrifice to the Office gods may make your computing life a bit easier, a template isn't a place of worship. Rather, a *template* is a special type of document file that holds formatting information, boilerplate text, and other customized settings that you can use as the basis for new documents.

Three Office programs — Word, Excel, and PowerPoint — enable you to specify a template whenever you create a new document. When you create a new document in Word, Excel, or PowerPoint by choosing File⇨New, you see a dialog box that lets you choose a template for the new document.

Office comes with a set of templates for the most common types of documents. These templates are grouped under the various tabs that appear across the top of the New dialog box.

In addition to the templates that come with Office, you can create your own templates in Word, Excel, and PowerPoint. Creating your own templates is especially useful if you want to establish a consistent look for documents prepared by your network users. For example, you can create a Letter template that includes your company's letterhead or a Proposal template that includes a company logo.

Office enables you to store templates in two locations. Where you put them depends on what you want to do with them:

- **The User Templates folder on each user's local disk drive:** If a particular user needs a specialized template, put it here.

» **The Workgroup Templates folder on a shared network drive:** If you have templates that you want to make available to all network users on the network server, put them here. This arrangement still allows each user to create templates that aren't available to other network users.

When you use both a User Templates folder and a Workgroup Templates folder, Office combines the templates from both folders and lists them in alphabetical order in the New dialog box. For example, the User Templates folder may contain templates named Blank Document and Web Page, and the Workgroup Templates folder may contain a template named Company Letterhead. In this case, three templates appear in the New dialog box, in this order: Blank Document, Company Letterhead, and Web Page.

To set the location of the User Templates and Workgroup Templates folders, follow these steps in Microsoft Word:

1. **In Word, create a new document or open an existing document.**

   It doesn't matter which document you open. This step is required simply because Word doesn't let you access the template folder locations unless a document is open.

2. **Choose File ⇨ Options.**

   The Word Options dialog box opens.

3. **Click the Advanced tab.**

   The Advanced options appear.

4. **Scroll down to the General section and then click the File Locations button.**

   The File Locations dialog box appears, as shown in Figure 4-6.

5. **Double-click the Workgroup Templates item.**

   This step opens a dialog box that lets you browse to the location of your template files.

6. **Browse to the template files and then click OK.**

   You return to the File Locations dialog box.

7. **Click OK to dismiss the File Locations dialog box.**

   You return to the Word Options dialog box.

8. **Click OK again.**

   The Word Options dialog box is dismissed.

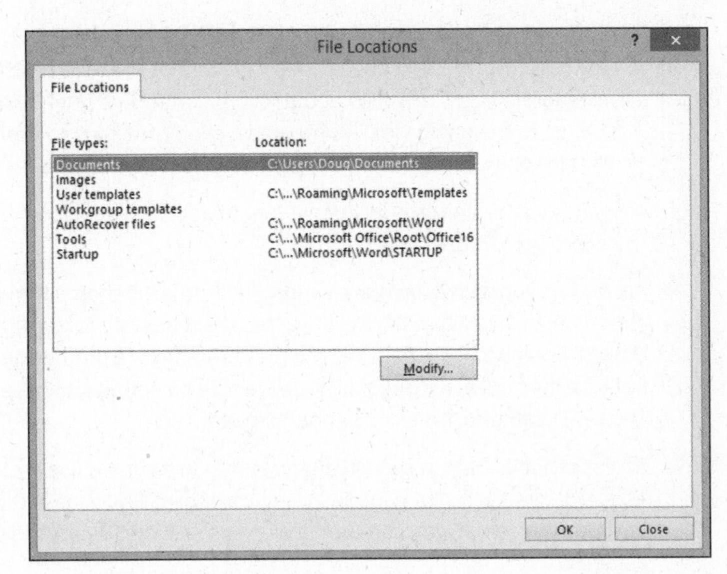

FIGURE 4-6: Setting the file locations.

TIP

Although the User Templates and Workgroup Templates settings affect Word, Excel, and PowerPoint, you can change these settings only from Word. The Options dialog boxes in Excel and PowerPoint don't show the User Templates or Workgroup Templates options.

When you install Office, the standard templates that come with Office are copied into a folder on the computer's local disk drive, and the User Templates option is set to this folder. The Workgroup Templates option is left blank. You can set the Workgroup Templates folder to a shared network folder by clicking Network Templates, clicking the Modify button, and specifying a shared network folder that contains your workgroup templates.

Networking an Access database

If you want to share a Microsoft Access database among several network users, be aware of a few special considerations. Here are the most important ones:

- When you share a database, more than one user may try to access the same record at the same time. This situation can lead to problems if two or more users try to update the record. To handle this potential traffic snarl, Access locks the record so that only one user at a time can update it. Access uses one of three methods to lock records:
  - *Edited Record:* This method locks a record whenever a user begins to edit a record. For example, if a user retrieves a record in a form that allows the record to be updated, Access locks the record while the user edits it so that other users can't edit the record until the first record is finished.

- *No Locks:* This method doesn't really mean that the record isn't locked. Instead, No Locks means that the record isn't locked until a user writes a change to the database. This method can be confusing to users because it enables one user to overwrite changes made by another user.
- *All Records:* All Records locks an entire table whenever a user edits any record in the table.

» Access lets you split a database so that the forms, queries, and reports are stored on each user's local disk drive, but the data itself is stored on a network drive. This feature can make the database run more efficiently on a network, but it's a little more difficult to set up. (To split a database, choose Tools ⇨ Database Utilities ⇨ Database Splitter.)

» Access includes built-in security features that you should use if you share an Access database from a Windows client computer. If you store the database on a domain server, you can use the server's security features to protect the database.

» Access automatically refreshes forms and datasheets every 60 seconds. That way, if one user opens a form or datasheet and another user changes the data a few seconds later, the first user sees the changes within one minute. If 60 seconds is too long (or too short) an interval, you can change the refresh rate by using the Advanced tab in the Options dialog box.

Working with Offline Files

Desktop computers are by nature stationary beasts. As a result, they're almost always connected to their networks. Laptop computers, however, are more transitory. If you have a laptop computer, you're likely to tote it around from place to place. If you have a network at work, you probably connect to the network when you're at work. But then you take the laptop computer home for the weekend, and you aren't connected to your network.

Of course, your boss wants you to spend your weekends working, so you need a way to access your important network files while you're away from the office and disconnected from the network. That's where the offline files feature comes in. It lets you access your network files even while you're disconnected from the network.

It sounds like magic, but it isn't really. Imagine how you'd work away from the network without this feature. You simply copy the files you need to work on to your laptop computer's local hard disk. Then, when you take the computer home, you work on the local copies. When you get back to the office, you connect to the network and copy the modified files back to the network server.

That's essentially how the offline files feature works, except that Windows does all the copying automatically. Windows also uses smoke and mirrors to make it look like the copies are actually on the network even though you're not connected to the network. For example, if you map a drive (drive `M:`, for example) and make it available offline, you can still access the offline copies of the file on the `M:` drive. That's because Windows knows that when you aren't connected to the network, it should redirect drive `M:` to its local copy of the drive `M:` files.

WARNING

The main complication of working with offline files is what happens when two or more users want to access the same offline files. Windows can attempt to straighten that mess out for you, but it doesn't do a great job of it. Your best bet is to not use the offline files feature with network resources that other users may want available offline, too. In other words, it's okay to make your home drive available offline because that drive is accessible only to you. I don't recommend making shared network resources available offline, though, unless they're read-only resources that don't contain files you intend to modify.

Before you can use offline files, you must first enable the Offline Files feature. To do that, open the Control Panel, double-click the Sync Center icon, and click Manage Offline Files. This brings up the Offline Files dialog box, shown in Figure 4-7. Next, click Enable Offline Files and then click OK.

After you've enabled offline files, using the offline files feature is easy: Just open the Computer folder, right-click the mapped network drive you want to make available offline, and choose Always Available Offline.

If you don't want to designate an entire mapped drive for offline access, you can designate individual folders within a mapped drive by using the same technique: Right-click the folder and then choose Always Available Offline.

When you first designate a drive or folder as available offline, Windows copies all the files on the drive or folder to local storage. Depending on how many files are involved, this process can take a while, so plan accordingly.

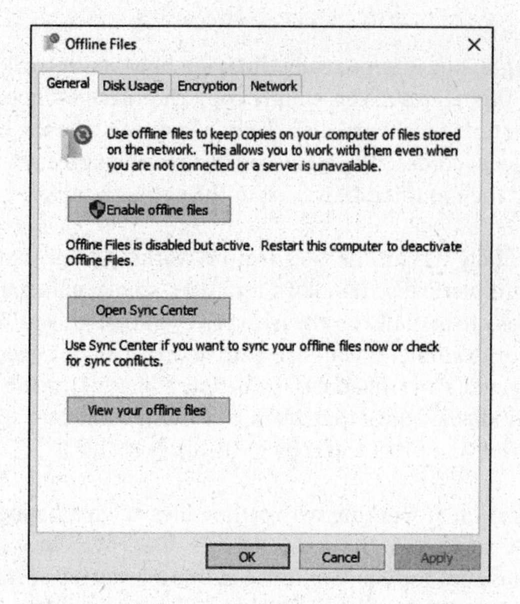

FIGURE 4-7: Enabling offline files.

After you designate a drive as available offline, Windows takes care of the rest. Each time you log on or out of the network, Windows synchronizes your offline files. Windows compares the time stamp on each file on both the server and the local copy and then copies any files that have changed.

Here are a few other thoughts to consider about offline files:

- If you want, you can force Windows to synchronize your offline files by right-clicking the drive or folder and choosing Sync.
- Make sure that no files in the folder are currently open at the time you set the Make Available Offline option. If any files are open, you'll receive an error message. You have to close the open files before you can designate the folder for offline access.
- The Properties dialog box for mapped drives includes an Offline Files tab, as shown in Figure 4-8.
- Employers love the offline files feature because it encourages their employees to work at home during evenings and weekends. In fact, every time you use the offline files feature to work at home, your boss sends Bill Gates a nickel. That's how he got so rich.

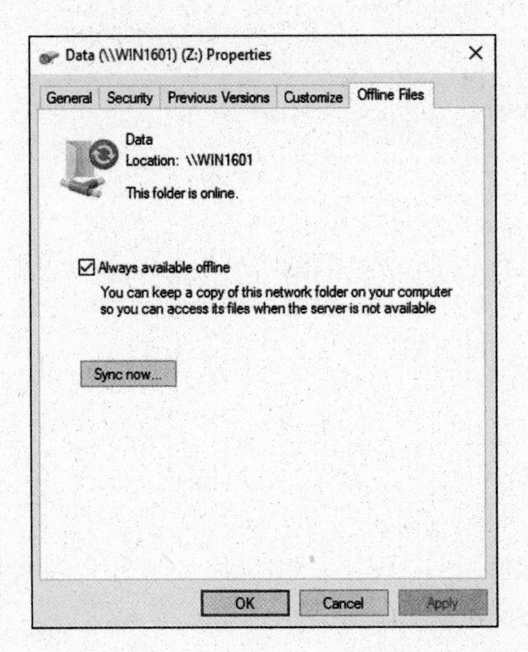

FIGURE 4-8: Offline file properties.

2 Designing Your Network

IN THIS PART . . .

Create a network plan.

Understand and use TCP/IP.

Work with cables, network adapters, switches, and other important networking components.

Use wireless devices in your network.

Get connected to the Internet.

IN THIS CHAPTER

» Making a network plan

» Taking stock of your computer stock

» Making sure that you know why you need a network

» Identifying basic network decisions that you can't avoid

Chapter 5 Planning a Network

Okay, so you're convinced that you need to network your computers. What now? Do you stop by Computers-R-Us on the way to work, install the network before drinking your morning coffee, and expect the network to be fully operational by noon?

I don't think so.

Networking your computers is just like any other worthwhile endeavor: Doing it right requires a bit of planning. This chapter helps you to think through your network before you start spending money. It shows you how to come up with a networking plan that's every bit as good as the plan that a network consultant would charge thousands of dollars for. See? This book is already saving you money!

Making a Network Plan

Before you begin any networking project, whether a new network installation or an upgrade of an existing network, start with a detailed plan. If you make technical decisions too quickly before studying all the issues that affect the project, you'll regret it. You'll discover too late that a key application won't run over the network, the network has unacceptably slow performance, or key components of the network don't work together.

Here are some general thoughts to keep in mind while you create your network plan:

- **Don't rush the plan.** The costliest networking mistakes are the ones that you make before you buy anything. Think things through and consider alternatives.
- **Write down the network plan.** The plan doesn't have to be a fancy, 500-page document. If you want to make it look good, pick up a small three-ring binder. This binder will be big enough to hold your network plan with room to spare.
- **Ask someone else to read your network plan before you buy anything.** Preferably, ask someone who knows more about computers than you do.
- **Keep the plan up to date.** If you add to the network, dig up the plan, dust it off, and update it.

TIP

"The best laid schemes of mice and men gang aft agley, and leave us naught but grief and pain for promised joy." Robert Burns lived a few hundred years before computer networks, but his famous words ring true. A network plan isn't chiseled in stone. If you discover that something doesn't work how you thought it would, that's okay. Just change your plan.

Being Purposeful

One of the first steps in planning your network is making sure that you understand why you want the network in the first place. Here are some of the more common reasons for creating or upgrading a network, all of them quite valid:

- Everyone in the office needs access to the Internet. Probably the most common reason for setting up a small network is to share an Internet connection. And even in larger networks, shared Internet access is one of the primary benefits of the network.
- My co-worker and I exchange files using flash drives just about every day. With a network, it would be easier to trade files.
- I don't want to buy everyone a color laser printer when I know the one we have now just sits there taking up space most of the day. So wouldn't investing in a network be better than buying a color laser printer for every computer?
- Business is so good that one person typing in orders eight hours each day can't keep up. If the sales and accounting data existed on a network server, I could hire another person to help, and I won't have to pay overtime to either person.

- My sister-in-law just upgraded the network at her office, and I don't want her to think that I'm behind the times.
- My existing network performs like it's made of kite string and tin cans. I should have upgraded it five years ago to speed up access to shared files, provide better security, and easier management.

Make sure that you identify all the reasons why you think you need a network and then write them down. Don't worry about winning the Pulitzer Prize for your stunning prose. Just make sure that you write down what you expect a network to do for you.

If you were making a 500-page networking proposal, you'd place the description of why a network is needed in a tabbed section labeled "Justification." In your network binder, file the description under "Purpose."

TIP

As you consider the reasons why you need a network, you may conclude that you don't need a network after all. That's okay. You can always use the binder for your stamp collection.

Taking Stock

One of the most challenging parts of planning a network is figuring out how to work with the computers that you already have. In other words, how do you get from here to there? Before you can plan how to get "there," you have to know where "here" is. In other words, you have to take a thorough inventory of your current computers.

What you need to know

You need to know the following information about each of your computers. Don't sweat it right now if some of these terms don't make sense. They're all just pieces of the puzzle.

- **The processor type and, if possible, its clock speed:** It would be nice if each of your computers had a shiny new Core i7 eight-core processor. In most cases, though, you find a mixture of computers: some new, some old, some borrowed, some blue. You may even find a few archaic Pentium computers.

  You can't usually tell what kind of processor that a computer has just by looking at the computer's case. The easiest way to find your computer's processor model is to open Windows Explorer, right-click This PC, and then choose Properties. A Properties page appears that includes the type of processor the computer has. For example, Figure 5-1 shows the Properties page for a computer that uses a Core i7 processor.

- **The amount of memory:** This information can also be found on the computer's Properties page, as shown in Figure 5-1.
- **The size of the hard drive and the arrangement of its partitions:** To find out the size of your computer's hard drive in Windows 10, open the File Explorer (found in the desktop taskbar), and then right-click the drive icon and choose the Properties command from the shortcut menu that appears. (The procedure for earlier versions of Windows is similar.) Figure 5-1 shows the Properties dialog box for a 126GB disk drive that has about 115GB of free space.

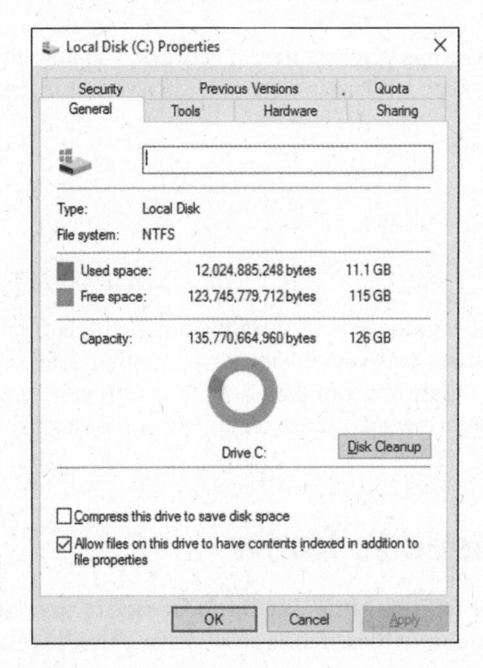

FIGURE 5-1: The Properties dialog box for a disk drive.

If your computer has more than one hard drive, Windows lists an icon for each drive in the Computer window. Jot down the size and amount of free space available on each drive.

- **The operating system version:** This you can also deduce from the System Properties dialog box. For example, the Properties page shown in Figure 5-2 indicates that the computer is running Windows 10 Pro.

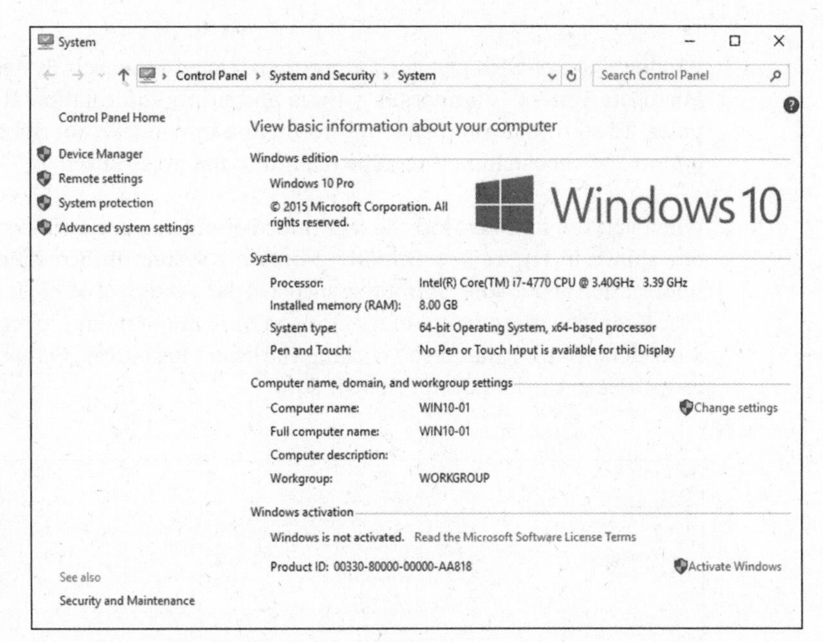

FIGURE 5-2: The Properties page for a computer with Core i7 processor and 8GB of RAM.

- **What kind of printer, if any, is attached to the computer:** Usually, you can tell just by looking at the printer. You can also tell by double-clicking the Devices and Printers icon in Control Panel.
- **Any other devices connected to the computer:** A DVD or Blu-ray drive? Scanner? External disk or tape drive? Video camera? Battle droid? Hot tub?
- **What software is used on the computer:** Microsoft Office? AutoCAD? QuickBooks? Make a complete list and include version numbers.
- **Does the computer have wireless capability?** Nearly all laptops do. Most desktops do not, but you can always add an inexpensive USB wireless adapter if you want your network to be entirely wireless.

Programs that gather information for you

Gathering information about your computers is a lot of work if you have more than a few computers to network. Fortunately, several software programs are available that can automatically gather the information for you. These programs inspect various aspects of a computer, such as the CPU type and speed, amount of RAM, and the size of the computer's hard drives. Then they show the information on the screen and give you the option of saving the information to a hard drive file or printing it.

Windows comes with just such a program, called Microsoft System Information. Microsoft System Information gathers and prints information about your computer. To start Microsoft System Information in Windows 10, right-click the Start button and choose Run, then type msinfo32 and press Enter.

When you fire up Microsoft System Information, you see a window similar to the one shown in Figure 5-3. Initially, Microsoft System Information displays basic information about your computer, such as your version of Microsoft Windows, the processor type, the amount of memory on the computer, and so on. You can obtain more detailed information by clicking Hardware Resources, Components, or other categories in the left side of the window.

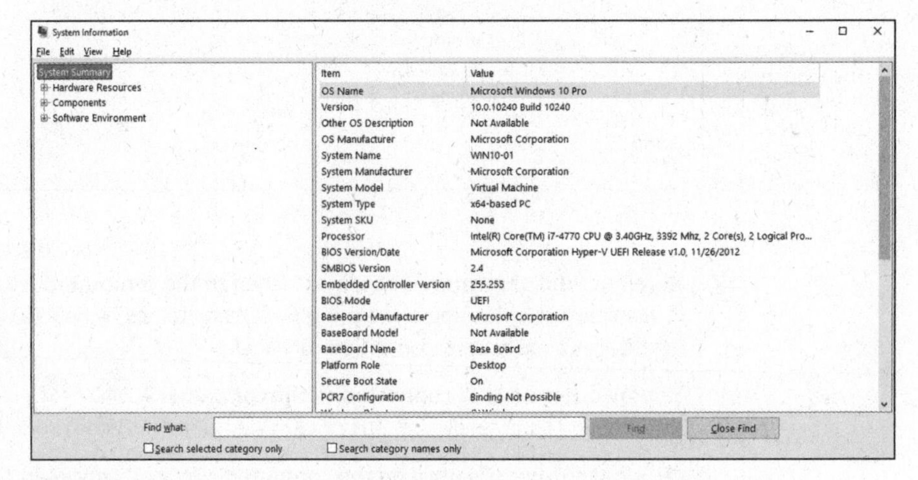

FIGURE 5-3: Let the System Information program gather the data you need.

To Dedicate or Not to Dedicate: That Is the Question

One of the most basic questions that a network plan must answer is whether the network will have one or more dedicated servers or rely completely on peer-to-peer networking. If the only reason for purchasing your network is to share a printer and exchange an occasional file, you may not need a dedicated server computer. In that case, you can create a peer-to-peer network by using the computers that you already have. However, all but the smallest networks will benefit from having a separate, dedicated server computer.

- **Using a dedicated server computer makes the network faster, easier to work with, and more reliable.** Consider what happens, though, when the user of a server computer that doubles as a workstation decides to turn off the computer, not realizing that someone else is accessing files on his hard drive.

- **You don't necessarily have to use your biggest and fastest computer as your server computer.** I've seen networks where the slowest computer on the network is the server. This advice is especially true when the server is mostly used to share a printer or to store a small number of shared files. So if you need to buy a computer for your network, consider promoting one of your older computers to be the server and using the new computer as a client.

Assuming that your network will require one or more dedicated servers, you should next consider what types of servers the network will need. In some cases, a single server computer can fill one or more of these roles. Whenever possible, limit each server computer to a single server function.

File servers

File servers provide centralized disk storage that can be conveniently shared by client computers on the network. The most common task of a file server is to store shared files and programs. For example, members of a small workgroup can use disk space on a file server to store their Microsoft Office documents.

File servers must ensure that two users don't try to update the same file at the same time. The file servers do this by *locking* a file while a user updates the file so that other users can't access the file until the first user finishes. For document files (for example, word processing or spreadsheet files), the whole file is locked. For database files, the lock can be applied just to the portion of the file that contains the record or records being updated.

Print servers

Sharing printers is one of the main reasons that many small networks exist. Although it isn't necessary, a server computer can be dedicated for use as a *print server*, whose sole purpose is to collect information being sent to a shared printer by client computers and print it in an orderly fashion.

- A single computer may double as both a file server and a print server, but performance is better if you use separate print and file server computers.

- With inexpensive inkjet printers running about $100 each, just giving each user his or her own printer is tempting. However, you get what you pay for. Instead of buying $100 printers for 15 users, you may be better off buying one high-speed $1,500 laser printer and sharing it. The $1,500 laser printer will be much faster, will probably produce better-looking output, and will be less expensive to operate.

Web servers

A *web server* is a server computer that runs software that enables the computer to host an Internet website. The two most popular web server programs are Microsoft's IIS (Internet Information Services) and Apache, an open source web server managed by the Apache Software Foundation.

Mail servers

A *mail server* is a server that handles the network's email needs. It is configured with email server software, such as Microsoft Exchange Server. Exchange Server is designed to work with Microsoft Outlook, the email client software that comes with Microsoft Office.

Most mail servers actually do much more than just send and receive email. For example, here are some of the features that Exchange Server offers beyond simple email:

- Collaboration features that simplify the management of collaborative projects
- Audio and videoconferencing
- Chat rooms and instant messaging (IM) services
- Microsoft Exchange Forms Designer, which lets you develop customized forms for applications, such as vacation requests or purchase orders

TIP

An increasingly popular alternative to setting up your own mail server is to subscribe to Microsoft's Exchange Online service. With Exchange Online, Microsoft takes care of all the details of setting up and maintaining an Exchange server. Your Microsoft Outlook clients are configured to connect to the Exchange Online service through the Internet rather than a local Exchange server within your network.

Database servers

A *database server* is a server computer that runs database software, such as Microsoft's SQL Server 2019. Database servers are usually used along with customized business applications, such as accounting or marketing systems.

Application servers

An *application server* is a server computer that runs a specific application. For example, you might use an accounting application that requires its own server. In that case, you'll need to dedicate a server to the accounting application.

License servers

Some organizations use software that requires licenses that are distributed from a centralized license server. For example, engineering firms often use computer-aided design (CAD) software such as AutoCAD that requires a license server. In that case, you'll need to set up a server to handle the licensing function.

Choosing a Server Operating System

If you determine that your network will require one or more dedicated servers, the next step is to determine what network operating system those servers should use. If possible, all the servers should use the same network operating system (NOS) so that you don't find yourself supporting different operating systems.

Although you can choose from many network operating systems, from a practical point of view, your choices are limited to the following:

- Windows Server 2019 or 2016
- Linux or another version of Unix

For more information, see Chapter 11.

Planning the Infrastructure

You also need to plan the details of how you will connect the computers in the network. This task includes determining whether you'll use a wired or wireless network (or both), what networking devices you need to service your network, and how you'll connect your network to the Internet.

If you chose to use a wired network, you can choose between Cat5e and Cat6 cabling. Cat6 is a bit more expensive than Cat5e but supports faster network speeds. Beyond that basic choice, you have many additional decisions to make:

- » Will you use inexpensive consumer-grade network switches such as those you can buy at a consumer electronics store or an office supply store, or will you want professional-grade switches, which are more expensive but provide advanced management features?
- » Where will you place the switch — on a desktop somewhere within the group or in a central wiring closet?
- » How many client computers and other devices will you place on each switch, and how many switches will you need to support all of these computers and other devices?
- » If you need more than one switch, what type of cabling will you use to connect the switches to one another?

For more information about network cabling, see Chapter 7.

TIP

If you're installing new network cable, don't scrimp on the cable itself. Because installing network cable is a labor-intensive task, the cost of the cable itself is a small part of the total cable installation cost. Consider using Cat6 cable instead of Cat5e. And in many cases, you should consider installing more cable than you actually need. For example, if one area of your office has four desks, don't pull just four cables to that location — pull six or even eight cables. The cost of labor for pulling the cables is more than the cost of the cables themselves, and it doesn't take much more labor to pull a few additional cables to the same location.

Drawing Diagrams

One of the most helpful techniques for creating a network plan is to draw a picture of it. The diagram can be a detailed floor plan, showing the actual location of each network component. This type of diagram is sometimes called a "physical map." If you prefer, the diagram can be a *logical map*, which is more abstract and

Picasso-like. Any time you change the network layout, update the diagram. Also include a detailed description of the change, the date that the change was made, and the reason for the change.

You can diagram very small networks on the back of a napkin, but if the network has more than a few computers, you'll want to use a drawing program to help you create the diagram. You can use a professional drawing program such as Microsoft Visio, but you can also find simpler and less expensive online drawing tools to help you document your network. Figure 5-4 shows a network diagram drawn with an inexpensive online tool called Lucidchart (`www.lucidchart.com`).

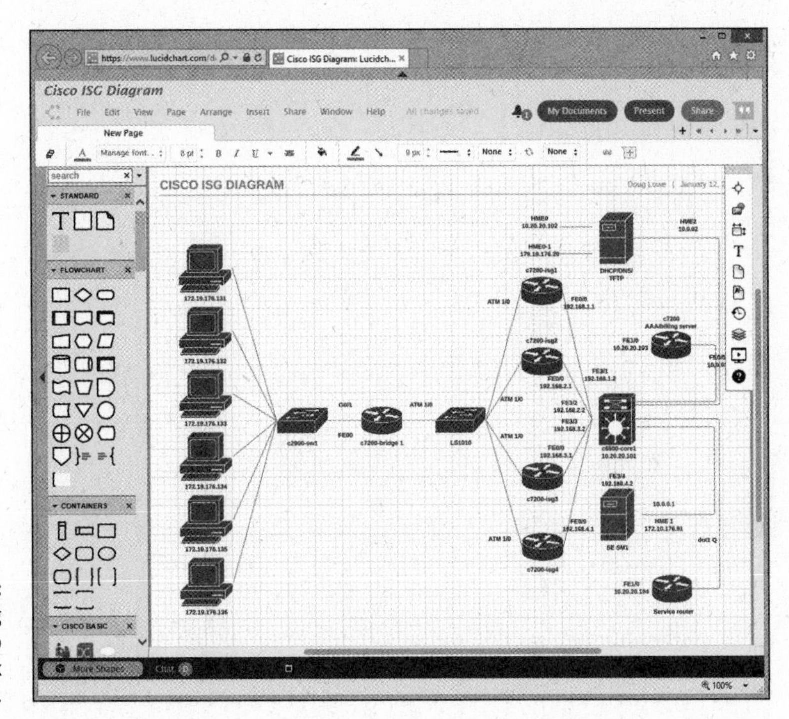

FIGURE 5-4: Using Lucidchart to draw a network diagram.

- » Digging into IP addresses
- » Finding out how subnetting works
- » Understanding private and public IP addresses
- » Looking at network address translation
- » Finding out how DHCP works
- » Understanding how DNS works

Chapter 6
Dealing with TCP/IP

Transmission Control Protocol/Internet Protocol — TCP/IP — is the basic protocol by which computers on a network talk to each other. Without TCP/IP, networks wouldn't work. In this chapter, I introduce you to the most important concepts of TCP/IP.

WARNING

This chapter is far and away the most technical chapter in this book. It helps you examine the binary system, the details of how IP addresses are constructed, how subnetting works, and how two of the most important TCP/IP services — Dynamic Host Configuration Protocol (DHCP) and Domain Name System (DNS) — work. You don't need to understand every detail in this chapter to set up a simple TCP/IP network. However, the more you understand the information in this chapter, the more TCP/IP will start to make sense. Be brave.

Understanding Binary

Before you can understand the details of how TCP/IP — in particular, IP — addressing works, you need to understand how the binary numbering system works because binary is the basis of IP addressing. If you already understand binary, please skip right over this section to the next main section, "Introducing IP Addresses." I don't want to bore you with stuff that's too basic.

Counting by ones

The *binary* counting system uses only two numerals: 0 and 1. In the decimal system to which most people are accustomed, you use ten numerals: 0 through 9. In an ordinary decimal number, such as 3,482, the rightmost digit represents ones; the next digit to the left, tens; the next, hundreds; the next, thousands; and so on. These digits represent powers of ten: first 10^0 (which is 1); next, 10^1 (10); then 10^2 (100); then 10^3 (1,000); and so on.

In binary, you have only two numerals rather than ten, which is why binary numbers look somewhat monotonous, as in 110011, 101111, and 100001.

The positions in a binary number (called *bits* rather than *digits*) represent powers of two rather than powers of ten — working from right to left, each bit represents the decimal values 1, 2, 4, 8, 16, 32, and so on. To figure the decimal value of a binary number, you multiply each bit by its corresponding power of two and then add the results. The decimal value of binary 10111, for example, is calculated as follows:

$$
\begin{array}{l}
1 \times 2^0 = 1 \times 1 = 1 \\
+\,1 \times 2^1 = 1 \times 2 = 2 \\
+\,1 \times 2^2 = 1 \times 4 = 4 \\
+\,0 \times 2^3 = 0 \times 8 = 0 \\
+\,1 \times 2^4 = 1 \times 16 = \underline{16} \\
\qquad\qquad 23
\end{array}
$$

Fortunately, a computer is good at converting a number between binary and decimal — so good, in fact, that you're unlikely ever to need to do any conversions yourself. The point of knowing binary isn't to be able to look at a number, such as 1110110110110, and say instantly, "Ah! Decimal 7,606!" (If you could do that, you would probably be interviewed on the *Today* show, and they would even make a movie about you.)

Instead, the point is to have a basic understanding of how computers store information and — most important — to understand how the hexadecimal counting system works (which I describe in the following section).

Here are some of the more interesting characteristics of binary and how the system is similar to and differs from the decimal system:

- » The number of bits allotted for a binary number determines how large that number can be. If you allot eight bits, the largest value that number can store is 11111111, which happens to be 255 in decimal.
- » To quickly determine how many different values you can store in a binary number of a given length, use the number of bits as an exponent of two. An eight-bit binary number, for example, can hold 2^8 values. Because 2^8 is 256, an 8-bit number can have any of 256 different values, which is why a byte, which is eight bits, can have 256 different values.

TIP

- » This powers-of-two concept is why computers don't use nice, even, round numbers in measuring such values as memory or disk space. A value of 1K, for example, isn't an even 1,000 bytes — it's 1,024 bytes because 1,024 is 2^{10}. Similarly, 1MB isn't an even 1,000,000 bytes but rather is 1,048,576 bytes, which happens to be 2^{20}.

Doing the logic thing

One of the great things about binary is that it's very efficient at handling special operations called *logical operations.* Four basic logical operations exist, although additional operations are derived from the basic four operations. Three of the operations — AND, OR, and XOR — compare two binary digits (bits). The fourth (NOT) works on just a single bit.

The following list summarizes the basic logical operations:

- » AND: An AND operation compares two binary values. If both values are 1, the result of the AND operation is 1. If one value is 0 or both of the values are 0, the result is 0.
- » OR: An OR operation compares two binary values. If at least one of the values is 1, the result of the OR operation is 1. If both values are 0, the result is 0.
- » XOR: An XOR operation compares two binary values. If exactly one of them is 1, the result is 1. If both values are 0 or if both values are 1, the result is 0.

» NOT: The NOT operation doesn't compare two values. Instead, it simply changes the value of a single binary value. If the original value is 1, NOT returns 0. If the original value is 0, NOT returns 1.

TIP

Logical operations are applied to binary numbers that have more than one binary digit by applying the operation one bit at a time. The easiest way to do this manually is to

1. *Line one of the two binary numbers on top of the other.*
2. *Write the result of the operation beneath each binary digit.*

The following example shows how you calculate 10010100 AND 11001101:

```
10010100
AND 11001101
10000100
```

As you can see, the result is 10000100.

Introducing IP Addresses

An *IP address* is a number that uniquely identifies every host on an IP network. IP addresses operate at the Network layer of the TCP/IP protocol stack, so they're independent of lower-level addresses, such as MAC addresses. (MAC stands for *Media Access Control.*)

IP addresses are 32-bit binary numbers, which means that theoretically, a maximum of something in the neighborhood of 4 billion unique host addresses can exist throughout the Internet. You'd think that'd be enough, but TCP/IP places certain restrictions on how IP addresses are allocated. These restrictions severely limit the total number of usable IP addresses, and about half of the total available IP addresses have already been assigned. However, new techniques for working with IP addresses have helped to alleviate this problem, and a new standard for 128-bit IP addresses (known as *IPv6*) is on the verge of winning acceptance.

Networks and hosts

The primary purpose of Internet Protocol (IP) is to enable communications between networks. As a result, a 32-bit IP address consists of two parts:

» **The network ID (or network address):** Identifies the network on which a host computer can be found

» **The host ID (or host address):** Identifies a specific device on the network indicated by the network ID

Most of the complexity of working with IP addresses has to do with figuring out which part of the complete 32-bit IP address is the network ID and which part is the host ID. The original IP specification uses the *address classes* system to determine which part of the IP address is the network ID and which part is the host ID. A newer system — classless IP addresses — is rapidly taking over the address classes system. You come to grips with both systems later in this chapter.

The dotted-decimal dance

IP addresses are usually represented in a format known as *dotted-decimal notation.* In dotted-decimal notation, each group of eight bits — an *octet* — is represented by its decimal equivalent. For example, consider the following binary IP address:

```
11000000101010001000100000011100
```

The dotted-decimal equivalent to this address is

```
192.168.136.28
```

Here, 192 represents the first eight bits (`11000000`); 168, the second set of eight bits (`10101000`); 136, the third set of eight bits (`10001000`); and 28, the last set of eight bits (`00011100`). This is the format in which you usually see IP addresses represented.

Classifying IP Addresses

When the original designers of the IP protocol created the IP addressing scheme, they could have assigned an arbitrary number of IP address bits for the network ID. The remaining bits would then be used for the host ID. For example, the designers may have decided that half of the address (16 bits) would be used for the network and the remaining 16 bits would be used for the host ID. The result of that scheme would be that the Internet could have a total of 65,536 networks, and each of those networks could have 65,536 hosts.

In the early days of the Internet, this scheme probably seemed like several orders of magnitude more than would ever be needed. However, the IP designers realized from the start that few networks would actually have tens of thousands of hosts. Suppose that a network of 1,000 computers joins the Internet and is assigned one of these hypothetical network IDs. Because that network uses only 1,000 of its 65,536 host addresses, more than 64,000 IP addresses would be wasted.

As a solution to this problem, the idea of IP address classes was introduced. The IP protocol defines five different address classes: A, B, C, D, and E. Each of the first three classes, A–C, uses a different size for the network ID and host ID portion of the address. Class D is for a special type of address called a *multicast address.* Class E is an experimental address class that isn't used.

The first four bits of the IP address are used to determine into which class a particular address fits:

- » If the first bit is 0, the address is a Class A address.
- » If the first bit is 1 and the second bit is 0, the address is a Class B address.
- » If the first two bits are both 1 and the third bit is 0, the address is a Class C address.
- » If the first three bits are all 1 and the fourth bit is 0, the address is a Class D address.
- » If the first four bits are all 1, the address is a Class E address.

Because Class D and E addresses are reserved for special purposes, I focus the rest of this discussion on Class A, B, and C addresses. Table 6-1 summarizes the details of each address class.

TABLE 6-1 IP Address Classes

Class	Address Range	Starting Bits	Length of Network ID	Number of Networks	Number of Hosts
A	1–126.*x*.*y*.*z*	0	8	126	16,777,214
B	128–191.*x*.*y*.*z*	10	16	16,384	65,534
C	192–223.*x*.*y*.*z*	110	24	2,097,152	254

Class A addresses

Class A addresses are designed for very large networks. In a Class A address, the first octet of the address is the network ID, and the remaining three octets are the host ID. Because only eight bits are allocated to the network ID and the first of these bits is used to indicate that the address is a Class A address, only 126 Class A networks can exist in the entire Internet. However, each Class A network can accommodate more than 16 million hosts.

Only about 40 Class A addresses are assigned to companies or organizations. The rest are either reserved for use by the Internet Assigned Numbers Authority (IANA) or are assigned to organizations that manage IP assignments for geographic regions, such as Europe, Asia, and Latin America.

Class B addresses

In a Class B address, the first two octets of the IP address are used as the network ID, and the second two octets are used as the host ID. Thus, a Class B address comes close to my hypothetical scheme of splitting the address down the middle, using half for the network ID and half for the host ID. It isn't identical to this scheme, however, because the first two bits of the first octet are required to be 10, to indicate that the address is a Class B address. Thus, a total of 16,384 Class B networks can exist. All Class B addresses fall within the range `128.x.y.z` to `191.x.y.z`. Each Class B address can accommodate more than 65,000 hosts.

The problem with Class B networks is that even though they're much smaller than Class A networks, they still allocate far too many host IDs. Very few networks have tens of thousands of hosts. Thus, the careless assignment of Class B addresses can lead to a large percentage of the available host addresses being wasted on organizations that don't need them.

Class C addresses

In a Class C address, the first three octets are used for the network ID, and the fourth octet is used for the host ID. With only eight bits for the host ID, each Class C network can accommodate only 254 hosts. However, with 24 network ID bits, Class C addresses allow for more than 2 million networks.

The problem with Class C networks is that they're too small. Although few organizations need the tens of thousands of host addresses provided by a Class B address, many organizations need more than a few hundred. The large discrepancy between Class B networks and Class C networks led to the development of subnetting, which I describe in the next section.

WHAT ABOUT IPV6?

Most of the current Internet is based on version 4 of the Internet Protocol, also known as IPv4. IPv4 has served the Internet well for more than 20 years. However, the growth of the Internet has put a lot of pressure on IPv4's limited 32-bit address space. This chapter describes how IPv4 has evolved to make the best possible use of 32-bit addresses, but eventually all the addresses will be assigned; the IPv4 address space will be filled to capacity. When that happens, the Internet will have to migrate to the next version of IP, known as IPv6.

IPv6 is also called *IP next generation,* or *IPng,* in honor of the favorite television show of most Internet gurus, *Star Trek: The Next Generation.*

IPv6 offers several advantages over IPv4, but the most important is that it uses 128 bits for Internet addresses rather than 32 bits. The number of host addresses possible with 128 bits is a number so large that it would make Carl Sagan proud. It doesn't just double or triple the number of available addresses. Just for the fun of it, here's the number of unique Internet addresses provided by IPv6:

340,282,366,920,938,463,463,374,607,431,768,211,456

This number is so large that it defies understanding. If the IANA had been around at the creation of the universe and started handing out IPv6 addresses at a rate of one per millisecond, it would now, 15 billion years later, have not yet allocated even 1 percent of the available addresses.

Unfortunately, the transition from IPv4 to IPv6 has been a slow one. Thus, the Internet will continue to be driven by IPv4 for at least a few more years.

Subnetting

Subnetting is a technique that lets network administrators use the 32 bits available in an IP address more efficiently by creating networks that aren't limited to the scales provided by Class A, B, and C IP addresses. With subnetting, you can create networks with more realistic host limits.

Subnetting provides a more flexible way to designate which portion of an IP address represents the network ID and which portion represents the host ID. With standard IP address classes, only three possible network ID sizes exist: 8 bits for

Class A, 16 bits for Class B, and 24 bits for Class C. Subnetting lets you select an arbitrary number of bits to use for the network ID.

Two reasons compel me to use subnetting. The first is to allocate the limited IP address space more efficiently. If the Internet were limited to Class A, B, or C addresses, every network would be allocated 254, 65,000, or 16 million IP addresses for host devices. Although many networks with more than 254 devices exist, few (if any) exist with 65,000, let alone 16 million. Unfortunately, any network with more than 254 devices would need a Class B allocation and probably waste tens of thousands of IP addresses.

The second reason for subnetting is that even if a single organization has thousands of network devices, operating all those devices with the same network ID would slow the network to a crawl. The way TCP/IP works dictates that all the computers with the same network ID must be on the same physical network. The physical network comprises a single *broadcast domain*, which means that a single network medium must carry all the traffic for the network. For performance reasons, networks are usually segmented into broadcast domains that are smaller than even Class C addresses provide.

Subnets

A *subnet* is a network that falls within another (Class A, B, or C) network. Subnets are created by using one or more of the Class A, B, or C host bits to extend the network ID. Thus, rather than the standard 8-, 15-, or 24-bit network ID, subnets can have network IDs of any length.

Figure 6-1 shows an example of a network before and after subnetting has been applied. In the unsubnetted network, the network has been assigned the Class B address `144.28.0.0`. All the devices on this network must share the same broadcast domain.

In the second network, the first four bits of the host ID are used to divide the network into two small networks, identified as subnets 16 and 32. To the outside world (that is, on the other side of the router), these two networks still appear to be a single network identified as `144.28.0.0`. For example, the outside world considers the device at `144.28.16.22` to belong to the `144.28.0.0` network. As a result, a packet sent to this device is delivered to the router at `144.28.0.0`. The router then considers the subnet portion of the host ID to decide whether to route the packet to subnet 16 or subnet 32.

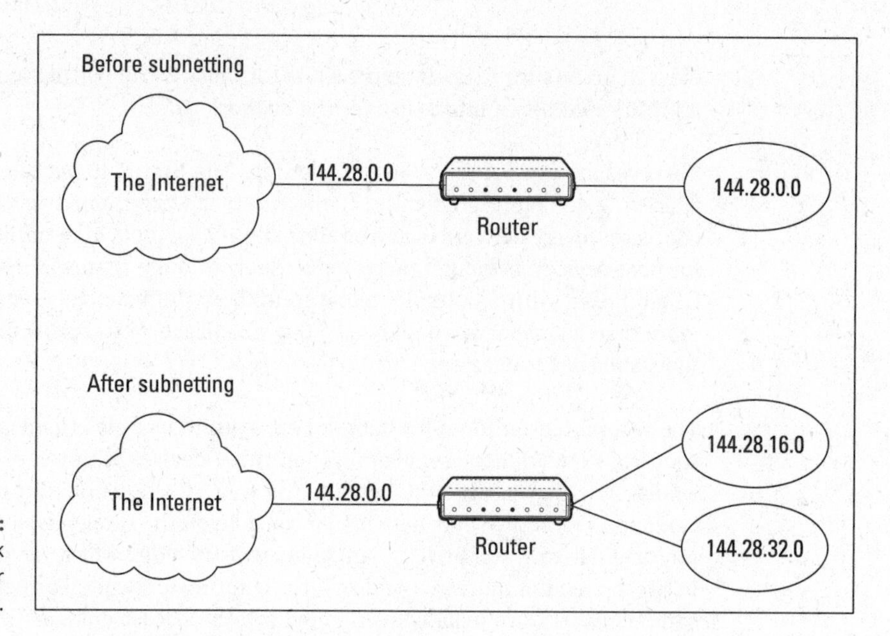

FIGURE 6-1: A network before and after subnetting.

Subnet masks

For subnetting to work, the router must be told which portion of the host ID to use for the subnet's network ID. This little sleight of hand is accomplished by using another 32-bit number, known as a *subnet mask.* Those IP address bits that represent the network ID are represented by a 1 in the mask, and those bits that represent the host ID appear as a 0 in the mask. As a result, a subnet mask always has a consecutive string of ones on the left, followed by a string of zeros.

For example, the subnet mask for the subnet, as shown in Figure 6-1, in which the network ID consists of the 16-bit network ID plus an additional 4-bit subnet ID, would look like this:

```
11111111 11111111 11110000 00000000
```

In other words, the first 20 bits are ones; the remaining 12 bits are zeros. Thus, the complete network ID is 20 bits in length, and the actual host ID portion of the subnetted address is 12 bits in length.

To determine the network ID of an IP address, the router must have both the IP address and the subnet mask. The router then performs a bitwise operation called a *logical AND* on the IP address to extract the network ID. To perform a logical AND, each bit in the IP address is compared to the corresponding bit in the subnet mask. If both bits are 1, the resulting bit in the network ID is set to 1. If either of the bits is 0, the resulting bit is set to 0.

For example, here's how the network address is extracted from an IP address using the 20-bit subnet mask from the previous example:

```
144 . 28 . 16 . 17
IP address: 10010000 00011100 00100000 00001001
Subnet mask: 11111111 11111111 11110000 00000000
Network ID: 10010000 00011100 00100000 00000000
144 . 28 . 16 . 0
```

Thus, the network ID for this subnet is `144.28.16.0`.

The subnet mask itself is usually represented in dotted-decimal notation. As a result, the 20-bit subnet mask used in the previous example would be represented as `255.255.240.0`:

```
Subnet mask: 11111111 11111111 11110000 00000000
255 . 255 . 240 . 0
```

TIP

Don't confuse a subnet mask with an IP address. A subnet mask doesn't represent any device or network on the Internet. It's just a way of indicating which portion of an IP address should be used to determine the network ID. (You can spot a subnet mask right away because the first octet is always `255`, and `255` isn't a valid first octet for any class of IP address.)

The great subnet roundup

You should know about a few additional restrictions that are placed on subnet masks — in particular:

- The minimum number of network ID bits is eight. As a result, the first octet of a subnet mask is always 255.
- The maximum number of network ID bits is 30. You have to leave at least two bits for the host ID portion of the address, to allow for at least two hosts. If you used all 32 bits for the network ID, that would leave no bits for the host ID. Obviously, that doesn't work. Leaving just one bit for the host ID doesn't work, either. That's because a host ID of all ones is reserved for a broadcast address — and all zeros refers to the network itself. Thus, if you used 31 bits for the network ID and left only one for the host ID, host ID 1 would be used for the broadcast address and host ID 0 would be the network itself, leaving no room for actual hosts. That's why the maximum network ID size is 30 bits.
- Because the network ID is always composed of consecutive bits set to 1, only nine values are possible for each octet of a subnet mask (including counting 0). For your reference, these values are listed in Table 6-2.

TABLE 6-2 The Eight Subnet Octet Values

Binary Octet	Decimal	Binary Octet	Decimal
00000000	0	11111000	248
10000000	128	11111100	252
11000000	192	11111110	254
11100000	224	11111111	255
11110000	240		

Private and public addresses

Any host with a direct connection to the Internet must have a globally unique IP address. However, not all hosts are connected directly to the Internet. Some are on networks that aren't connected to the Internet. Some hosts are hidden behind firewalls, so their Internet connection is indirect.

Several blocks of IP addresses are set aside just for this purpose — for use on private networks that aren't connected to the Internet or to use on networks hidden behind a firewall. Three such ranges of addresses exist, as summarized in Table 6-3. Whenever you create a private TCP/IP network, use IP addresses from one of these ranges.

TABLE 6-3 Private Address Spaces

Subnet Mask	Address Range
`255.0.0.0`	`10.0.0.1-10.255.255.254`
`255.255.240.0`	`172.16.1.1-172.31.255.254`
`255.255.0.0`	`192.168.0.1-192.168.255.254`

Understanding Network Address Translation

Many firewalls use a technique called *network address translation* (NAT) to hide the actual IP address of a host from the outside world. When that's the case, the NAT device must use a globally unique IP to represent the host to the Internet; behind the firewall, however, the host can use any IP address it wants. As packets cross the firewall, the NAT device translates the private IP address to the public IP address, and vice versa.

One of the benefits of NAT is that it helps slow down the rate at which the IP address space is assigned because a NAT device can use a single public IP address for more than one host. It does this by keeping track of outgoing packets so that it can match up incoming packets with the correct host. To understand how this process works, consider this sequence of steps:

1. A host whose private address is `192.168.1.100` sends a request to `216.58.192.4`, which happens to be `www.google.com`. The NAT device changes the source IP address of the packet to `208.23.110.22`, the IP address of the firewall. That way, Google will send its reply back to the firewall router. The NAT records that `192.168.1.100` sent a request to `216.58.192.4`.
2. Now another host, at address `192.168.1.107`, sends a request to `17.172.224.47`, which happens to be `www.apple.com`. The NAT device changes the source of this request to `208.23.110.22` so that Microsoft will reply to the firewall router. The NAT records that `192.168.1.107` sent a request to `17.172.224.47`.
3. A few seconds later, the firewall receives a reply from `216.58.192.4`. The destination address in the reply is `208.23.110.22`, the address of the firewall. To determine to whom to forward the reply, the firewall checks its records to see who's waiting for a reply from `216.58.192.4`. It discovers that `192.168.1.100` is waiting for that reply, so it changes the destination address to `192.168.1.100` and sends the packet on.

Actually, the process is a little more complicated than that because it's very likely that two or more users may have pending requests from the same public IP. In that case, the NAT device uses other techniques to figure out to which user each incoming packet should be delivered.

Configuring Your Network for DHCP

Every host on a TCP/IP network must have a unique IP address. Each host must be properly configured so that it knows its IP address. When a new host comes online, it must be assigned an IP address within the correct range of addresses for the subnet — one that's not already in use. Although you can manually assign IP addresses to each computer on your network, that task quickly becomes overwhelming if the network has more than a few computers.

That's where Dynamic Host Configuration Protocol (DHCP) comes into play. DHCP automatically configures the IP address for every host on a network, thus ensuring that each host has a valid, unique IP address. DHCP even automatically reconfigures IP addresses as hosts come and go. As you can imagine, DHCP can save a network administrator many hours of tedious configuration work.

In this section, you discover the ins and outs of DHCP: what it is, how it works, and how to set it up.

Understanding DHCP

DHCP allows individual computers on a TCP/IP network to obtain their configuration information — in particular, their IP addresses — from a server. The DHCP server keeps track of which IP addresses have already been assigned so that when a computer requests an IP address, the DHCP servers offer it an IP address that isn't already in use.

The alternative to DHCP is to assign each computer on your network a static IP address, which can be good or problematic:

WARNING

- Static IP addresses are okay for networks with a handful of computers.
- For networks with more than a few computers, using static IP addresses is a huge mistake. Eventually, some poor, harried administrator (guess who?) will make the mistake of assigning two computers the same IP address. Then you have to manually check each computer's IP address to find the conflict. DHCP is a must for any but the smallest networks.

Although the primary job of DHCP is to assign IP addresses, DHCP provides more configuration information than just the IP address to its clients. The additional configuration information is referred to as *DHCP options.* The following list describes some common DHCP options that can be configured by the server:

- Router address, also known as the default gateway address
- Expiration time for the configuration information
- Domain name
- DNS server address
- Windows Internet Name Service (WINS) server address

DHCP servers

A DHCP server can be a server computer located on the TCP/IP network. Fortunately, all modern server operating systems have a built-in DHCP server capability. To set up DHCP on a network server, all you have to do is enable the server's

DHCP function and configure its settings. In the section "Managing a Windows Server 2019 DHCP Server," later in this chapter, I show you how to configure a DHCP server for Windows 2019.

A server computer running DHCP doesn't have to be devoted entirely to DHCP unless the network is very large. For most networks, a file server can share duty as a DHCP server, especially if you provide long leases for your IP addresses. (I explain the idea of leases later in this chapter.)

Most multifunction routers also have built-in DHCP servers. So if you don't want to burden one of your network servers with the DHCP function, you can enable the router's built-in DHCP server. An advantage of allowing the router to be your network's DHCP server is that you rarely need to power down a router. By contrast, you occasionally need to restart or power down a file server to perform system maintenance, to apply upgrades, or to do some needed troubleshooting.

TIP

Most networks require only one DHCP server. Setting up two or more servers on the same network requires that you carefully coordinate the IP address ranges (known as *scopes*) for which each server is responsible. If you accidentally set up two DHCP servers for the same scope, you may end up with duplicate address assignments if the servers attempt to assign the same IP address to two different hosts. To prevent this situation from happening, set up just one DHCP server unless your network is so large that one server can't handle the workload.

Understanding scopes

A *scope* is simply a range of IP addresses that a DHCP server is configured to distribute. In the simplest case, in which a single DHCP server oversees IP configuration for an entire subnet, the scope corresponds to the subnet. However, if you set up two DHCP servers for a subnet, you can configure each one with a scope that allocates only one part of the complete subnet range. In addition, a single DHCP server can serve more than one scope.

You must create a scope before you can enable a DHCP server. When you create a scope, you can provide it these properties:

- A **scope name,** which helps you identify the scope and its purpose.
- A **scope description,** which lets you provide additional details about the scope and its purpose.
- A **starting IP address** for the scope.
- An **ending IP address** for the scope.

- A **subnet mask** for the scope. You can specify the subnet mask with dotted decimal notation or with Classless Inter Domain Routing (CIDR) notation.
- **One or more ranges of excluded addresses.** These addresses aren't assigned to clients. (For more information, see the next section, "Feeling excluded?")
- **One or more reserved addresses.** These addresses are always assigned to particular host devices. (For more information, see the section "Reservations suggested," later in this chapter.)
- The **lease duration,** which indicates how long the host is allowed to use the IP address. The client attempts to renew the lease when half of the lease duration has elapsed. For example, if you specify a lease duration of eight days, the client attempts to renew the lease after four days have passed. The host then has plenty of time to renew the lease before the address is reassigned to some other host.
- The **router address** for the subnet.

  This value is also known as the *default gateway address.*

- The **domain name and the IP address** of the network's DNS servers and WINS servers.

Feeling excluded?

Everyone feels excluded once in a while. With a wife and three daughters, I know how that feels. Sometimes, however, being excluded is a good thing. In the case of DHCP scopes, exclusions can help you prevent IP address conflicts and can enable you to divide the DHCP workload for a single subnet among two or more DHCP servers.

An *exclusion* is a range of addresses not included in a scope but falling within the range of the scope's starting and ending addresses. In effect, an exclusion range lets you punch a hole in a scope: The IP addresses that fall within the hole aren't assigned.

Here are a couple of reasons to exclude IP addresses from a scope:

- **The computer that runs the DHCP service itself must usually have a static IP address assignment.** As a result, the address of the DHCP server should be listed as an exclusion.
- **You may want to assign static IP addresses to your other servers.** In that case, each server IP address should be listed as an exclusion.

TIP

Reservations are often better solutions to this problem, as I describe in the next section.

Reservations suggested

In some cases, you may want to assign a specific IP address to a particular host. One way to do this is to configure the host with a static IP address so that the host doesn't use DHCP to obtain its IP configuration. However, two major disadvantages to that approach exist:

- **TCP/IP configuration supplies more than just the IP address.** If you use static configuration, you must manually specify the subnet mask, default gateway address, DNS server address, and other configuration information required by the host. If this information changes, you have to change it not only at the DHCP server, but also at each host that you configured statically.
- **You must remember to exclude the static IP address from the DHCP server's scope.** Otherwise, the DHCP server doesn't know about the static address and may assign it to another host. Then comes the problem: You have two hosts with the same address on your network.

TIP

A better way to assign a fixed IP address to a particular host is to create a DHCP reservation. A *reservation* simply indicates that whenever a particular host requests an IP address from the DHCP server, the server should provide it the address that you specify in the reservation. The host doesn't receive the IP address until the host requests it from the DHCP server, but whenever the host does request IP configuration, it always receives the same address.

To create a reservation, you associate the IP address that you want assigned to the host with the host's MAC address. Accordingly, you need to get the MAC address from the host before you create the reservation:

- Usually, you can get the MAC address by running the command `ipconfig /all` from a command prompt.
- If TCP/IP has not yet been configured on the computer, you can get the MAC address by choosing Start ⇨ All Programs ⇨ Accessories ⇨ System Tools ⇨ System Information.

TIP

If you set up more than one DHCP server, be sure to specify the same reservations on each server. If you forget to repeat a reservation on one of the servers, that server may assign the address to another host.

How long to lease?

One of the most important decisions that you make when you configure a DHCP server is the length of time to specify for the lease duration. The default value is eight days, which is appropriate in many cases. However, you may encounter situations in which a longer or shorter interval may be appropriate:

- The more stable your network, the longer the lease duration can safely exist. If you only periodically add new computers to your network (or replace existing computers), you can safely increase the lease duration past eight days.
- The more volatile the network, the shorter the lease duration should be. For example, you may have a wireless network in a university library, used by students who bring their laptop computers into the library to work for a few hours at a time. For this network, a duration as short as one hour may be appropriate.

Don't configure your network to allow leases of infinite duration. Although some administrators feel that this duration cuts down the workload for the DHCP server on stable networks, no network is permanently stable. Whenever you find a DHCP server that's configured with infinite leases, look at the active leases. I guarantee that you'll find IP leases assigned to computers that no longer exist.

Managing a Windows Server 2019 DHCP Server

The exact steps to follow when you configure and manage a DHCP server depend on the network operating system or router you're using. The following paragraphs describe how to work with a DHCP server in Windows Server 2019. The procedures for other operating systems are similar.

If you haven't already installed the DHCP server on the server, install it using the Server Manager (click Server Manager in the task bar, and then use Add Roles and Features to add the DHCP role). Once the DHCP server role is installed, you can manage it by opening the DHCP management console, as shown in Figure 6-2. To open this console, open System Manager and choose Tools ⇨ DHCP.

To get started with a DHCP server, you must create at least one scope. You can create a scope by using the New Scope Wizard, which you start by selecting the server you want to create the scope on and then clicking New Scope. The wizard asks for the essential information required to define the scope, including the scope's name, its starting and ending IP addresses, and the subnet mask. You can also specify any IP addresses you want to exclude from the scope, the lease duration (the default is eight days), the IP address of your gateway router, the domain name for your network, and the IP addresses for the DNS servers you want the client computers to use. Figure 6-3 shows the New Scope Wizard in action.

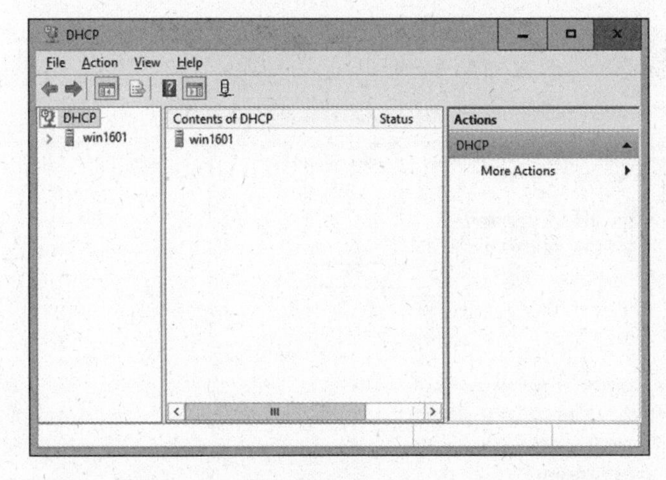

FIGURE 6-2: The DHCP management console.

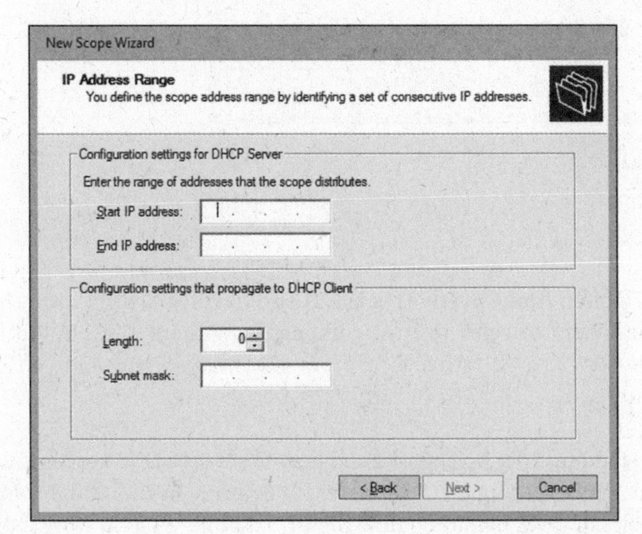

FIGURE 6-3: The New Scope Wizard.

Configuring a Windows DHCP Client

Configuring a Windows client for DHCP is easy. The DHCP client is included automatically when you install the TCP/IP protocol, so all you have to do is configure TCP/IP to use DHCP. To do this, open the Network Properties dialog box by choosing Network or Network Connections in the Control Panel (depending on which version of Windows the client is running). Then select the TCP/IP protocol and click the Properties button. This action opens the TCP/IP Properties dialog box, as shown in Figure 6-4. To configure the computer to use DHCP, select the Obtain an IP Address Automatically and Obtain DNS Server Address Automatically options. Click OK, and you're done.

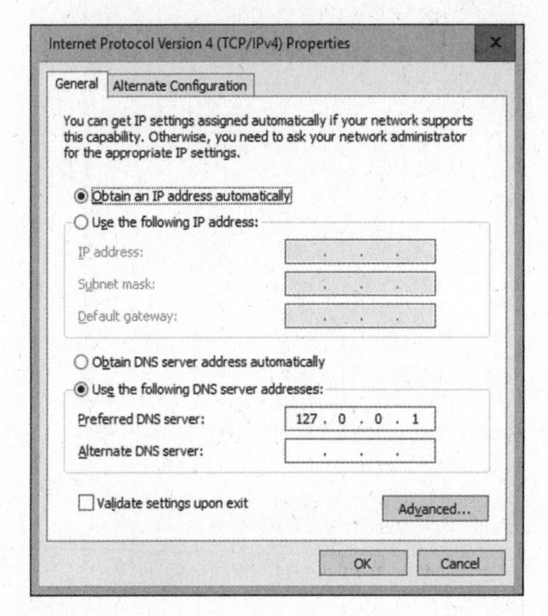

FIGURE 6-4: Configuring a Windows client to use DHCP.

Using DNS

DNS (Domain Name System) is the TCP/IP facility that lets you use names rather than numbers to refer to host computers. Without DNS, you'd buy your books from 99.84.233.99 rather than from www.amazon.com and you'd sell your used furniture at 104.78.128.48 rather than on www.ebay.com.

Understanding how DNS works and how to set up a DNS server is crucial to setting up and administering a TCP/IP network. The rest of this chapter introduces you to the basics of DNS, including how the DNS naming system works and how to set up a DNS server.

Domains and domain names

To provide a unique DNS name for every host computer on the Internet, DNS uses a time-tested technique: divide and conquer. DNS uses a hierarchical naming system that's similar to the way folders are organized hierarchically on a Windows computer. Instead of folders, however, DNS organizes its names into *domains.* Each domain includes all the names that appear directly beneath it in the DNS hierarchy.

For example, Figure 6-5 shows a small portion of the DNS domain tree. At the top of the tree is the *root domain,* which is the anchor point for all domains. Directly beneath the root domain are four *top-level domains,* named `edu`, `com`, `org`, and `gov`.

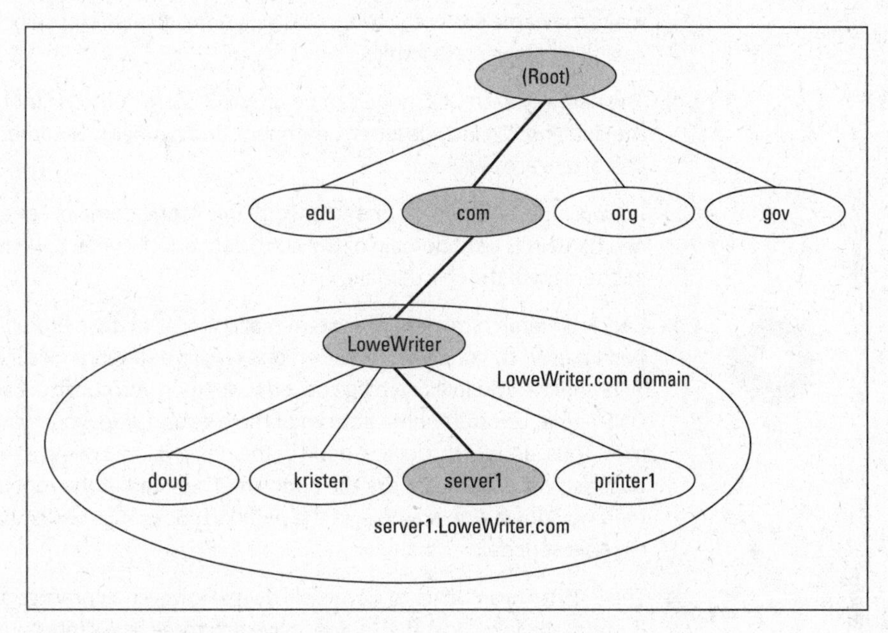

FIGURE 6-5: DNS names.

In reality, many more top-level domains than this exist in the Internet's root domain. In fact, at the time I wrote this, there were about a thousand top-level domains in use.

Beneath the `com` domain in Figure 6-5 is another domain named `LoweWriter`, which happens to be my own personal domain. (Pretty clever, eh?) To completely identify this domain, you have to combine it with the name of its *parent domain* (in this case, `com`) to create the complete domain name: `LoweWriter.com`. Notice that the parts of the domain name are separated from each other by periods, which are pronounced "dot." As a result, when you read this domain name, you should pronounce it "LoweWriter dot com."

Beneath the `LoweWriter` node are four host nodes, named `doug`, `kristen`, `server1`, and `printer1`. These nodes correspond to three computers and a printer on my home network. You can combine the host name with the domain name to get the complete DNS name for each of my network's hosts. For example, the complete

DNS name for my server is `server1.LoweWriter.com`. Likewise, my printer is `printer1.LoweWriter.com`.

Here are a few additional details that you need to remember about DNS names:

TIP

- DNS names aren't case-sensitive. As a result, `LoweWriter` and `Lowewriter` are treated as the same name, as are `LOWEWRITER`, `LOWEwriter`, and `LoWeWrItEr`. When you use a domain name, you can use capitalization to make the name easier to read, but DNS ignores the difference between capital and lowercase letters.
- The name of each DNS node can be up to 63 characters long (not including the dot) and can include letters, numbers, and hyphens. No other special characters are allowed.
- A *subdomain* is a domain that's beneath an existing domain. For example, the `com` domain is a subdomain of the `root` domain. Likewise, `LoweWriter` is a subdomain of the `com` domain.
- DNS is a hierarchical naming system that's similar to the hierarchical folder system used by Windows. However, one crucial difference exists between DNS and the Windows naming convention. When you construct a complete DNS name, you start at the bottom of the tree and work your way up to the root. Thus, `doug` is the lowest node in the name `doug.LoweWriter.com`. By contrast, Windows paths are the opposite: They start at the root and work their way down. For example, in the path `\Windows\System32\dns`, `dns` is the lowest node.
- The DNS tree can be up to 127 levels deep. However, in practice, the DNS tree is pretty shallow. Most DNS names have just three levels (not counting the root), and although you sometimes see names with four or five levels, you rarely see more levels than that.
- Although the DNS tree is shallow, it's very broad. In other words, each of the top-level domains has a huge number of second-level domains immediately beneath it. For example, at the time I wrote this book, the `com` domain had more than two million second-level domains beneath it.

Fully qualified domain names

If a domain name ends with a trailing dot, that trailing dot represents the root domain, and the domain name is said to be a *fully qualified domain name* (FQDN). A fully qualified domain name — also called an *absolute name* — is unambiguous because it identifies itself all the way back to the root domain. In contrast, if a domain name doesn't end with a trailing dot, the name may be interpreted in the context of some other domain. Thus, DNS names that don't end with a trailing dot are *relative names.*

This concept is similar to the way relative and absolute paths work in Windows. For example, if a path begins with a backslash, such as `\Windows\System32\dns`, the path is absolute. However, a path that doesn't begin with a backslash, such as `System32\dns`, uses the current folder as its starting point. If the current folder happens to be `\Windows`, `\Windows\System32\dns` and `System32\dns` refer to the same location.

In many cases, relative and fully qualified domain names are interchangeable because the software that interprets them always interprets relative names in the context of the root domain. That's why, for example, you can type `www.wiley.com` — without the trailing dot — rather than `www.wiley.com.` to go to the Wiley home page in a web browser. Some applications, such as DNS servers, may interpret relative names in the context of a domain other than the root.

Working with the Windows DNS Server

The procedure for installing and managing a DNS server depends on the network operating system you're using. This section is specific to working with a DNS server in Windows 2019. Working with a DNS server in a Linux or Unix environment is similar but without the help of a graphical user interface.

You can install the DNS server on Windows Server 2019 from the Server Manager (choose Server Manager on the taskbar). After you install the DNS server, you can manage it from the DNS management console. Here, you can perform common administrative tasks, such as adding additional zones, changing zone settings, or adding new records an existing zone. The DNS management console hides the details of the resource records from you, thus allowing you to work with a friendly graphical user interface instead.

To add a new host (which is defined by a DNS record called an A record) to a zone, right-click the zone in the DNS management console and choose the Add New Host command. This action opens the New Host dialog box, as shown in Figure 6-6.

Here, you specify the following information:

- **Name:** The host name for the new host.
- **IP Address:** The host's IP address.
- **Create Associated Pointer (PTR) Record:** Automatically creates a PTR record in the reverse lookup zone file. Select this option if you want to allow reverse lookups for the host. (A *reverse lookup* determines the domain name for a

given IP address. It's called that because the normal type of DNS lookup determines the IP address for a given domain name.)

- **Allow Any Authenticated User to Update:** Select this option if you want to allow other users to update this record or other records with the same host name. You should usually leave this option deselected.
- **Time to Live:** The TTL value for this record, which indicates how long (in seconds) the data should be cached.

You can add other records, such as MX records, in the same way.

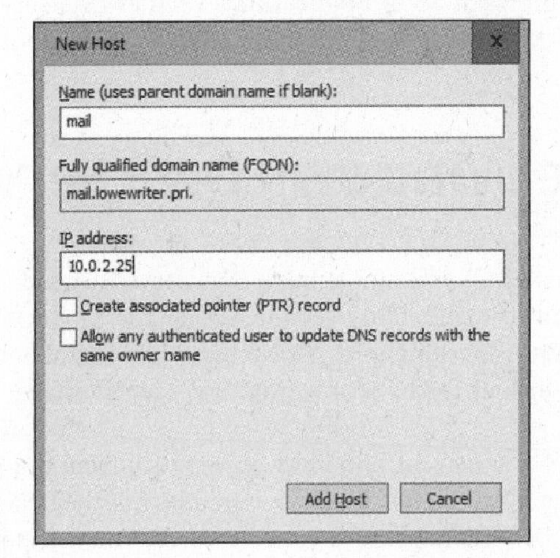

FIGURE 6-6: The New Host dialog box.

Configuring a Windows DNS Client

Client computers don't need much configuration to work properly with DNS. The client must have the address of at least one DNS server. Usually, this address is supplied by DHCP, so if the client is configured to obtain its IP address from a DHCP server, it also obtains the DNS server address from DHCP.

To configure a client computer to obtain the DNS server location from DHCP, open the Network Properties dialog box by choosing Network or Network Connections in the Control Panel (depending on which version of Windows the client is running). Then select the TCP/IP protocol and click the Properties button. This action summons the TCP/IP Properties dialog box, which is shown back in Figure 6-4. To configure the computer to use DHCP, select the Obtain an IP Address Automatically and the Obtain DNS Server Address Automatically options. Click OK, and you're done.

» Checking out the different types of network cable

» Installing twisted-pair cable

» Working with switches

Chapter 7

Oh, What a Tangled Web We Weave: Cables and Switches

Cable is the plumbing of your network. In fact, working with network cable is a lot like working with pipe: You have to use the right pipe (cable) along with the right valves and connectors (switches).

And network cables have an advantage over pipes: You don't get wet when they leak.

This chapter tells you far more about network cables than you probably need to know. I introduce you to *Ethernet*, the most common system of network cabling for small networks. Then you find out how to work with the cables used to wire up an Ethernet network. Finally, I show you how to work with switches, which create the necessary interconnections so that all your computers can talk to each other.

What Is Ethernet?

Ethernet is a standardized way of connecting computers to create a network.

You can think of Ethernet as a kind of municipal building code for networks: It specifies what kind of cables to use, how to connect the cables, how long the cables can be, how computers transmit data to one another by using the cables, and more.

Although Ethernet is now the overwhelming choice for networking, that wasn't always the case. In ye olde days, Ethernet had two significant competitors:

- **Token Ring:** This IBM standard for networking is still used in some organizations (especially where IBM mainframe or midrange systems are in use).
- **ARCnet:** This standard is still sometimes used for industrial network applications, such as building automation and factory robot control.

But these older networks are now pretty much obsolete, so you don't need to worry about them. Ethernet is now the only real networking choice.

OBLIGATORY INFORMATION ABOUT NETWORK TOPOLOGY

TECHNICAL STUFF

A networking book wouldn't be complete without the usual textbook description of the three basic network topologies: bus, ring, and star.

In a bus topology, network nodes (that is, computers) are strung together in a line, like this:

A *bus* is the simplest type of topology, but it has some drawbacks. If the cable breaks somewhere in the middle, the whole network breaks.

A second type of topology is the ring:

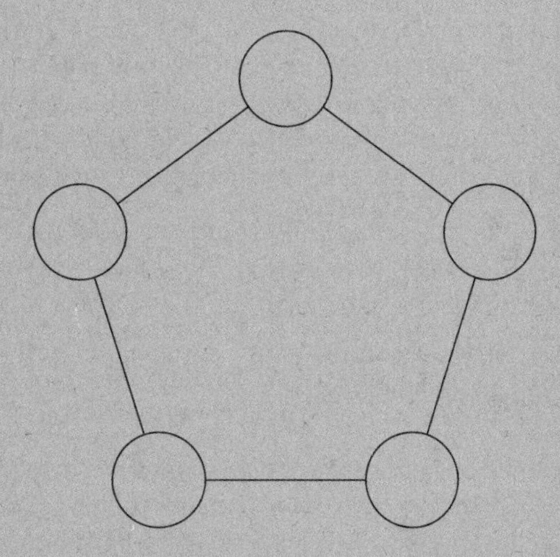

A ring is very much like a bus except with no end to the line: The last node on the line is connected to the first node, forming an endless loop.

A third type of topology is a star:

In a star network, all the nodes are connected to a central hub. In effect, each node has an independent connection to the network, so a break in one cable doesn't affect the others.

Ethernet networks are based on a bus design. However, fancy cabling tricks make an Ethernet network look more like a star.

Here are a few tidbits you're likely to run into at parties where the conversation is about Ethernet standards:

- Ethernet is a set of standards for the infrastructure on which a network is built. All modern operating systems can operate on an Ethernet network. If you build your network on a solid Ethernet base, you can intermix different operating systems: Windows, macOS, and Linux.
- Ethernet is often referred to by network gurus as 802.3 (pronounced "eight-oh-two-dot-three"), which is the official designation used by the *IEEE* (pronounced "eye-triple-ee," not "ai*eeee!*"), a group of electrical engineers who wear bow ties and love to argue about things like inductance and cross-talk — and it's a good thing they do. If not for them, you couldn't mix and match Ethernet components made by different companies.
- The original vintage Ethernet transmits data at a paltry rate of 10 million bits per second, or 10 Mbps. (*Mbps* is usually pronounced "megabits per second.") Because 8 bits are in a byte, that translates into roughly 1.2 million bytes per second. In practice, Ethernet can't move information that fast because data must be transmitted in packages of no more than 1,500 bytes, called *packets.* So 150KB of information has to be split into 100 packets.

  Ethernet's transmission speed has nothing to do with how fast electrical signals move on the cable. The electrical signals travel at about 70 percent of the speed of light, or as Captain Kirk would say, "Warp factor point-seven-oh."

- A faster version of Ethernet, called *100 Mbps Ethernet* or *Fast Ethernet,* moves data ten times as fast as normal Ethernet.
- The most common version of Ethernet today is *gigabit Ethernet,* which moves data at 1,000 Mbps.
- Most networking components that you can buy these days support 10, 100 Mbps, and 1,000 Mbps. These components are called *10/100/1000 Mbps components.*
- Some networking components support 10 gigabit Ethernet, which moves data at 10,000 Mbps (or 10 Gbps). Ten Gbps Ethernet is usually used for high-speed connections between servers and network switches.

All about Cable

Although you can use wireless technology to create networks without cables, most networks still use cables to physically connect each computer to the network. Over the years, various types of cables have been used with Ethernet networks.

Almost all networks are now built with twisted-pair cable. In this type of cable, pairs of wires are twisted around each other to reduce electrical interference. (You almost need a PhD in physics to understand why twisting the wires helps to reduce interference, so don't feel bad if this concept doesn't make sense.)

Although unlikely, you might encounter older types of cable on ancient networks. Before twisted-pair cable became the standard, *coaxial* cable (also known as *coax*, pronounced "COE-ax") was used. The most common type of coax cable, called RG-58, resembled television cable. The second, even older type, was called 10base5, also known as *thicknet* because of its unwieldy thickness.

You may also encounter fiber optic cables that span long distances at high speeds or thick twisted-pair bundles that carry multiple sets of twisted-pair cable between wiring closets in a large building. Most networks, however, use simple twisted-pair cable.

Twisted-pair cable is sometimes called *UTP.* (The *U* stands for *u*nshielded, but "twisted-pair" is the standard name.) Figure 7-1 shows a twisted-pair cable.

FIGURE 7-1: Twisted-pair cable.

When you use UTP cable to construct an Ethernet network, you connect the computers in a star arrangement, as Figure 7-2 illustrates. In the center of this star is a *switch.* Depending on the model, Ethernet switches enable you to connect 4 to 48 computers (or more) by using twisted-pair cable.

TIP

In the UTP star arrangement, if one cable goes bad, only the computer attached to that cable is affected. The rest of the network continues to chug along.

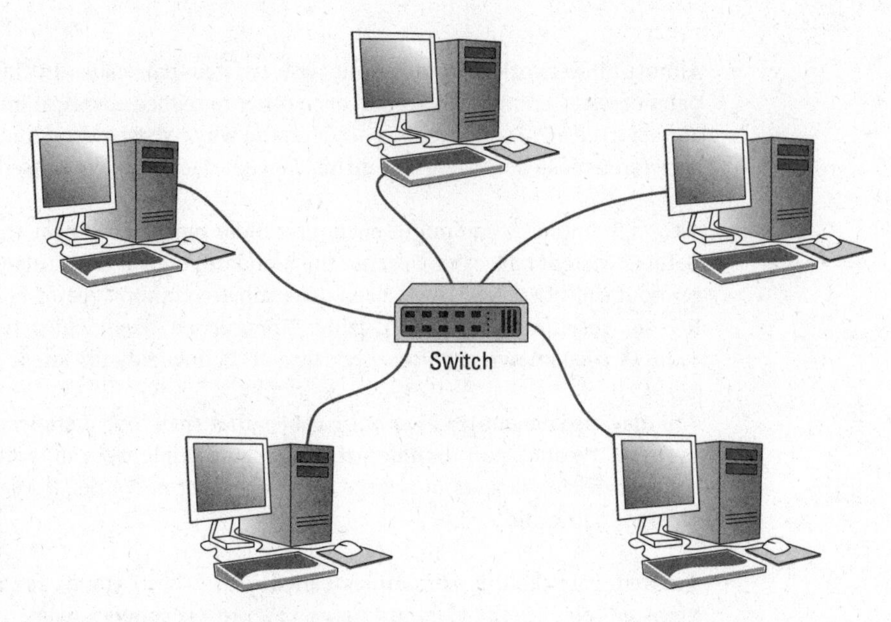

FIGURE 7-2: A network cabled with twisted-pair cable.

Cable categories

Twisted-pair cable comes in various grades called *categories.* These categories are specified by the ANSI/EIA Standard 568. (*ANSI* stands for American National Standards Institute; *EIA* stands for Electronic Industries Association). The standards indicate the data capacity — bandwidth — of the cable. Table 7-1 lists the various categories of twisted-pair cable that are currently available. (Higher categories, including Cat-7 and Cat-8, are in the design stage but are not yet commercially available.)

TABLE 7-1 **Twisted-Pair Cable Categories**

Category	Maximum Data Rate	Intended Use
1	1 Mbps	Voice only
2	4 Mbps	4 Mbps Token Ring
3	16 Mbps	10BaseT Ethernet
4	20 Mbps	16 Mbps Token Ring
5	100 Mbps (2-pair)	100BaseT Ethernet
	1000 Mbps (4-pair)	1000BaseTX
5e	1000 Mbps (2-pair)	1000BaseT
6	1000 Mbps (2-pair)	1000BaseT and faster broadband applications
6a	10000 Mbps (2-pair)	Provides for 10 Gbps Ethernet

Although higher-category cables are more expensive, the real cost of installing Ethernet cabling is the labor required to pull the cables through the walls. You should never install anything less than Category (Cat) 5e cable. And if at all possible, invest in Cat 6 cable to allow for upgrades to your network.

To sound like the cool kids, say "Cat 6" rather than "Category 6."

What's with the pairs?

Most twisted-pair cables have four pairs of wires, for a total of eight wires. Standard Ethernet uses only two of the pairs, so the other two pairs are unused. You may be tempted to save money by purchasing cable with just two pairs of wires, but that's a bad idea. If a network cable develops a problem, you can sometimes fix it by switching over to one of the extra pairs. If you use two-pair cable, though, you don't have any spare pairs to use.

Don't use the extra pairs for some other purpose, such as a voice line or a second data line. The electrical "noise" in the extra wires can interfere with your network.

To shield or not to shield

Unshielded twisted-pair cable (UTP) is designed for normal office environments. When you use UTP cable, you must be careful not to route cable close to fluorescent light fixtures, air conditioners, or electric motors (such as automatic door motors or elevator motors). UTP is the least expensive type of cable.

In environments that have a lot of electrical interference (such as factories), you may want to use shielded twisted-pair cable (STP). Because STP can be as much as three times more expensive than regular UTP, you don't want to use STP unless you have to. With a little care, UTP can withstand the amount of electrical interference found in a normal office environment. But for harsh environments, where cable will be placed outdoors, buried in the ground, or near industrial equipment, you should use STP cable instead.

Most STP cable is shielded by a layer of aluminum foil. For buildings with unusually high amounts of electrical interference, the more expensive braided-copper shielding offers even more protection.

When to use plenum cable

The outer sheath of shielded and unshielded twisted-pair cable comes in two kinds:

- **PVC:** The most common and least expensive type.
- **Plenum:** A special type of fire-retardant cable designed for use in the plenum space (definition coming right up) of a building. Plenum cable has a special Teflon coating that not only resists heat, but also gives off fewer toxic fumes if it does burn. Unfortunately, plenum cable costs more than twice as much as ordinary PVC cable.

Most local building codes require plenum cable when the wiring is installed in the building's *plenum space* (a compartment that's part of the building's air-distribution system, usually the space above a suspended ceiling or under a raised floor).

The area above a suspended ceiling is *not* a plenum space if both the delivery and return lines of the air-conditioning and heating systems are ducted. Plenum cable is required only if the air-conditioning and heating systems aren't ducted. When in doubt, have the local inspector look at your facility before you install cable.

Sometimes solid, sometimes stranded

The actual copper wire that makes up the cable comes in two varieties: solid and stranded. Your network will have some of each:

- **Stranded:** Each conductor is made from a bunch of very small wires that are twisted together. Stranded cable is more flexible than solid cable, so it doesn't break as easily. However, stranded cable is more expensive than solid cable and isn't very good at transmitting signals over long distances. Stranded cable is best used for *patch cables* (such as patch panels to hubs and switches).

  Strictly speaking, the cable that connects your computer to the wall jack is called *station cable* — not patch cable — but it's an appropriate use for stranded cable. (Although not technically correct, most people refer to the cable that connects a computer to a wall jack as a "patch cable.")
- **Solid:** Each conductor is a single, solid strand of wire. Solid cable is less expensive than stranded cable and carries signals farther, but it isn't very flexible. If you bend it too many times, it breaks. Typically, you find solid cable in use as permanent wiring within the walls and ceilings of a building.

Installation guidelines

The hardest part of installing network cable is the physical task of pulling the cable through ceilings, walls, and floors. This job is just tricky enough that I recommend you don't attempt it yourself, except for small offices. For large jobs, hire a professional cable installer. You may even want to hire a professional for small jobs if the ceiling and wall spaces are difficult to access.

Keep these pointers in mind if you install cable yourself:

- You can purchase twisted-pair cable in prefabricated lengths, such as 10, 15, or 20 feet. Longer lengths, such as 50 feet or 100 feet, are also available.
- Alternatively, you can purchase cable in bulk rolls, cut them to length, and attach the connectors yourself.
- Always use a bit more cable than you need, especially if you're running cable through walls. For example, when you run a cable up a wall, leave a few feet of slack in the ceiling above the wall. That way, you have plenty of cable if you need to make a repair.
- When running cable, avoid sources of interference, such as fluorescent lights, big motors, X-ray machines, nuclear reactors, cyclotrons, flux capacitors, or other gadgets you may have hidden in behind closed doors in your office.

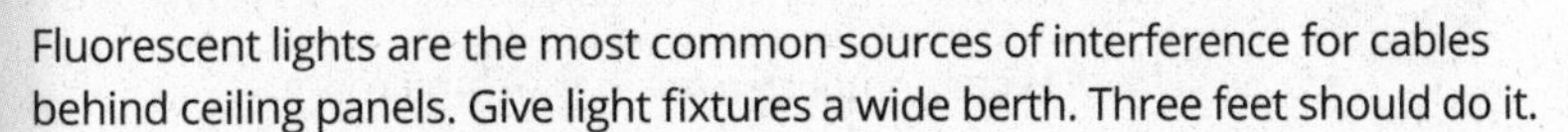

  Fluorescent lights are the most common sources of interference for cables behind ceiling panels. Give light fixtures a wide berth. Three feet should do it.

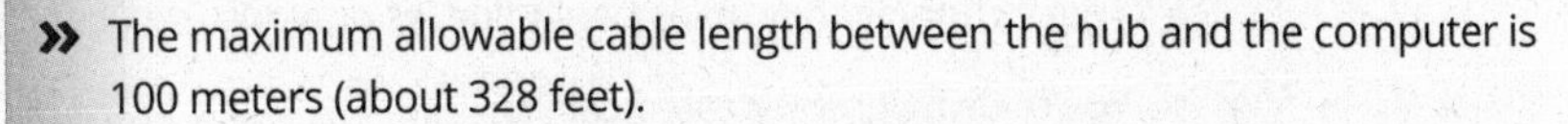

- The maximum allowable cable length between the hub and the computer is 100 meters (about 328 feet).
- If you must run cable across the floor where people walk, cover the cable so no one trips over it. Cable protectors are available at most hardware stores.
- When running cables through walls, label each cable at both ends. Most electrical supply stores carry pads of cable labels that are perfect for the job. These pads contain 50 sheets or so of precut labels with letters and numbers. They look much more professional than wrapping a loop of masking tape around the cable and writing on the tape with a marker.

  You can also purchase label makers that can print labels designed to attach to cables. As a last resort, you can always write directly on the label with a permanent marker.

- If you're installing cable in new construction, label each end of the cable at least three times, leaving about a foot of space between the labels. The drywallers or painters will probably spray mud or paint all over your cables, making the labels difficult to find.

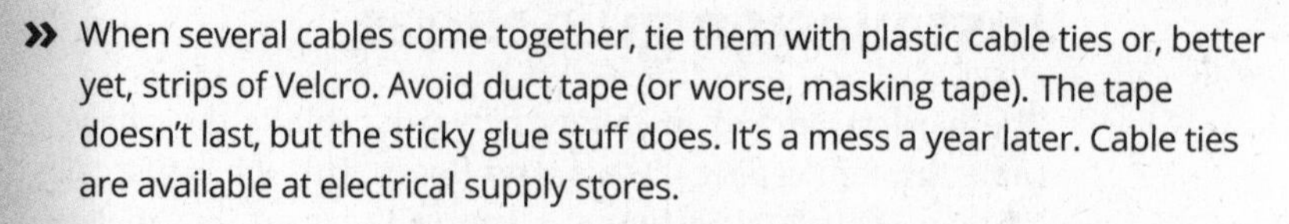

» When several cables come together, tie them with plastic cable ties or, better yet, strips of Velcro. Avoid duct tape (or worse, masking tape). The tape doesn't last, but the sticky glue stuff does. It's a mess a year later. Cable ties are available at electrical supply stores.

TIP

» Cable ties have all sorts of useful purposes. Once, on a backpacking trip, I used a pair of cable ties to attach an unsuspecting buddy's hat to a high tree limb. He wasn't impressed with my innovative use of the cable ties, but my other hiking companions were.

» When you run cable above suspended ceiling panels, use cable ties, hooks, or clamps to secure the cable to the ceiling or to the metal frame that supports the ceiling tiles. Don't just lay the cable on top of the panels.

The tools you need

Of course, to do a job right, you must have the right tools:

» Start with a basic set of computer tools, which you can get for about $15 from any computer store and most office-supply stores. These kits include socket wrenches and screwdrivers to open your computers and insert adapter cards.

TIP

The computer tool kit probably contains everything you need if

- All your computers are in the same room.
- You're running the cables behind desks or along the floor.
- You're using prefabricated cables.

TIP

If you don't have a computer tool kit, make sure that you have several flat-head and Phillips screwdrivers of various sizes.

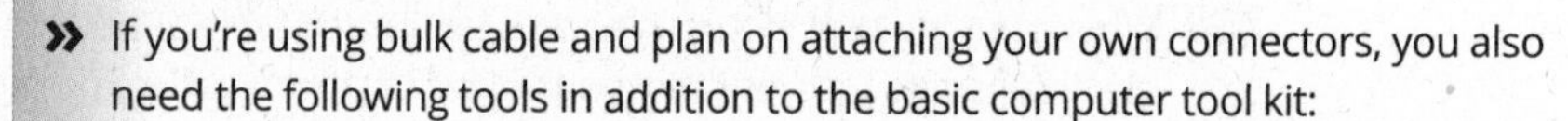

» If you're using bulk cable and plan on attaching your own connectors, you also need the following tools in addition to the basic computer tool kit:

- *Wire cutters:* You need big ones for coax; smaller ones work for twisted-pair cable. For yellow cable, you need the Jaws of Life.
- *A crimp tool:* You need the crimp tool to attach the connectors to the cable. Don't use a cheap $25 crimp tool. A good crimp tool costs $100 and will save you many headaches in the long run.

REMEMBER

When you crimp, you mustn't scrimp.

- *Wire stripper:* You need this tool only if the crimp tool doesn't include a wire stripper.
- *A cable tester,* which lets you determine whether the cable will work.

» If you plan on running cables through walls, you need these additional tools:
  - *A hammer*
  - *A keyhole saw:* This one is useful if you plan on cutting holes through walls to route your cable.
  - *A flashlight*
  - *A ladder*
  - *Someone to hold the ladder*
  - *Fish tape:* Possibly. A *fish tape* is a coiled-up length of stiff metal tape. To use it, you feed the tape into one wall opening and fish it toward the other opening, where a partner is ready to grab it when the tape arrives. Next, your partner attaches the cable to the fish tape and yells something like "Let 'er rip!" or "Bombs away!" Then you reel in the fish tape and the cable along with it. (You can find fish tape in the electrical section of most well-stocked hardware stores.)

WARNING

If you plan on routing cable through a concrete subfloor, you need to rent a jackhammer and a backhoe and then hire someone to hold a yellow flag while you work. Better yet, find some other route for the cable.

Pinouts for twisted-pair cables

Each pair of wires in a twisted-pair cable is one of four colors: orange, green, blue, or brown. The two wires that make up each pair are complementary: one is white with a colored stripe; the other is colored with a white stripe. For example, the orange pair has an orange wire with a white stripe (the *orange wire*) and a white wire with an orange stripe (the *white/orange wire*). Likewise, the blue pair has a blue wire with a white stripe (the *blue wire*) and a white wire with a blue stripe (the *white/blue wire*).

When you attach a twisted-pair cable to a modular connector or jack, you must match up the right wires to the right pins. It's harder than it sounds; you can use any of several different standards to wire the connectors. To confuse matters further, you can use one of the two popular standard ways of hooking up the wires: EIA/TIA 568A or EIA/TIA 568B, also known as AT&T 258A. Both of these wiring schemes are shown in Table 7-2.

WARNING

It doesn't matter which of these wiring schemes you use, but pick one and stick with it. If you use one wiring standard on one end of a cable and the other standard on the other end, the cable doesn't work.

TABLE 7-2 **Pin Connections for Twisted-Pair Cable**

Pin	Function	EIA/TIA 568A	EIA/TIA568B AT&T 258A
1	Transmit +	White/Green	White/orange wire
2	Transmit –	Green	Orange wire
3	Receive +	White/Orange	White/green wire
4	Unused	Blue	Blue wire
5	Unused	White/Blue	White/blue wire
6	Receive –	Orange	Green wire
7	Unused	White/Brown	White/brown wire
8	Unused	Brown	Brown wire

Ethernet only uses two of the four pairs, connected to Pins 1, 2, 3, and 6. One pair transmits data; the other receives data. The only difference between the two wiring standards is which pair transmits and which receives. In the EIA/TIA 568A standard, the green pair is used for transmit, and the orange pair is used for receive. In the EIA/TIA 568B and AT&T 258A standards, the orange pair is used for transmit and the green pair for receive.

Don't be tempted to just connect Pins 1, 2, 3, and 6. Connect all four pairs as indicated in Table 7-2.

RJ-45 connectors

RJ-45 connectors for twisted-pair cables aren't too difficult to attach if you have the right crimping tool. The only trick is making sure that you attach each wire to the correct pin and then press the tool hard enough to ensure a good connection.

Here's the procedure for attaching an RJ-45 connector:

1. **Cut the end of the cable to the desired length.**

   Make sure that you make a square cut — not a diagonal cut.

2. **Insert the cable into the stripper portion of the crimp tool so that the end of the cable is against the stop.**

   Squeeze the handles and slowly pull out the cable, keeping it square. This strips off the correct length of outer insulation without puncturing the insulation on the inner wires.

3. **Arrange the wires so that they lie flat and line up according to Table 7-2.**

   You have to play with the wires a little bit to get them to lay out in the right sequence.

4. **Slide the wires into the pinholes on the connector.**

   Double-check to make sure all the wires are slipped into the correct pinholes.

5. **Insert the plug and wire into the crimping portion of the tool and then squeeze the handles to crimp the plug.**

   Squeeze it tight!

6. **Remove the plug from the tool and double-check the connection.**

   You're done!

Here are a few other points to remember when dealing with RJ-45 connectors and twisted-pair cable:

TIP

» The pins on the RJ-45 connectors aren't numbered.

  You can tell which is Pin 1 by holding the connector so that the metal conductors are facing up, as shown in Figure 7-3. Pin 1 is on the left.

» Some people wire the cable differently — using the green-and-white pair for Pins 1 and 2, and the orange-and-white pair for Pins 3 and 6. Doing it this way doesn't affect the operation of the network (the network is color blind) *as long as the connectors on both ends of the cable are wired the same way!*

» When you attach the connectors, don't untwist more than half an inch of cable. And, don't try to stretch the cable runs beyond the 100-meter maximum. When in doubt, have the cable professionally installed.

FIGURE 7-3: Attaching an RJ-45 connector to twisted-pair cable.

Crossover cables

A *crossover cable* can directly connect two devices without a switch. You can use a crossover cable to connect two computers directly to each other, but crossover cables are more often used to daisy-chain switches to each other.

If you want to create your own crossover cable, you must reverse the wires on one end of the cable, as shown in Table 7-3. This table shows how you should wire both ends of the cable to create a crossover cable. Connect one of the ends according to the Connector A column and the other according to the Connector B column.

TABLE 7-3 Creating a Crossover Cable

Pin	Connector A	Connector B
1	White/green	White/orange
2	Green	Orange
3	White/orange	White/green
4	Blue	Blue
5	White/blue	White/blue
6	Orange	Green
7	White/brown	White/brown
8	Brown	Brown

Crossover cables aren't as widely necessary as they used to be, because most switches can now automatically detect whether a crossover cable is necessary and adjust internally to allow you to use a standard cable instead of a crossover cable.

TIP

If you study Table 7-3 long enough and then compare it with Table 7-2, you may notice that a crossover cable is a cable that's wired according to the 568A standard on one end and the 568B standard on the other end.

Wall jacks and patch panels

If you want, you can run a single length of cable from a network hub or switch in a wiring closet through a hole in the wall, up the wall to the space above the ceiling, through the ceiling space to the wall in an office, down the wall, through a hole, and all the way to a desktop computer. That's not a good idea. For example, every time someone moves the computer or even cleans behind it, the cable will get moved a little bit. Eventually, the connection will fail, and the RJ-45 plug will have to be replaced. Then the cables in the wiring closet will quickly become a tangled mess.

The alternative is to put a wall jack in the wall at the user's end of the cable and connect the other end of the cable to a patch panel. Then the cable itself is completely contained within the walls and ceiling spaces. To connect a computer to the network, you plug one end of a patch cable (properly called a *station cable*) into the wall jack and plug the other end into the computer's network interface. In the wiring closet, you use a patch cable to connect the wall jack to the network hubs or switches. Figure 7-4 shows how this arrangement works.

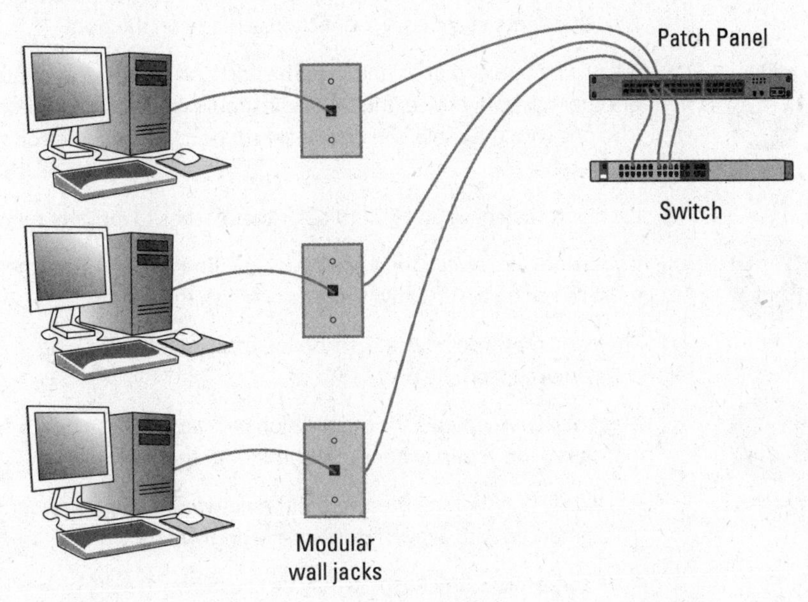

FIGURE 7-4: Using wall jacks and patch panels.

Connecting a twisted-pair cable to a wall jack or a patch panel is similar to connecting it to an RJ-45 plug. However, you don't usually need any special tools. Instead, the back of the jack has a set of slots that you lay each wire across. You then snap a removable cap over the top of the slots and press it down. This action forces the wires into the slots, where little metal blades pierce the insulation and establish the electrical contact.

TIP

When you connect the wire to a jack or a patch panel, be sure to untwist as little of the wire as possible. If you untwist too much of the wire, the signals that pass through the wire may become unreliable.

Understanding Switches

When you use twisted-pair cable to wire a network, you don't plug the computers into each other. Instead, each computer plugs into a separate device called a *switch.*

You need to know only a few details when working with switches. Here they are:

- Installing a switch is usually very simple. Just plug in the power cord and then plug in patch cables to connect the network.
- Each port on the switch has an RJ-45 jack and a single LED indicator, labeled *Link,* that lights up when a connection is made on the port.

TIP

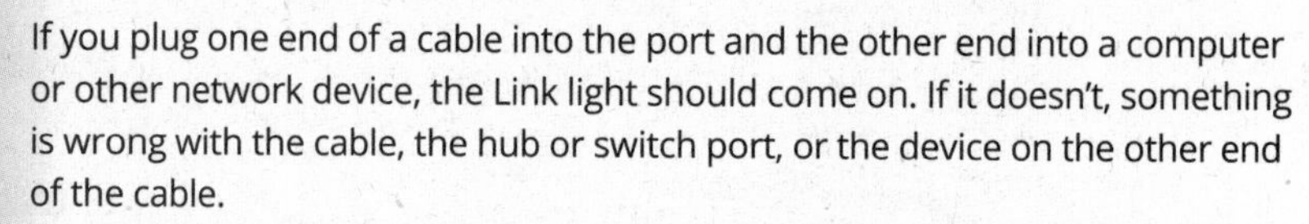

  If you plug one end of a cable into the port and the other end into a computer or other network device, the Link light should come on. If it doesn't, something is wrong with the cable, the hub or switch port, or the device on the other end of the cable.

- Each port may have an LED indicator that flashes to indicate network activity.

TIP

  If you stare at a switch for a while, you can find out who uses the network most by noting which activity indicators flash the most.

- The ports may also have a collision indicator that flashes whenever a packet collision occurs on the port.

WARNING

  It's perfectly acceptable for the collision indicator to flash now and then, but if it flashes a lot, you may have a problem with the network:

  - Usually, the flashing means that the network is overloaded and should be segmented with a switch to improve performance.
  - In some cases, the flashing may be caused by a faulty network node that clogs the network with bad packets.

Comparing managed and unmanaged switches

Not all switches are created equal. Some switches are designed for very small networks in homes or single-office businesses. Small networks are so simple to manage that the switch itself doesn't require any management or configuration of its own. You simply plug all the computers into the switch, and the network takes care of itself.

WHY IS IT CALLED A SWITCH?

You might be wondering why a switch is called a switch. After all, in your everyday experience, a switch is used to turn something on and off. But network switches don't turn networks on and off.

In networking, a switch is a device that receives incoming packets of information from the network and determines where each packet should be sent. In that sense, a network switch is more like a railroad track switch than a light switch. Instead of turning something on or off, a network switch determines which of several tracks a particular packet of information should be sent to.

Consider a small switch with eight ports, numbered 1 through 8. When the switch is powered on, it pays attention to the devices that it can connect to on each of its eight ports. It does this by studying the Ethernet packets that arrive on each port and taking note of the sender's address contained in each packet. The switch keeps track of which device is attached to each of its ports. When a packet arrives on a port, the switch looks at the recipient's address contained in the packet. The switch then determines which port the recipient is on and sends the packet to that port.

Thus, switches efficiently manage the travel of packets throughout a network by switching each packet traveling on the network through the correct cables, ensuring that each packet arrives at its destination.

In larger networks, however, switches have a much more complicated job to do. In these environments, you need the ability to monitor and configure the behavior of each of the switches in the network. Switches that provide this capability are called *managed switches.* Switches that do not provide this capability are called *unmanaged switches.*

A managed switch has an IP address of its own and provides a web-based management console that you can access by pointing your favorite web browser to the IP address of the switch. After you've logged in to the management console, you can do things such as configure each port for different types of network traffic, view the amount of traffic on each port, and monitor each port's performance, as well as the overall performance of the switch.

TIP

As a general rule, if your network requires more than a single switch, you should use managed switches. Managed switches are more expensive than unmanaged switches, but when your network grows large enough to require more than one switch, you'll appreciate the benefits that managed switches provide.

Daisy-chaining switches

If a single switch doesn't have enough ports for your entire network, you can connect switches by *daisy-chaining* them, as shown in Figure 7-5. (Note that although you can daisy-chain unmanaged switches, I recommend you use managed switches if your network is large enough to require more than one switch.)

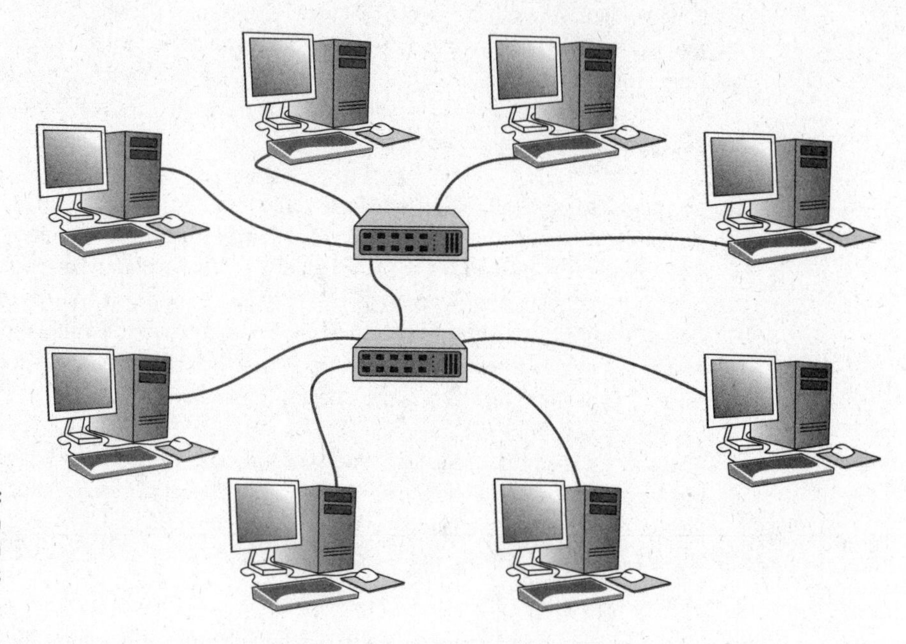

FIGURE 7-5: You can daisy-chain switches together.

On most switches, it doesn't matter which port you use to daisy chain to another switch. Just pick a port on both switches and use a patch cable to connect the switches to each other via these ports. And if your switch has ports with differing speeds, use the fastest ports to connect the switches to one another.

TIP

If your building is prewired and has a network jack near each desk, you can use a small switch to connect two or more computers to the network by using a single jack. Just use one cable to plug the switch into the wall jack and then plug each computer into one of the switch's ports.

Stacking switches

Some switches, called *stackable switches*, can be expanded by adding additional switch modules that add additional ports to the switch. The additional modules may be in the form of physically separate switches that are connected via special interconnected ports, or they may be modules that can be inserted into a larger chassis. Stackable switches are by nature managed switches, and the defining characteristic of a stackable switch is that all the switches connected together into a single stack are managed as if they're a single switch.

For example, a stackable switch may be initially configured with just a single module that provides 48 switch ports. If you need additional ports, you can add a second module with 48 additional ports, creating a single switch with 96 ports.

Stackable switches are more expensive than non-stacking switches, but the simplicity of managing one large switch rather than managing multiple smaller switches may justify the added cost.

Looking at distribution switches and access switches

A network large enough to require more than one switch may also be large enough to require several distinct types of switches:

- **Access switches:** An *access switch* is a switch that typically has a large number of 1 Gigabit (Gb) ports whose job is to connect individual devices such as computers and printers to the network.

  For example, if your company has 100 employees, and you have wired two Cat-5e cables to each user's desk, you'll need at least 200 1 Gb switch ports to support these users. You'll also need a few extra ports for things like printers and Wi-Fi access points.

  Your network design may end up with a total of five 48-port access switches, providing a total of 240 1 Gb ports to support the 100 users.

- **Distribution switches:** A *distribution switch* is a switch that isn't designed to directly support end users. Instead, it's designed to connect the access switches to each other and to your servers. Because the purpose of distribution switches is to manage the aggregate traffic from all the access switches, distribution switches are sometimes called *aggregation switches.*

  Often, a distribution switch uses 10 Gbps ports rather than 1 Gbps ports. The added speed is helpful because each port on the distribution switch carries much more data than each port on the access switches. The access port switches should also be configured with a few 10 Gbps ports, which are used to connect to the distribution switch.

- **Core switches:** The largest networks may also utilize separate *core switches,* which are used to connect the distribution switches. Core switches manage traffic between distribution switches.

Figure 7-6 shows a network that has four separate access switches that are linked together via a distribution switch.

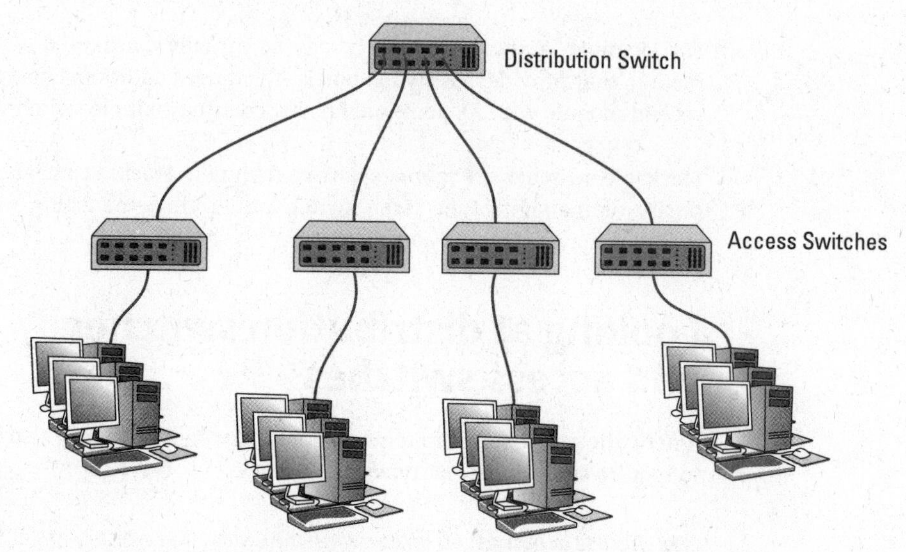

FIGURE 7-6: Distribution and access switches.

Powering Up with Power over Ethernet

In addition to delivering data, Ethernet can also be used to deliver power to devices that don't consume a lot of electrical power. Ethernet circuits that provide power are called *Power over Ethernet* (PoE). Using PoE requires that you use special switches that are designed to provide PoE.

Three types of devices are commonly used on PoE networks:

- **IP phones:** Phone systems that use TCP/IP to transmit voice conversations often use PoE to provide power to the phones at users' desks. This eliminates the need for a separate power supply to power the phone.
- **Wi-Fi access points:** Wi-Fi access points often use PoE rather than a separate power supply. Because Wi-Fi access points are often placed in the ceiling, getting power to them can be tricky. Using PoE eliminates the need to have an electrician provide an electrical outlet next to each Wi-Fi access point.
- **Security cameras:** Surveillance cameras often use PoE for the same reason that Wi-Fi access points do: to eliminate the need for power outlets at each camera location.

PoE does not require any special cabling, but special PoE switches are required. PoE switches are more expensive than non-PoE switches. But if you plan on deploying IP phones, Wi-Fi access points, or surveillance cameras, you should consider using PoE switches to power them.

Looking at Three Types of Network Rooms

As a final topic for this chapter, let's have a brief look at three distinct types of rooms where you might put your network's switches:

- **Main point of entry (MPOE):** Every office building has a main electrical room in which services from the outside world, including not just electrical power but also data connections, enter the building. In networking parlance, this room is called the MPOE.

  The MPOE is often the room where your network's connection to the Internet will arrive. As a rule, you want to avoid placing other networking equipment such as switches and routers in this room, because it's usually not air-conditioned.

- **Main distribution facility (MDF):** Your primary computer equipment room, which you might refer to as the *server room,* is usually the room where you'll place your distribution switches. This room is called the MDF. Access switches may also be placed in the MDF if the MDF is close to the users served by the access switches.

- **Intermediate distribution facility (IDF):** For groups of users who are not close to the MDF, you can consider a separate *IDF* that is closer to those users. This room will contain access switches for those users and will be connected via one or more Ethernet cables to a distribution switch in the MDF. (An IDF also commonly called a *wiring closet.*)

  In a multi-story building, it's common to have an IDF on each floor. And if your network spans several buildings on a campus, it's also common to have an IDF in each building.

- Reviewing basic radio terms
- Considering infrastructure and ad-hoc networks
- Working with a wireless access point
- Configuring Windows for wireless networking
- Securing a wireless network

Chapter 8

Setting Up a Wireless Network

Since the beginning of Ethernet networking, cable has been getting smaller and easier to work with. The original Ethernet cable was about as thick as your thumb, weighed a ton, and was difficult to bend around tight corners. Then came coaxial cable, which was lighter and easier to work with. Coaxial cable was supplanted by unshielded twisted-pair (UTP) cable, which is the cable used for most networks today.

Although cable through the years has become smaller, cheaper, and easier to work with, it is still *cable.* So you have to drill holes in walls, pull cable through ceilings, and get insulation in your hair to wire your entire home or office.

The alternative to networking with cables is, of course, networking *without* cables . . . also known as *wireless networking.* Wireless networks use radio waves to send and receive network signals. As a result, a computer can connect to a wireless network at any location in your home or office.

Wireless networks are especially useful for notebook computers. After all, the main benefit of a notebook computer is that you can carry it around with you wherever you go. At work, you can use your notebook computer at your desk, in the conference room, in the break room, or even out in the parking lot. At home, you can use it in the bedroom, kitchen, den, or game room, or out by the pool. With wireless networking, your notebook computer can be connected to the network no matter where you take it.

Wireless networks have also become extremely useful for other types of mobile devices, such as smartphones and tablet computers. Sure, these devices can connect via a cell network, but that can get real pricey real quick. With a wireless network, though, you can connect your smartphone or tablet without having to pay your cellphone company for the connection time.

This chapter introduces you to the ins and outs of setting up a wireless network. I tell you what you need to know about wireless networking standards, how to plan your wireless network, and how to install and configure wireless network components. And if you end up with a hybrid network of wired and wireless, I show you how to create that, too.

Diving into Wireless Networking

As I mention earlier, a wireless network is just a network that uses radio signals rather than direct cable connections to exchange information. Simple as that. A computer with a wireless network connection is like a cellphone. Just as you don't have to be connected (tethered) to a phone line to use a cellphone, you don't have to be connected to a network cable to use a wireless networked computer.

Here are the key concepts and terms you need to understand to set up and use a basic wireless network:

- **WLAN:** A wireless network is often referred to as a wireless local area network (WLAN). Some people prefer to switch the acronym around to local area wireless network, or LAWN.
- **Wi-Fi:** The term *Wi-Fi* is often used to describe wireless networks although it technically refers to just one form of wireless network: the 802.11 standard. (See the section "Eight-Oh-Two-Dot-Eleventy Something: Understanding Wireless Standards," later in this chapter for more information.)
- **SSID:** A wireless network has a name, known as a SSID. *SSID* stands for *service set identifier.* (Wouldn't that make a great *Jeopardy!* question? I'll take obscure four-letter acronyms for $400, please!) All the computers that belong to a single wireless network must have the same SSID.

- **Channels:** Wireless networks can transmit over any of several channels. For computers to talk to one another, though, they must be configured to transmit on the same channel.
- **Ad-hoc:** The simplest type of wireless network consists of two or more computers with wireless network adapters. This type of network is an *ad-hoc mode network.*
- **Infrastructure mode:** A more complex type of network is an infrastructure mode network. All this really means is that a group of wireless computers can be connected not only to one another, but also to an existing cabled network via a device called a *wireless access point* (WAP). (I tell you more about ad-hoc and infrastructure networks later in this chapter.)

A Little High School Electronics

I was a real nerd in high school: I took three years of electronics. The electronics class at my school was right next door to the auto shop. All the cool kids took auto shop, of course, and only nerds like me took electronics. We hung in there, though, and learned all about capacitors and diodes while the cool kids were learning how to raise their cars and install 2-gigawatt stereo systems.

It turns out that a little of that high school electronics information proves useful when it comes to wireless networking — not much, but a little. You'll understand wireless networking much better if you know the meanings of some basic radio terms.

Waves and frequencies

For starters, radio consists of electromagnetic waves sent through the atmosphere. You can't see or hear them, but radio receivers can pick them up and convert them to sounds, images, or — in the case of wireless networks — data. Radio waves are actually cyclical waves of electronic energy that repeat at a particular rate: the *frequency.*

Figure 8-1 shows two frequencies of radio waves. The first is one cycle per second; the second is two cycles per second. (Real radio doesn't operate at that low a frequency, but I figured that one and two cycles per second would be easier to draw than 680,000 and 2.4 million cycles per second.)

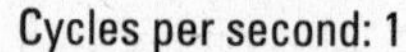

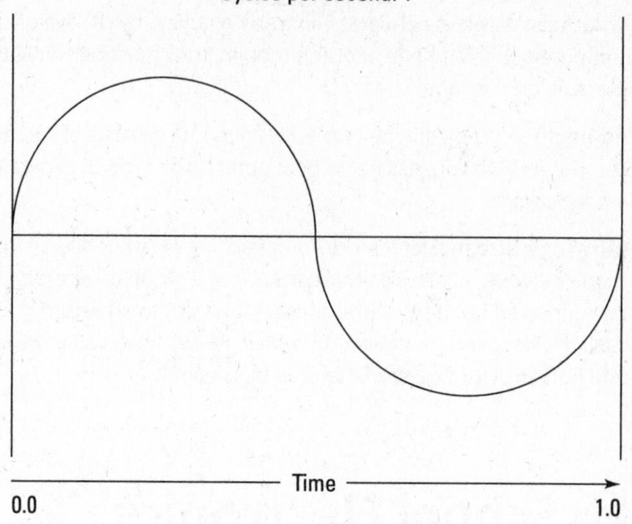

Cycles per second: 2

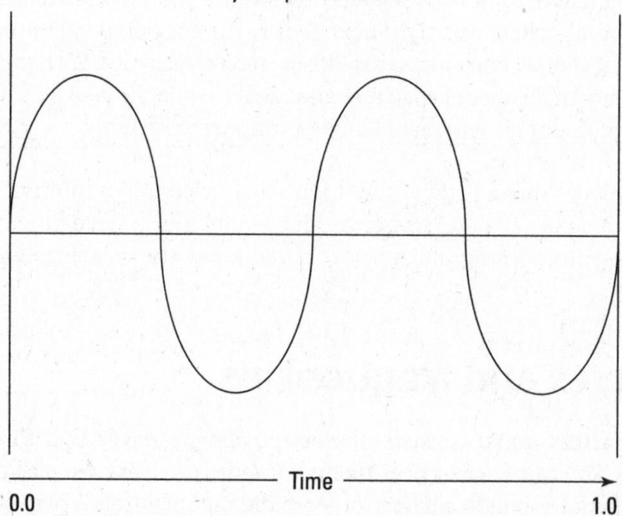

FIGURE 8-1: Radio waves frequently have frequency.

TIP

The measure of a frequency is *cycles per second*, which indicates how many complete cycles the wave makes in 1 second (duh). In honor of Heinrich Hertz — who was the first person to successfully send and receive radio waves (it happened in the 1880s) — cycles per second is usually referred to as *Hertz*, abbreviated *Hz*. Thus, 1 Hz is one cycle per second.

Incidentally, when the prefix *K* (for *kilo*, or 1,000), *M* (for *mega*, 1 million), or *G* (for *giga*, 1 billion) is added to the front of Hz, the *H* is still capitalized. Thus, 2.4 MHz is correct (not 2.4 Mhz).

So the beauty of radio frequencies is that transmitters can be tuned to broadcast radio waves at a precise frequency. Likewise, receivers can be tuned to receive radio waves at a precise frequency, ignoring waves at other frequencies. That's why you can tune the radio in your car to listen to dozens of radio stations: Each station broadcasts at its own frequency.

Wavelength and antennas

A term related to frequency is *wavelength.* Radio waves travel at the speed of light, and *wavelength* refers to how far the radio signal travels with each cycle. Or, put another way, *wavelength* refers to the physical distance between the crest of each wave. Because the speed of light is roughly 182,282 miles per second, for example, the wavelength of a 1 Hz radio wave is about 182,282 miles. The wavelength of a 2 Hz signal is about half that: a mere 91,141 miles.

As you can see, the wavelength decreases as the frequency increases. The wavelength of a typical AM radio station broadcasting at 580 KHz is about 522 yards. For a TV station broadcasting at 100 MHz, it's about 3 yards. For a wireless network broadcasting at 2.4 GHz, the wavelength is just shorter than 5 inches.

And the shorter the wavelength, the smaller the antenna needs to be to adequately receive the signal. As a result, higher-frequency transmissions need smaller antennas. You may have noticed that AM radio stations usually have huge antennas mounted on top of tall towers, but cellphone transmitters are much smaller, and their towers aren't nearly as tall because cellphones operate on a higher frequency than AM radio stations do. So who decides what type of radio gets to use specific frequencies? That's where spectrums and the FCC come in.

Spectrums and the FCC

Spectrum refers to a continuous range of frequencies on which radio can operate. In the United States, the Federal Communications Commission (FCC) regulates not only how much of Janet Jackson can be shown at the Super Bowl, but also how various portions of the radio spectrum can be used. Essentially, the FCC has divided the radio spectrum into dozens of small ranges — *bands* — and restricted certain uses to certain bands. AM radio, for example, operates in the band from 535 KHz to 1,700 KHz.

AND NOW, A WORD FROM THE IRONY DEPARTMENT

I was an English-literature major in college, so I like to use literary devices such as irony. I don't get to use it much in the computer books I write, so when I get the chance to use irony, I like to jump on it like a hog out of water.

So here's my juicy bit of irony for today: The very first Ethernet system was actually a wireless network. Ethernet traces its roots to a network called AlohaNet, developed at the University of Hawaii in 1970. This network transmitted its data by using small radios. If two computers tried to broadcast data at the same time, the computers detected the collision and tried again after a short, random delay. This technique was the inspiration for the basic technique of Ethernet, now called "carrier sense multiple access with collision detection" or CSMA/CD. The wireless AlohaNet was the network that inspired Robert Metcalfe to develop his cabled network, which he called Ethernet, as his doctoral thesis at Harvard in 1973.

For the next 20 years or so, Ethernet was pretty much a cable-only network. It wasn't until the mid-1990s that Ethernet finally returned to its wireless roots.

Table 8-1 lists some of the most popular bands. Note that some of these bands are wide — UHF television begins at 470 MHz and ends at 806 MHz — but other bands are restricted to a specific frequency. The difference between the lowest and highest frequency within a band is the *bandwidth.*

TABLE 8-1 **Popular Bands of the Radio Spectrum**

Band	Use
535 KHz–1,700 KHz	AM radio
5.9 MHz–26.1 MHz	Shortwave radio
26.96 MHz–27.41 MHz	Citizens Band (CB) radio
54 MHz–88 MHz	Television (VHF channels 2–6)
88 MHz–108 MHz	FM radio
174 MHz–220 MHz	Television (VHF channels 7–13)
470 MHz–806 MHz	Television (UHF channels)
806 MHz–890 MHz	Cellular networks
900 MHz	Cordless phones and wireless networks (802.11ah)

Band	Use
1850 MHz–1990 MHz	PCS cellular
2.4 GHz–2.4835 GHz	Cordless phones and wireless networks (802.11b and 802.11n)
4 GHz–5 GHz	Large-dish satellite TV
5 GHz	Wireless networks (802.11a)
11.7 GHz–12.7 GHz	Small-dish satellite TV

Two of the bands in the spectrum are allocated for use by wireless networks: 2.4 GHz and 5 GHz. Note that these bands aren't devoted exclusively to wireless networks. In particular, the 2.4 GHz band shares its space with cordless phones. As a result, cordless phones sometimes interfere with wireless networks. Note also that, as of 2016, some wireless networks can also operate in the 900 MHz spectrum.

Eight-Oh-Two-Dot-Eleventy Something: Understanding Wireless Standards

The most popular standards for wireless networks are the IEEE 802.11 standards. These standards are essential wireless Ethernet standards and use many of the same networking techniques that the cabled Ethernet standards (in other words, 802.3) use. Most notably, 802.11 networks use the same CSMA/CD technique as cabled Ethernet to recover from network collisions.

The 802.11 standards address the bottom two layers of the IEEE seven-layer model: the Physical layer and the Media Access Control (MAC) layer. Note that TCP/IP protocols apply to higher layers of the model. As a result, TCP/IP runs just fine on 802.11 networks.

The original 802.11 standard was adopted in 1997. Two additions to the standard, 802.11a and 802.11b, were adopted in 1999. Then came 802.11g in 2003 and 802.11n in 2009.

802.11n ruled the roost for a few years, until the latest to gain widespread acceptance came out in 2014: 802.11ac. Still more variations are in the works, including 802.11ah, which will operate in the 900 MHz spectrum.

Table 8-2 summarizes the basic characteristics of the five most popular variants of 802.11 as of early 2016. Currently, most wireless networks are based on the 802.11n and 802.11ac standards.

TABLE 8-2 **802.11 Variations**

Standard	Speeds	Frequency	Typical Range (Indoors)
802.11a	Up to 54 Mbps	5 GHz	150 feet
802.11b	Up to 11 Mbps	2.4 GHz	300 feet
802.11g	Up to 54 Mbps	2.4 GHz	300 feet
802.11n	Up to 600 Mbps (but most devices are in the 100 Mbps range)	2.4 GHz	230 feet
802.11ac	Up to 1,300 Mbps	5 GHz	230 feet

Home on the Range

The maximum range of an 802.11ac wireless device indoors is about 230 feet. This can have an interesting effect when you get a bunch of wireless computers together such that some of them are in range of one another but others are not. Suppose that Bart, Homer, and Lisa all have wireless laptops. Bart's computer is 150 feet away from Homer's computer, and Bart's computer is 150 feet away from Lisa's in the opposite direction (see Figure 8-2). In this case, Bart can access both Homer's and Lisa's computers, but Homer can access only Bart's computer, and Lisa can access only Bart's computer. In other words, Homer and Lisa won't be able to access each other's computers because they're 300 feet away from each other, well beyond the 230-foot range limit. (This is starting to sound suspiciously like an algebra problem. Now suppose that Homer starts walking toward Bart at 2 miles per hour, and Lisa starts running toward Bart at 4 miles per hour. . . .)

Note: Although the normal range for 802.11ac is 230 feet, the range may be less in actual practice. Obstacles — solid walls, bad weather, cordless phones, microwave ovens, backyard nuclear reactors, and so on — can all conspire to reduce the effective range of a wireless adapter. If you're having trouble connecting to the network, sometimes just adjusting the antenna helps.

Also, wireless networks tend to slow down when the distance increases. 802.11ac network devices claim to operate at 1,300 Mbps, but they usually achieve that speed only at close range. The farther out you get, the slower the actual speed becomes. At maximum distance, you may be able to connect, but your effective connection speed will be slow. You should also realize that when you're at the edge of the wireless device's range, you're more likely to lose your connection suddenly due to bad weather.

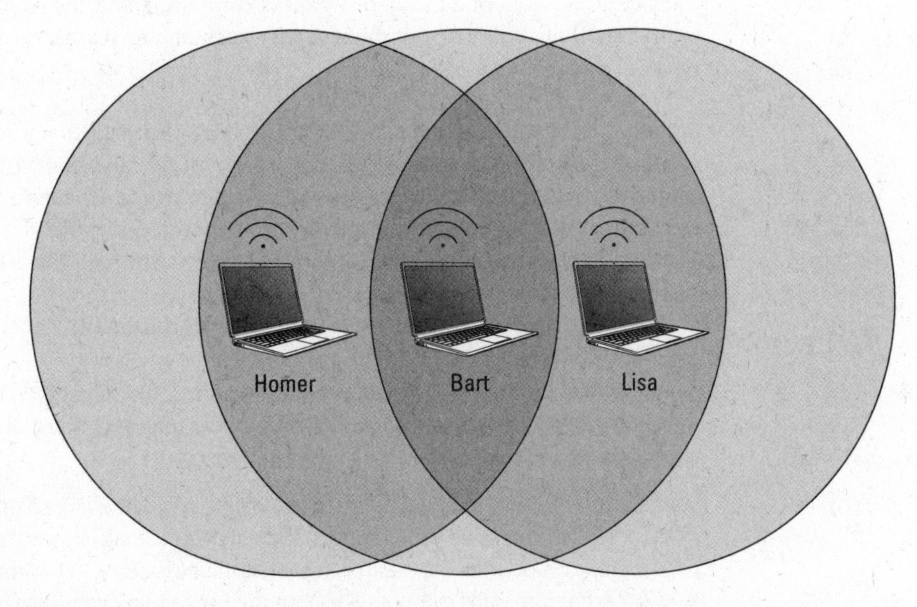

FIGURE 8-2: Homer, Bart, and Lisa playing with their wireless network.

Using Wireless Network Adapters

Each computer that will connect to your wireless network needs a wireless network adapter, which is similar to the network interface card (NIC) used for a standard Ethernet connection. Instead of having a cable connector on the back, however, a wireless network adapter has an antenna.

Just about all portable computers, such as notebooks and tablets, come with wireless networking built in, so you don't have to add a separate wireless network adapter to a portable computer. Desktop computers, though, are a different story. They often don't have built-in wireless networking. If yours doesn't, you'll have to purchase one of two types of wireless adapters:

- **A wireless PCI card:** You install this wireless network adapter in an available slot inside your desktop computer. Yup, you need to take your computer apart, so use this type of card only if you have the expertise and the nerves to dig into your computer's guts. A typical 802.11ac version costs about $75.
- **A wireless USB adapter:** This gizmo is a separate device that plugs into a USB port on your computer. A USB adapter should cost about the same as a PCI adapter. And you can install it without taking your computer apart.

Setting Wireless Access Points

Unlike cabled networks, wireless networks don't need a separate switch. If all you want to do is network a group of wireless computers and the computers are close to one another, you just purchase a wireless adapter for each computer and — *voilà!* — instant network.

But what if you already have an existing cabled network? Suppose that you work at an office with 15 computers all cabled up nicely, and you just want to add a couple of wireless notebook computers to the network. Or suppose that you have two computers in your den connected with network cable, but you want to link up a computer in your bedroom without pulling cable through the attic.

That's where a WAP comes in. A WAP actually performs two functions:

- It acts as a central connection point for all your computers that have wireless network adapters. In effect, the WAP performs the same function that a network switch performs for a wired network.
- It links your wireless network to your existing wired network so that your wired computer and your wireless computers get along like one big happy family. This sounds like the makings of a Dr. Seuss story. ("Now the wireless sneeches had hubs without wires. But the twisted-pair sneeches had cables to thires. . . .")

TIP

Wireless access points are sometimes just called access points (APs), which is basically a box with an antenna (or often a pair of antennae) and an RJ-45 Ethernet port. You plug the AP into a network cable and then plug the other end of the cable into a hub or switch, and your wireless network should be able to connect to your cabled network.

Figure 8-3 shows how an access point acts as a central connection point for wireless computers and also how it bridges your wireless network to your wired network.

Infrastructure mode

When you set up a wireless network with an AP, you're creating an *infrastructure mode network:* The AP provides a permanent infrastructure for the network. The APs are installed at fixed physical locations, so the network has relatively stable boundaries. Whenever a mobile computer wanders into the range of one of the APs, it has come into the sphere of the network and can connect.

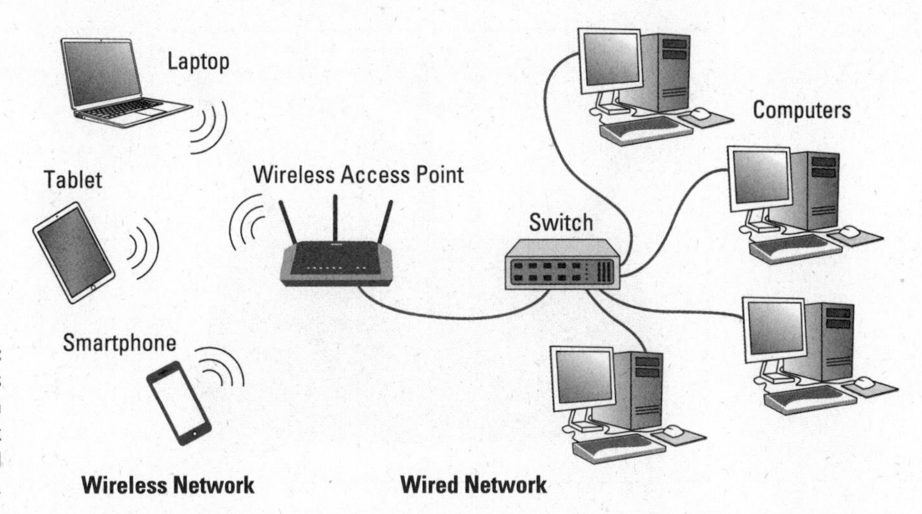

FIGURE 8-3: A wireless access point connects a wireless network to a cabled network.

An access point and all the wireless computers that are connected to it are called a *Basic Service Set* (BSS). Each BSS is identified by a *Service Set Identifier* (SSID). When you configure an access point, you specify the SSID that you want to use. The default SSID on most access points is a generic name that often includes the name of the manufacturer of the device. However, you can easily change the SSID to something more meaningful.

Multifunction WAPs

TIP

Wireless access points often include other built-in features. Some access points double as Ethernet switches, sporting a bank of RJ-45 ports that you can plug other computers or devices into. In addition, some access ports include broadband cable or DSL firewall routers that enable you to connect to the Internet. Figure 8-4 shows a typical 802.11ac wireless router intended for home or small office use. A device of this type would typically include the following:

- An 802.11ac wireless access point that can support multiple wireless devices
- A router with firewall capabilities that can be connected directly to the Internet output from a cable or DSL router
- A four-port gigabit Ethernet switch to connect cabled computers or other devices
- One or more USB ports that enable you to connect USB printers or disk drives to your network

TIP

A multifunction access point designed to serve as an Internet gateway for home networks is sometimes called a *wireless router* or a *residential gateway.*

FIGURE 8-4: A typical wireless router.

Roaming Capabilities

You can use two or more wireless access points to create a large wireless network in which computer users can roam from area to area and still be connected to the wireless network. As the user moves out of the range of one access point, another access point automatically picks up the user and takes over without interrupting the user's network service.

To set up two or more access points for roaming, you must carefully place the access points so that all areas of the office or building that are being networked are in range of at least one of the access points. Then just make sure that all the computers and access points use the same SSID.

Two or more access points joined for roaming, along with all the wireless computers connected to any of the access points, form an *Extended Service Set* (ESS). The access points in the ESS are usually connected to a wired network.

One limitation of roaming is that each access point in an ESS must be on the same TCP/IP subnet. That way, a computer that roams from one access point to another within the ESS retains the same IP address. If the access points had a different subnet, a roaming computer would have to change IP addresses when it moved from one access point to another.

Wireless bridging

Another use for wireless APs is to bridge separate subnets that can't easily be connected by cable. Suppose that you have two office buildings that are only about

50 feet apart. To run cable from one building to the other, you'd have to bury conduit — a potentially expensive job. Because the buildings are so close, though, you can probably connect them with a pair of wireless access points that function as a *wireless bridge* between the two networks. Connect one of the access points to the first network and the other access points to the second network. Then configure both APs to use the same SSID and channel.

Ad-hoc networks

A WAP isn't necessary to set up a wireless network. Any time two or more wireless devices come within range of each other, they can link up to form an ad-hoc network. If you and a few of your friends all have notebook computers with wireless adapters, for example, you can meet anywhere and form an ad-hoc network.

All the computers within range of one another in an ad-hoc network are an *Independent Basic Service Set* (IBSS).

Configuring a Wireless Access Point

The physical setup for a wireless access point is pretty simple: You take it out of the box, put it on a shelf or on top of a bookcase near a network jack and a power outlet, plug in the power cable, and plug in the network cable.

The software configuration for an access is a little more involved but still not very complicated. It's usually done via a web interface. To get to the configuration page for the access, you need to know its IP address. Then you just type that address in the address bar of a browser on any computer on the network.

Multifunction access points usually provide DHCP and NAT services for the networks and double as the network's gateway router. As a result, they typically have a private IP address that's either at the beginning of one of the Internet's private IP address ranges, typically `192.168.0.1` or `10.0.0.1`. Consult the documentation that came with the AP to find out more.

If you use a multifunction AP that serves as both your wireless AP and your Internet router, and you can't remember the IP address, run the `IPCONFIG` command at a command prompt on any computer on the network. The Default Gateway IP address should be the IP address of the access point.

Basic configuration options

Figure 8-5 shows the main configuration screen for a typical router. I called up this configuration page by entering `192.168.0.1` in the address bar of a web browser and then supplying the login password when prompted.

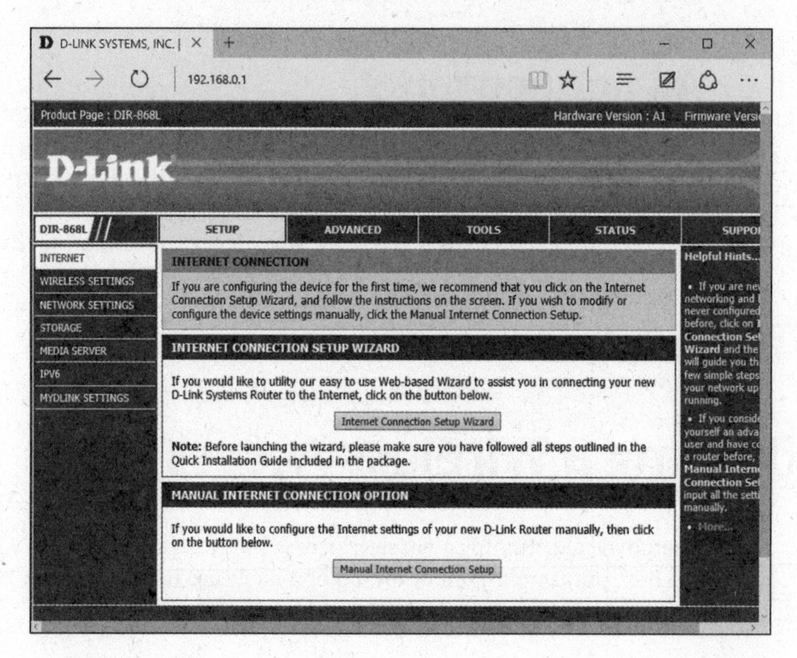

FIGURE 8-5: The main configuration page for a typical wireless router.

On the main setup page of this router, you configure information such as the hostname and IP address of the router and whether the router's DHCP server should be enabled. Options found on additional tabs allow you to configure wireless settings, such as the network name (SSID), the type of security to enforce, and a variety of other settings.

DHCP configuration

You can configure most multifunction access points to operate as a DHCP server. For small networks, the access point is commonly the DHCP server for the entire network. In that case, you need to configure the access point's DHCP server.

Figure 8-6 shows the DHCP configuration page for a typical wireless router. To enable DHCP, you select the Enable option and then specify the other configuration options to use for the DHCP server.

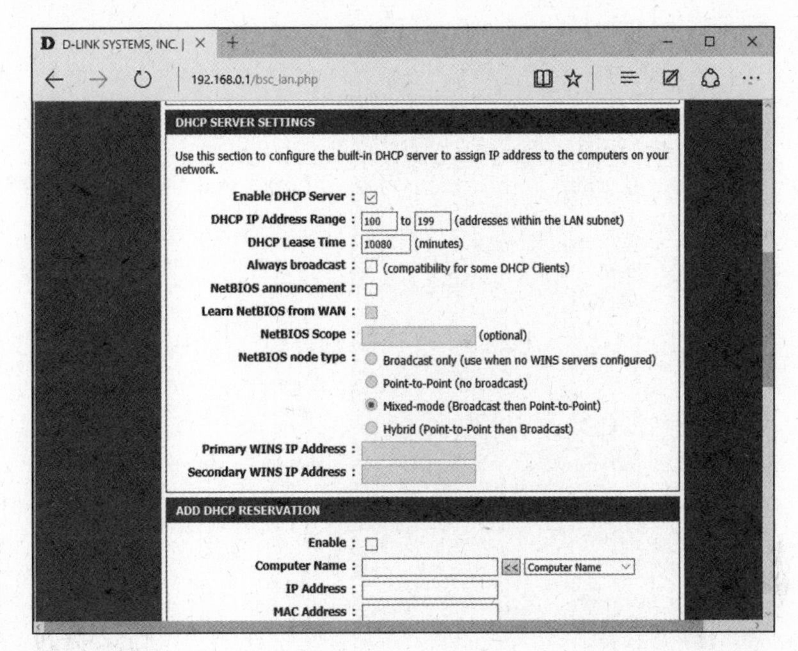

FIGURE 8-6: Configuring DHCP for a typical wireless router.

Larger networks with more demanding DHCP requirements are likely to have a separate DHCP server running on another computer. In that case, you can defer to the existing server by disabling the DHCP server in the access point.

For more information on configuring a DHCP server, refer to Chapter 6.

Connecting to a Wireless Network

Connecting to a wireless network on a Windows computer is straightforward. Windows automatically detects any wireless networks that are in range and displays them in a list when you tap the Wireless icon at the bottom of the screen, as shown in Figure 8-7.

To connect to a network, just tap it, and then enter the security key when prompted. If the key is correct, you'll be connected.

At the time you connect, you can choose to connect to the network automatically whenever it's in range. If you select this option, you won't have to select the network manually or enter the security key; you'll just be connected automatically.

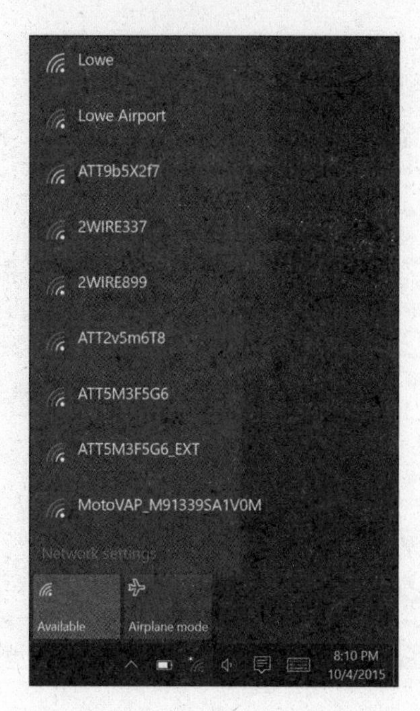

FIGURE 8-7: Choosing a wireless network.

Windows remembers every network you connect to, which is a plus for networks you frequently use but a drawback for networks you'll likely never use again. To tell Windows to forget a network, follow these steps:

1. **Click Start, and then click Settings.**

   The Settings window appears.

2. **Click Network & Internet.**

   This brings up the Network & Internet page, which lists the known networks.

3. **Scroll to the bottom of the list of known networks, and then click Manage Wi-Fi Settings.**

   This brings up the Manage Wi-Fi Settings page, which includes a section titled Manage Known Networks.

4. **In the Manage Known Networks section, click the network you want to forget.**

   The network is selected, as shown in Figure 8-8.

5. **Click Forget.**

   The network will be forgotten. To log into this network again, you'll have to enter the security key.

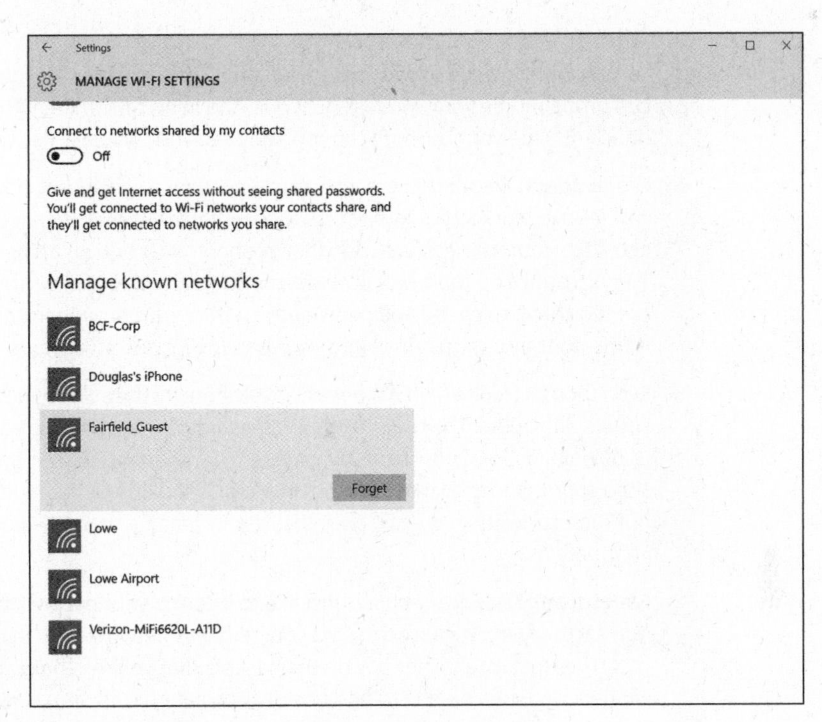

FIGURE 8-8: Forgetting a wireless network in Windows 10.

Paying Attention to Wireless Network Security

Before you unleash a wireless access point on your network, you should first consider the inherent security risks that come along with wireless networking. Unless you take a few basic precautions, adding a wireless access point will expose the innards of your network to anyone within range.

Refer to Chapter 19 for details about network security in general. The information in this section will help you prevent unwanted visitors from gaining access to your network via your wireless access point, but the security techniques outlined in Chapter 19 are a must regardless of whether you provide wireless access.

The following paragraphs describe the types of security threats that wireless networks are most likely to encounter. You should take each of these kinds of threats into consideration when you plan your network's security:

» **Intruders:** With a wired network, an intruder usually must gain access to your facility to physically connect to your network. That's not so with a wireless network. Anyone with a wireless device can gain access to your network if she can get within range of your network's radio signals. At home, your neighbors can probably see your wireless network. And in an office, kids sitting on a bench outside your building can probably see your wireless network.

- **Freeloaders:** *Freeloaders* are intruders who want to piggyback on your wireless network to get free access to the Internet. If they manage to gain access to your wireless network, they probably won't do anything malicious: They'll just fire up their web browsers and surf the web. These are folks who are too cheap to spend $40 per month on their own broadband connection at home, so they'd rather drive into your parking lot and steal yours.

  Even though freeloaders may be relatively benign, they can be a potential source of trouble. They suck up your bandwidth. They may use your network to download illegal pornography, or they may try to hijack your email server to send spam. And they may start out innocently looking for free Internet access, but their curiosity may grow once they get in, leading them to snoop around your network.

- **Eavesdroppers:** *Eavesdroppers* just like to listen to your network traffic. They don't actually try to gain access via your wireless network — at least, not at first. They just listen. They spy on the packets that you're sending over the wireless network, hoping to find useful information such as passwords or credit card numbers.

- **Spoilers:** A *spoiler* is a hacker who gets his kicks from jamming networks so that they become unusable. A spoiler usually accomplishes this act by flooding the network with meaningless traffic so that legitimate traffic gets lost in the flow. Spoilers may also try to place viruses or worm programs on your network via an unsecured wireless connection.

ROGUE ACCESS POINTS

One of the biggest problems that business networks face is the problem of *rogue access points,* which are access points that suddenly appears on your network out of nowhere. What usually happens is that an employee wants to connect his iPad or smartphone to your company network, but you won't give him the password. So the user stops at Walmart on the way home from work one day, buys a cheap wireless router, and plugs it into your network without asking permission.

Now, in spite of all the elaborate security precautions you've taken to fence in your network, this well-meaning user has opened the barn door. It's *very* unlikely that the user will enable the security features of the wireless access point; in fact, he probably isn't even aware that wireless access devices *have* security features.

Unless you take some kind of action to find it, a rogue access point can operate undetected on your network for months or even years. You may not discover it until you report to work one day and find that your network has been trashed by an intruder who found his way into your network via an unprotected wireless access point that you didn't even know existed.

Here are some steps you can take to reduce the risk of rogue access points appearing on your system:

- **Establish a policy prohibiting users from installing wireless access points on their own.** Then make sure that you inform all network users of the policy, and let them know why installing an access point on their own can be such a major problem.
- **If possible, establish a program that quickly and inexpensively grants wireless access to users who want it.** Rogue access points show up in the first place because users who want access can't get it. If you make it easier for users to get legitimate wireless access, you're less likely to find wireless access points hidden behind file cabinets or in flower pots.
- **Once in a while, take a walk through the premises, looking for rogue access points.** Take a look at every network outlet in the building and see what's connected to it.
- **Turn off all your wireless access points and then walk around the premises with a wireless-equipped mobile device such as a smartphone and look for wireless networks that pop up.** Just because you detect a wireless network, of course, doesn't mean you've found a rogue access point — you may have stumbled onto a wireless network in a nearby office or home. But knowing what wireless networks are available from within your office will help you determine whether any rogue access points exist.

Hopefully I've convinced you that wireless networks do, indeed, pose many security risks. Here are some steps you can take to help secure your wireless network:

- **Create a secure wireless password.** The first thing you should do when you set up a wireless network is change the default password required to access the network. Most manufacturers of wireless routers secure the SSID with a standard password that is well known. Make sure you change it to something that only you know, and then only share that password with those you want to grant access to.
- **Change the administrative password.** Most access points have a web-based setup page that you can access from any web browser to configure the access

point's settings. The setup page is protected by a username and password, but the username and password are initially set to default values that are easy to guess. Anyone who gains access to your network can then log in to the administrative page and take control of your network.

- **Hide the SSID.** A simple step you can take to secure your wireless network is to disable the automatic broadcast of your network's SSID. That way, only those who know of your network's existence will be able to access it. (Securing the SSID isn't a complete security solution, so you shouldn't rely on it as your only security mechanism.)
- **Disable guest mode.** Many access points have a guest-mode feature that enables client computers to specify a blank SSID or to specify "any" as the SSID. If you want to ensure that only clients that know the SSID can join the network, you must disable this feature.
- **Use MAC address filtering.** One of the most effective ways to secure a wireless network is to use a technique called *MAC address filtering.* MAC address filtering allows you to specify a list of MAC addresses for the devices that are allowed to access the network or are prohibited from accessing the network. If a computer with a different MAC address tries to join the network via the access point, the access point will deny access.

TIP

MAC address filtering is a great idea for wireless networks with a fixed number of clients. If you set up a wireless network at your office so that a few workers can connect their notebook computers, you can specify the MAC addresses of those computers in the MAC filtering table. Then other computers won't be able to access the network via the access point.

MAC address filtering isn't bulletproof, but it can go a long way toward keeping unwanted visitors off your network. Unfortunately, MAC address filtering is also pretty inconvenient. Whenever you want to grant access for a new device, you'll have to find out that device's MAC address and add it to the list of permitted devices.

- **Place your access points outside the firewall.** The most effective security technique for wireless networking is placing all your wireless access points *outside* your firewall. That way, all network traffic from wireless users will have to travel through the firewall to access the network.

As you can imagine, doing this can significantly limit network access for wireless users. To get around those limitations, you can enable a virtual private network (VPN) connection for your wireless users. The VPN will allow full network access to authorized wireless users.

Obviously, this solution requires a bit of work to set up and can be a little inconvenient for your users, but it's an excellent way to fully secure your wireless access points.

DON'T NEGLECT THE BASICS

The security techniques described in this chapter are specific to wireless networks. They should be used alongside the basic security techniques that are presented in Chapter 19. In other words, don't forget the basics, such as the following:

- Use strong passwords for your user accounts.
- Apply security patches to your servers.
- Change default server account information (especially the administrator password).
- Disable unnecessary services.
- Check your server logs regularly.
- Install virus protection.
- Back up!

» Examining high-speed private lines

» Using a router

» Securing your connection with a firewall

» Using the firewall that comes with Windows

Chapter 9

Connecting to the Internet

So you decided to connect your network to the Internet. All you have to do is call the cable company and have it send someone out, right? Wrong. Unfortunately, connecting to the Internet involves more than just calling the cable company. For starters, you have to make sure that cable is the right way to connect. Then you have to select and configure the software you use to access the Internet. Finally, you have to lie awake at night worrying whether hackers are breaking into your network via its Internet connection.

Not to worry. The advice in this chapter helps you decide how to connect to the Internet and, once the decision is made, how to do it safely.

Connecting to the Internet

Connecting to the Internet isn't free. For starters, you have to purchase the computer equipment necessary to make the connection. Then you have to obtain a connection from an Internet service provider (ISP). The ISP charges you a monthly fee that depends on the speed and capacity of the connection.

The following sections describe the most commonly used methods of connecting network users to the Internet.

Connecting with cable or DSL

For small and home offices, the two most popular methods of connecting to the Internet are cable and digital subscriber line (DSL). Cable and DSL connections are often called *broadband connections* for technical reasons you don't really want to know.

Cable Internet access works over the same cable that brings 40 billion TV channels into your home, whereas DSL is a digital phone service that works over a standard phone line. Both offer three major advantages over old-fashioned dialup connections:

- **Cable and DSL are much faster than dialup connections.** A cable connection can be anywhere from 10 to 200 times faster than a dialup connection, depending on the service you get. And the speed of a DSL line is comparable with cable. (Although DSL is a dedicated connection, cable connections are shared among several subscribers. The speed of a cable connection may slow down when several subscribers use the connection simultaneously.)
- **With cable and DSL, you're always connected to the Internet.** You don't have to connect and disconnect each time you want to go online like you would if you use a modem. No more waiting for the modem to dial your service provider and listening to the annoying modem shriek while it attempts to establish a connection.
- **Cable and DSL don't tie up a phone line while you're online.** With cable, your Internet connection works over TV cables, not over phone cables. With DSL, the phone company installs a separate phone line for the DSL service, so your regular phone line isn't affected.

Unfortunately, there's no such thing as a free lunch, and the high-speed, always-on connections offered by cable and DSL don't come without a price. For starters, you can expect to pay a higher monthly access fee for cable or DSL. In most areas of the United States, cable runs about $50 per month for residential users; business users can expect to pay more, especially if more than one user will be connected to the Internet via the cable.

The cost for DSL service depends on the access speed you choose. In some areas, residential users can get a relatively slow DSL connection for as little as $30 per month. For higher access speeds or for business users, DSL can cost substantially more.

Too, cable and DSL access aren't available everywhere. But if you live in an area where cable or DSL isn't available, you can still get high-speed Internet access by using a satellite hookup or a cellular network.

Connecting with high-speed private lines

If your network is large and high-speed Internet access is a high priority, contact your local phone company (or companies) about installing a dedicated high-speed digital line. These lines can cost you plenty (on the order of hundreds of dollars per month), so they're best suited for large networks in which 20 or more users are accessing the Internet simultaneously.

The following paragraphs describe three basic options for high-speed private lines:

- **T1 and T3 lines:** T1 and T3 lines run over standard copper phone lines. A T1 line has a connection speed of up to 1.544 Mbps. A T3 line is a bit faster: It transmits data at 44.184 Mbps. Of course, T3 lines are more expensive than T1 lines.

  If you don't have enough users to justify the expense of an entire T1 or T3 line, you can lease just a portion of the line. With a fractional T1 line, you can get connections with speeds of 128 Kbps to 768 Kbps; with a fractional T3 line, you can choose speeds ranging from 4.6 Mbps to 32 Mbps.

TIP

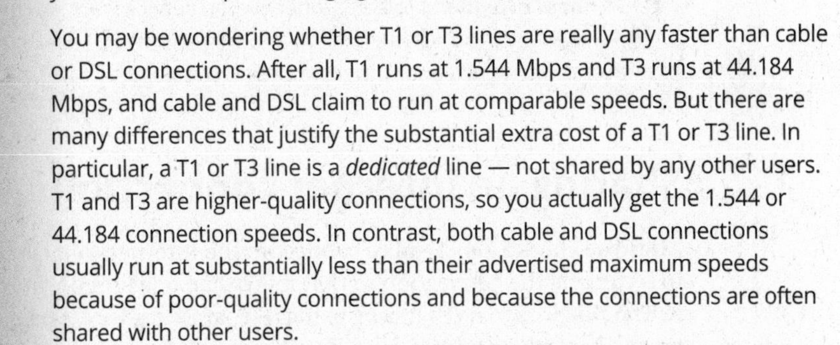

  You may be wondering whether T1 or T3 lines are really any faster than cable or DSL connections. After all, T1 runs at 1.544 Mbps and T3 runs at 44.184 Mbps, and cable and DSL claim to run at comparable speeds. But there are many differences that justify the substantial extra cost of a T1 or T3 line. In particular, a T1 or T3 line is a *dedicated* line — not shared by any other users. T1 and T3 are higher-quality connections, so you actually get the 1.544 or 44.184 connection speeds. In contrast, both cable and DSL connections usually run at substantially less than their advertised maximum speeds because of poor-quality connections and because the connections are often shared with other users.

  Also, both cable and DSL connections download data much faster than they upload data. So, while you may be able to download data at 100 Mbps over a cable connections, you'll be lucky to upload data at much more than 7 or 8 Mbps.

- **Business-class cable:** Cable TV providers (such as Comcast) offer business-class service on their cable network. The price and speed depends on your area. For example, where I live, I can get 100 Mbps service for about $400/month.

Like residential cable, the upload speed for business-class cable is usually much slower than the download speed. For example, a typical plan that allows 100 Mbps for downloads can support only 10 Mbps for uploads. Thus, if you need to upload large amounts of data, you'll notice the performance drop.

Another drawback of business-class cable service is that it is, well, cable service. Your Internet connection is service by the same people who service cable TV in your community. Although business-class customers get priority service over residential customers, business-class service usually does not include response-time guarantees the way T1/T3 or fiber service does. So if your connection goes down, you might find yourself down for hours or even a few days instead of minutes or, at worse, a few hours.

- **Fiber optic:** The fastest, most reliable, and most expensive form of Internet connection is fiber optic. Fiber optic cable uses strands of glass to transmit data over light signals at very high speeds. Because the light signals traveling within the fiber cables are not subject to electromagnetic interference, fiber connections are extremely reliable; about the only thing that can interrupt a fiber connection is if someone physically cuts the wire.

  Fiber can also be very expensive. In some locations, a 100 Mbps fiber connection can cost well over $1,000 per month. However, the connection is extremely reliable, and response time to service interruptions is measured in minutes instead of hours. And prices are coming down, especially in metropolitan areas as fiber carriers continue to build out their networks.

- **Wireless providers:** In areas where wired service (such as cable or fiber) is not available, you may be able to find wireless service, which provides Internet access using cellular or other wireless technology.

Sharing an Internet connection

After you choose a method to connect to the Internet, you can turn your attention to setting up the connection so that more than one user on your network can share it. The best way to do that is by using a separate device called a *router*. You can pick up an inexpensive router for a small network for less than $75. Routers suitable for larger networks will, naturally, cost a bit more.

Because all communications between your network and the Internet must go through the router, the router is a natural place to provide the security measures necessary to keep your network safe from the many perils of the Internet. As a result, a router used for Internet connections often doubles as a firewall, as described in the section "Using a firewall," later in this chapter.

Securing Your Connection with a Firewall

If your network is connected to the Internet, a whole host of security issues bubbles to the surface. You probably connected your network to the Internet so that your network's users could get out to the Internet. Unfortunately, however, your Internet connection is a two-way street. It not only enables your network's users to step outside the bounds of your network to access the Internet, but it also enables others to step in and access your network.

And step in they will. The world is filled with hackers who are looking for networks like yours to break into. They may do it just for the fun of it, or they may do it to steal your customers' credit card numbers or to coerce your mail server into sending thousands of spam messages on behalf of the bad guys. Whatever their motive, rest assured that your network will be broken into if you leave it unprotected.

Using a firewall

A *firewall* is a security-conscious router that sits between the Internet and your network with a single-minded task: preventing *them* from getting to *us*. The firewall acts as a security guard between the Internet and your local area network (LAN). All network traffic into and out of the LAN must pass through the firewall, which prevents unauthorized access to the network.

WARNING

Some type of firewall is a must-have if your network has a connection to the Internet, whether that connection is broadband (cable modem or DSL), T1, or some other high-speed connection. Without it, sooner or later a hacker will discover your unprotected network and tell his friends about it, and within a few hours, your network will be toast.

You can set up a firewall in two basic ways:

- **Firewall appliance:** The easiest way, and usually the best choice. A firewall appliance is basically a self-contained router with built-in firewall features.

  Most firewall appliances include web-based interfaces that enable you to connect to the firewall from any computer on your network by using a browser. You can then customize the firewall settings to suit your needs.

- **Server computer:** Can be set up to function as a firewall computer.

  The server can run just about any network operating system, but most dedicated firewall systems run Linux.

Whether you use a firewall appliance or a firewall computer, the firewall must be located between your network and the Internet, as shown in Figure 9-1. Here, one end of the firewall is connected to a network switch, which is, in turn, connected to the other computers on the network. The other end of the firewall is connected to the Internet. As a result, all traffic from the LAN to the Internet (and vice versa) must travel through the firewall.

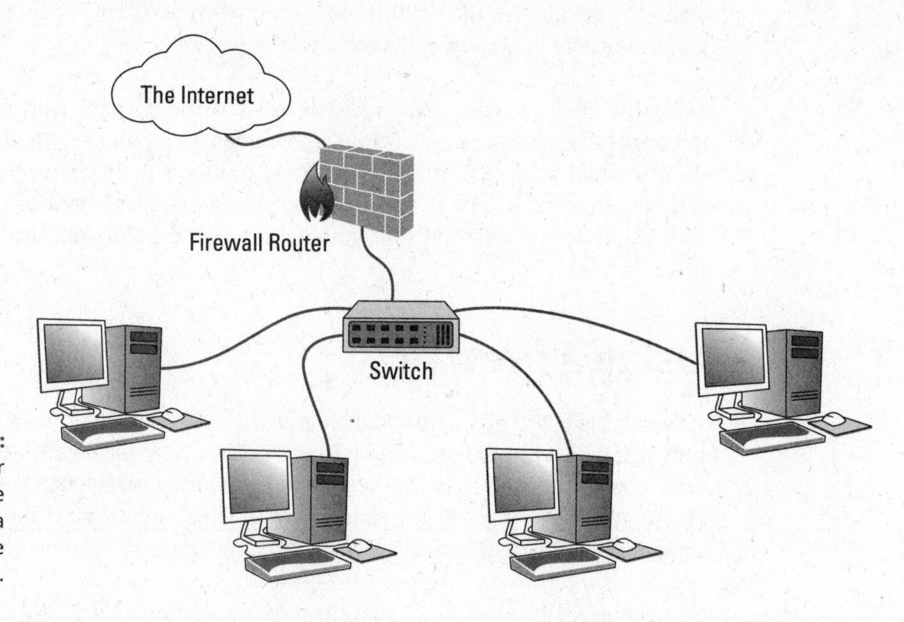

FIGURE 9-1: A firewall router creates a secure link between a network and the Internet.

The term *perimeter* is sometimes used to describe the location of a firewall on your network. In short, a firewall is like a perimeter fence that completely surrounds your property and forces all visitors to enter through the front gate.

WARNING

In large networks, figuring out exactly where the perimeter is located can be a little difficult. If your network has two or more Internet connections, make sure that every one of those connections connects to a firewall — and not directly to the network. You can do this by providing a separate firewall for each Internet connection or by using a firewall with more than one Internet port.

TIP

Some firewall routers can also enforce virus protection for your network. For more information about virus protection, see Chapter 20.

Comparing residential gateways to firewall routers

If you peruse the shelves of your local big-box electronics store, you'll see a variety of devices called *Wi-Fi routers.* Technically, these devices are actually called *residential gateways*, because they provide more than just Wi-Fi capability. A residential gateway typically combines four distinct components in one handy package:

- » A router that can connect a small network to the Internet.
- » A firewall to protect the internal network from hackers who would love nothing more than to compromise your network.
- » A small switch (typically four ports) to connect a few computers to the network. If you have more than four users, you can connect a larger switch to one of the gateway's switch ports. (For more information about switches, refer to Chapter 7.)
- » A Wireless Access Point, which allows wireless devices to connect. (For more information about wireless access, refer to Chapter 8.)

Residential gateways are fine for home networks or for very small businesses. However, if your network has more than a dozen computers, you should consider stepping up to a dedicated firewall router, which separates the firewall and router features from the switching and Wi-Fi features.

A true firewall router has just two types of ports:

- » A wide area network (WAN) port, which connects the firewall router to the Internet
- » A local area network (LAN) port, which connects the firewall router to your internal network

Some firewall routers contain more than one WAN port, allowing you to connect to two separate Internet connections. For more information about this capability, see the section "Providing a Backup Internet Connection" later in this chapter.

Looking at the built-in Windows firewall

Windows includes a built-in firewall that provides basic packet-filtering firewall protection. In most cases, you're better off using a dedicated firewall router because these devices provide better security features than the built-in Windows firewall does. Still, the built-in firewall is suitable for home networks or very small office networks.

Here are the steps that activate the built-in firewall in Windows:

1. **Open the Control Panel.**
2. **Click the System and Security link.**

   The System and Security page appears.
3. **Click the Windows Firewall link.**

   The Windows Firewall page appears.
4. **Click the Turn Windows Firewall On or Off link.**

   The page shown in Figure 9-2 appears.

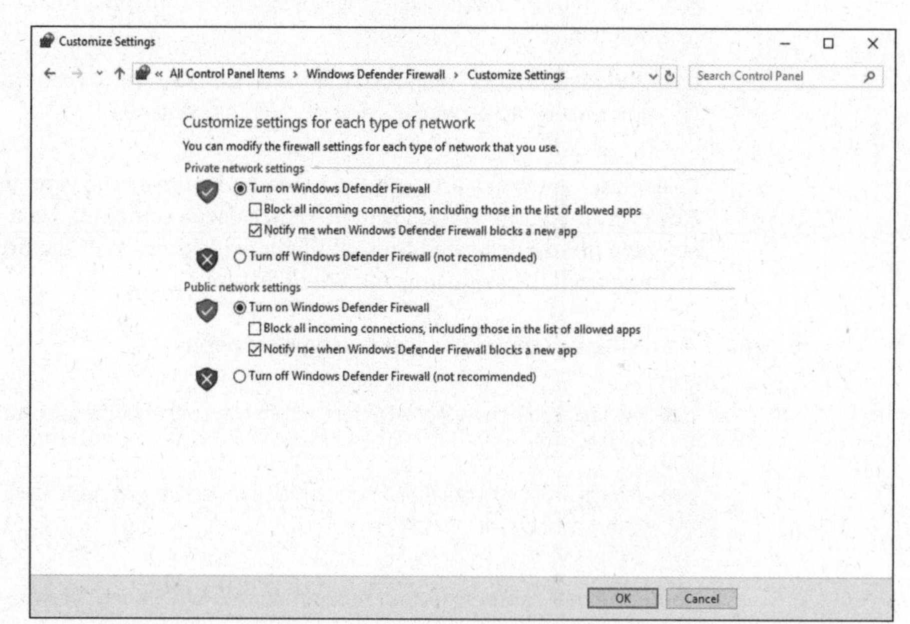

FIGURE 9-2: Activating the Windows firewall.

5. **Select the Turn on Windows Firewall option.**

   Note that you can independently turn the firewall on or off for public network — that is, for your connection to the Internet — and for your home or work network — that is, if you have a network that connects other computers in your home or office. I recommend you either turn the firewall on for both or turn it off for both. Turn the firewall off if you're using a separate firewall built into the router that connects your computer or home or work network to the Internet. Turn the firewall on if you don't have a separate firewall.

I also recommend leaving the Notify Me When Windows Firewall Blocks a New Program option enabled. That way, you'll be notified when the firewall blocks a suspicious program.

6. **Click OK.**

   The firewall is enabled.

Note that the Windows firewall includes additional options you can configure. However, I recommend against fiddling with those options unless you've taken an upper-division college course in computer security.

Do *not* enable the Windows firewall if you're using a separate firewall router to protect your network. Because the other computers on the network are connected directly to the router and not to your computer, the Windows firewall doesn't protect the rest of the network. Additionally, as an unwanted side effect, the rest of the network will lose the capability of accessing your computer.

Providing a Backup Internet Connection

For most businesses, a reliable connection to the Internet is absolutely vital. Consider the consequences of losing your Internet connection:

- **All email stops flowing.** Your users can neither send nor receive emails. Clients, customers, and other business partners can't communicate with you.
- **Access to the web stops.** Your users can't access any websites. If you subscribe to any vital web-based services, such as conferencing sites (like GoToMeeting or Zoom), online banking, or other vital services, those services will be unavailable while your Internet circuit is down.
- **If you use an IP-based phone service such as RingCentral or 8x8, your phones won't work.**
- **If you store data in the cloud, such as on OneDrive, Dropbox, or Google Drive, you won't be able to access your data.**

In short, without the Internet, your business may come to a standstill.

For that reason, it's a good idea to provide not just one connection to the Internet, but two. You can then configure your firewall router to automatically switch to the backup Internet circuit if your primary connection fails. This capability is called *failover*, and it's available in most dedicated firewall router devices. (Note that failover is usually available on residential gateways, commonly known as Wi-Fi routers.)

TIP

In most cases, you can get away with a slower Internet connection to use as your backup connection. For example, if your primary Internet connection is a 1 Gbps fiber connection, you could use a 100 Mbps cable connection for your backup. If your primary circuit goes down, performance will be noticeably slower, but at least you'll still be connected.

When selecting a backup Internet connection, it's usually a good idea to use a different provider than the company that provides your primary connection. That way, if a system-wide outage occurs at your primary connection's provider, the outage is less likely to affect your backup connection.

One common scenario is to use cellular service for your backup connection. A company called CradlePoint (`www.cradlepoint.com`) provides several devices designed specifically for this purpose. For example, CradlePoint's AER1600 devices include the following:

- Five Ethernet ports, each of which can be configured for Internet connections (WAN) or internal connections (LAN).
- Two LTE cellular modems to connect to a cell provider. All you need to do is provide a SIM card for your provider of choice. (If you want, you can provide two SIM cards to provide two backup cell providers.)

Many options are available for providing a backup Internet circuit. None is without cost, but compare that cost with the cost of being without an Internet connection altogether. For most businesses, it's worth a small additional expenditure per month to ensure that the Internet is always available.

3 Working with Servers

IN THIS PART . . .

Set up virtual servers.

Install and configure Windows Server 2019.

Understand and use Active Directory user accounts.

Manage storage on your network.

» Weighing the benefits of virtualization

» Looking at what a hypervisor does

» Considering how disks and networks are virtualized

» Activating Hyper-V

» Creating and using virtual machines

Chapter 10
Virtualizing Your Network

Virtualization is one of the hottest trends in networking today. According to some industry pundits, virtualization is the best thing to happen to computers since the invention of the transistor. If you haven't already begun to virtualize your network, you're standing on the platform watching as the train is pulling out.

This chapter is a brief introduction to virtualization, with an emphasis on using it to leverage your network server hardware to provide more servers using less hardware. In addition to the general concepts of virtualization, you find out how to experiment with virtualization by activating Microsoft's Hyper-V, which is a free virtualization product included with Windows.

Understanding Virtualization

The basic idea behind virtualization is to use software to simulate the existence of hardware. This powerful idea enables you to run more than one independent computer system on a single physical computer system. Suppose that your organization requires a total of 12 servers to meet its needs. You could run each of these 12 servers on a separate computer, in which case you would have 12 computers in your server room, or you could use virtualization to run these 12 servers on just 2 computers. In effect, each of those computers would simulate 6 separate computer systems, each running one of your servers.

Each of the simulated computers is called a *virtual machine* (VM). For all intents and purposes, each virtual machine appears to be a complete, self-contained computer system with its own processor (or, more likely, processors), memory, disk drives, CD-ROM/DVD drives, keyboard, mouse, monitor, network interfaces, USB ports, and so on.

Like a real computer, each virtual machine requires an operating system to do productive work. In a typical network server environment, each virtual machine runs its own copy of Windows Server. The operating system has no idea that it's running on a virtual machine rather than on a real machine.

Here are a few terms you need to be familiar with if you expect to discuss virtualization intelligently:

- **Host:** The actual physical computer on which one or more virtual machines run.
- **Bare metal:** Another term for the host computer that runs one or more virtual machines.
- **Guest:** Another term for a virtual machine running on a host.
- **Guest operating system:** An operating system that runs within a virtual machine. By itself, a guest is just a machine; it requires an operating system to run. The guest operating system is what brings the guest to life.

WARNING

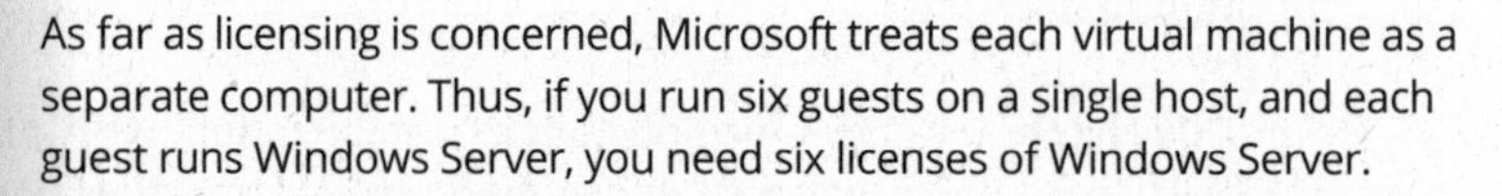
As far as licensing is concerned, Microsoft treats each virtual machine as a separate computer. Thus, if you run six guests on a single host, and each guest runs Windows Server, you need six licenses of Windows Server.

- **Hypervisor:** The virtualization operating system that creates and runs virtual machines. For more information about hypervisors, read the next section, "Understanding Hypervisors."
- **Hardware abstraction layer (HAL):** A layer of software that acts as a go-between to separate actual hardware from the software that interacts with it. An operating system provides a HAL, because it uses device drivers to communicate with actual hardware devices so that software running in the operating system doesn't have to know the details of the specific device it's interacting with. A hypervisor also provides a HAL that enables the guest operating systems in virtual machines to interact with virtualized hardware.

THE LONG TREK OF VIRTUALIZATION

Kids these days think they invented everything, including virtualization.

Little do they know.

Virtualization was developed for PC-based computers in the early 1990s, around the time Captain Picard was flying the Enterprise around in *Star Trek: The Next Generation*.

But the idea is much older than that.

The first virtualized server computers predate Captain Picard by about 20 years. In 1972, IBM released an operating system called simply VM, which had nearly all the basic features found in today's virtualization products.

VM allowed the administrators of IBM's System/370 mainframe computers to create multiple independent virtual machines, each of which was called (you guessed it) a virtual machine, or VM. This terminology is still in use today.

Each VM could run one of the various guest operating systems that were compatible with the System/370 and appeared to this guest operating system to be a complete, independent System/370 computer with its own processor cores, virtual memory, disk partitions, and input/output devices.

The core of the VM system itself was called the *hypervisor* — another term that persists to this day.

The VM product that IBM released in 1972 was actually based on an experimental product that IBM released on a limited basis in 1967.

So whenever someone tells you about this new technology called *virtualization*, you can tell them that it was invented when *Star Trek* was originally on TV. When they ask, "You mean the one with Picard?" you can say, "No, the one with Kirk."

Understanding Hypervisors

At the core of virtualization is a *hypervisor*, a layer of software that manages the creation and execution of virtual machines. A hypervisor provides several core functions:

- It provides a HAL, which virtualizes all the hardware resources of the host computer on which it runs. This includes processor cores, RAM, and I/O devices such as disk drives, keyboards, mice, monitors, USB devices, and so on.

- It creates pools of these abstracted hardware resources that can be allocated to virtual machines.
- It creates virtual machines. A virtual machine is a complete implementation of an idealized computer system that has the hardware resources of the host available to it. The hardware for each virtual machine is drawn from the pools of available hardware resources managed by the hypervisor.
- It manages the execution of its virtual machines, allocating host hardware resources as needed to each virtual machine and starting and stopping virtual machines when requested by users.
- It ensures that each virtual machine is completely isolated from all other virtual machines, so that if a problem develops in one virtual machine, none of the other virtual machines is affected.
- It manages communication among the virtual machines over virtual networks, enabling the virtual machines to connect with each other and with a physical network that reaches beyond the host.

There are two basic types of hypervisors you should know about:

- **Type-1:** A type-1 hypervisor runs directly on the host computer, with no intervening operating system. This is the most efficient type of hypervisor because it has direct access to the hardware resources of the host system.

  The two best-known examples of type-1 hypervisors are VMware's ESXi and Microsoft's Hyper-V. ESXi is part of a suite of popular virtualization products from VMware, and Hyper-V is the built-in virtualization platform that is included with recent versions of Windows Server.
- **Type-2:** A type-2 hypervisor runs as an application within an operating system that runs directly on the host computer. Type-2 hypervisors are less efficient than type-1 hypervisors because when you use a type-2 hypervisor, you add an additional layer of hardware abstraction — the first provided by the operating system that runs natively on the host, and the second by the hypervisor that runs as an application on the host operating system.

TIP

For production use, you should always use type-1 hypervisors because they're much more efficient than type-2 hypervisors. Type-1 hypervisors are considerably more expensive than type-2 hypervisors, however. As a result, many people use inexpensive or free type-2 hypervisors to experiment with virtualization before making a commitment to purchase an expensive type-1 hypervisor.

Understanding Virtual Disks

Computers aren't the only things that are virtualized in a virtual environment. In addition to creating virtual computers, virtualization also creates virtual disk storage. Disk virtualization lets you combine a variety of physical disk storage devices to create pools of disk storage that you can then parcel out to your virtual machines as needed.

Virtualization of disk storage is nothing new. In fact, there are actually several layers of virtualization involved in an actual storage environment. At the lowest level are the actual physical disk drives. Physical disk drives are usually bundled together in arrays of individual drives. This bundling is a type of virtualization in that it creates the image of a single large disk drive that isn't really there. For example, four 2TB disk drives might be combined in an array to create a single 8TB disk drive.

Note that disk arrays are usually used to provide data protection through redundancy. This is commonly called RAID, which stands for *redundant array of inexpensive disks.*

One common form of RAID, called RAID-10, lets you create mirrored pairs of disk drives so that data is always written to both of the drives in a mirror pair. So, if one of the drives in a mirror pair fails, the other drive can carry the load. With RAID-10, the usable capacity of the complete array is equal to one-half of the total capacity of the drives in the array. For example, a RAID-10 array consisting of four 2TB drives contains two pairs of mirrored 2TB disk drives, for a total usable capacity of 4TB.

Another common form of RAID is RAID-5, in which disk drives are combined and one of the drives in the group is used for redundancy. Then, if any one of the drives in the array fails, the remaining drives can be used to re-create the data that was on the drive that failed. The total capacity of a RAID-5 array is equal to the sum of the capacities of the individual drives, minus one of the drives. For example, an array of four 2TB drives in a RAID-5 configuration has a total usable capacity of 6TB.

In a typical virtual environment, the host computers can be connected to disk storage in several distinct ways:

- **Local disk storage:** In local disk storage, disk drives are mounted directly on the host computer and are connected to the host computer via its internal disk drive controllers. For example, a host computer might include four 1TB disk drives mounted within the same chassis as the computer itself. These four drives might be used to form a RAID-10 array with a usable capacity of 2TB.

The main drawbacks of local disk storage is that it's limited to the physical capacity of the host computers and is available only to the host computer that it's installed in.

- **Storage Area Network (SAN):** In a SAN, disk drives are contained in a separate device that is connected to the host via a high-speed controller. The difference between a SAN and local storage is that the SAN is a separate device. Its high-speed connection to the host is often just as fast as the internal connection of local disk storage, but the SAN includes a separate storage controller that is responsible for managing the disk drives.

  A typical SAN can hold a dozen or more disk drives and can allow high-speed connections to more than one host. A SAN can often be expanded by adding one or more expansion chassis, which can contain a dozen or more disk drives each. Thus, a single SAN can manage hundreds of terabytes of disk data.

- **Network attached storage (NAS):** This type of storage is similar to a SAN, but instead of connecting to the hosts via a high-speed controller, a NAS connects to the host computers via standard Ethernet connections and TCP/IP. NAS is the least expensive of all forms of disk storage, but it's also the slowest.

Regardless of the way the storage is attached to the host, the hypervisor consolidates its storage and creates virtual pools of disk storage typically called *data stores*. For example, a hypervisor that has access to three 2TB RAID-5 disk arrays might consolidate them to create a single 6TB data store.

From this data store, you can create *volumes*, which are essentially virtual disk drives that can be allocated to a particular virtual machine. Then, when an operating system is installed in a virtual machine, the operating system can mount the virtual machine's volumes to create drives that the operating system can access.

For example, let's consider a virtual machine that runs Windows Server. If you were to connect to the virtual machine, log in, and use Windows Explorer to look at the disk storage that's available to the machine, you might see a `C:` drive with a capacity of 100GB. That `C:` drive is actually a 100GB volume that is created by the hypervisor and attached to the virtual machine. The 100GB volume, in turn, is allocated from a data store, which might be 4TB in size. The data store is created from disk storage contained in a SAN attached to the host, which might be made up of a RAID-10 array consisting of four 2TB physical disk drives.

So, you can see that there are at least four layers of virtualization required to make the raw storage available on the physical disk drives available to the guest operating system:

- Physical disk drives are aggregated using RAID-10 to create a unified disk image that has built-in redundancy. RAID-10 is, in effect, the first layer of virtualization. This layer is managed entirely by the SAN.
- The storage available on the SAN is abstracted by the hypervisor to create data stores. This is, effectively, a second level of virtualization.
- Portions of a data store are used to create volumes that are then presented to virtual machines. Volumes represent a third layer of virtualization.
- The guest operating system sees the volumes as if they're physical devices, which can be mounted and then formatted to create usable disk storage accessible to the user. This is the fourth layer of virtualization.

Although it may seem overly complicated, these layers of virtualization give you a lot of flexibility when it comes to storage management. New disk arrays can be added to a SAN, or a new NAS can be added to the network, and then new data stores can be created from them without disrupting existing data stores. Volumes can be moved from one data store to another without disrupting the virtual machines they're attached to. In fact, you can increase the size of a volume on the fly, and the virtual machine will immediately see the increased storage capacity of its disk drives, without even requiring so much as a reboot.

Understanding Network Virtualization

When you create one or more virtual machines on a host system, you need to provide a way for those virtual machines to communicate not only with each other but also with the other physical computers on your network. To enable such connections, you must create a *virtual network* within your virtualization environment. The virtual network connects the virtual machines to each other and to the physical network.

To create a virtual network, you must create a *virtual switch,* which connects the virtual machines to each other and to a physical network via the host computer's network interfaces. Like a physical switch, a virtual switch has ports. When you create a virtual switch, you connect the virtual switch to one or more of the host computer's network interfaces. These interfaces are then connected with network cable to physical switches, which effectively connects the virtual switch to the physical network.

Then, when you create virtual machines, you connect each virtual machine to a port on the virtual switch. When all the virtual machines are connected to the switch, the VMs can communicate with each other via the switch. And they can communicate with devices on the physical network via the connections through the host computer's network interfaces.

Looking at the Benefits of Virtualization

You might suspect that virtualization is inefficient because a real computer is inherently faster than a simulated computer. Although it's true that real computers are faster than simulated computers, virtualization technology has become so advanced that the performance penalty for running on a virtualized machine rather than a real machine is only a few percent.

The small amount of overhead imposed by virtualization is usually more than made up for by the simple fact that even the most heavily used servers spend most of their time twiddling their digital thumbs, waiting for something to do. In fact, many servers spend nearly *all* their time doing nothing. As computers get faster and faster, they spend even more of their time with nothing to do.

Virtualization is a great way to put all this unused processing power to good use.

Besides this basic efficiency benefit, virtualization has several compelling benefits:

- **Hardware cost:** You typically can save a lot of money by reducing hardware costs when you use virtualization. Suppose that you replace ten servers that cost $4,000 each with one host server. Granted, you'll probably spend more than $4,000 on that server, because it needs to be maxed out with memory, processor cores, network interfaces, and so on. So you'll probably end up spending $10,000 or $15,000 for the host server. Also, you'll end up spending something like $5,000 for the hypervisor software. But that's still a lot less than the $40,000 you would have spent on ten separate computers at $4,000 each.
- **Energy costs:** Many organizations have found that going virtual has reduced their overall electricity consumption for server computers by 80 percent. This savings is a direct result of using less computer hardware to do more work. One host computer running ten virtual servers uses approximately one-tenth the energy that would be used if each of the ten servers ran on separate hardware.

- **Reduced downtime:** Virtual environments typically have less downtime than nonvirtual environments. For example, suppose you need to upgrade the BIOS on one of your server computers. With physical servers, this type of upgrade will ordinarily require that you shut down the operating system that runs on the server, upgrade the BIOS, and then restart the server. During the upgrade, the server will be unavailable.

  In a virtual environment, you don't need to shut down the servers to upgrade the BIOS on the host computer that runs the server. Instead, all you do is move the servers that run on the host that needs the upgrade to another host. When the servers are moved (an operation that can be done without shutting them down), you can shut down the host and upgrade its BIOS. Then, after you restart the host, you can move the servers back to the host — again, without shutting down the servers.

- **Recoverability:** One of the biggest benefits of virtualization isn't the cost savings, but the ability to recover quickly from hardware failures. Suppose that your organization has ten servers, each running on separate hardware. If any one of those servers goes down due to a hardware failure — say, a bad motherboard — that server will remain down until you can fix the computer. On the other hand, if those ten servers are running as virtual machines on two different hosts, and one of the hosts fails, the virtual machines that were running on the failed host can be brought up on the other host in a matter of minutes.

  Granted, the servers will run less efficiently on a single host than they would have on two hosts, but the point is that they'll all be running after only a short downtime.

  In fact, with the most advanced hypervisors available, the transfer from a failing host to another host can be done automatically and instantaneously, so downtime is all but eliminated.

- **Disaster recovery:** Besides the benefit of recoverability when hardware failures occur, an even bigger benefit of virtualization comes into play in a true disaster-recovery situation. Suppose that your organization's server infrastructure consists of 20 separate servers. In the case of a devastating disaster, such as a fire in the server room that destroys all hardware, how long will it take you to get all 20 of those servers back up and running on new hardware? Quite possibly, the recovery time will be measured in weeks.

  By contrast, virtual machines are actually nothing more than files that can be backed up onto tape. As a result, in a disaster-recovery situation, all you have to do is rebuild a single host computer and reinstall the hypervisor software. Then you can restore the virtual-machine backups from tape, restart the virtual machines, and get back up and running in a matter of days instead of weeks.

Choosing Virtualization Hosts

Having made the decision to virtualize your servers, you're next faced with the task of selecting the host computers on which you'll run your virtual servers. The good news is that you need to purchase fewer servers than if you use physical servers. The not-so-good news is that you need to purchase really good servers to act as hosts, because each host will support multiple virtual servers.

TIP

Here are some tips to get you started:

- **If possible, purchase at least two hosts, and make sure that each host is independently capable of running all your virtual servers.** That way, if one of the hosts goes down, you can temporarily move all your servers to the good host while the bad one is being repaired. When both hosts are up, you can spread the workload across the two hosts for better performance.
- **Add up the amount of memory you intend to allocate for each server to determine the amount of RAM for each host.** Then give yourself plenty of cushion. If your servers will require a total of 50GB of RAM, get 72GB on each host, for a total of 144GB if you have two hosts. That will give you plenty of room to grow.
- **Do a similar calculation for processor cores.** It's easier to oversubscribe processor cores on hosts than it is to oversubscribe memory. Like most computers, servers spend an enormous percentage of their time idling. Virtualization makes very efficient use of processor cores for a large number of servers.
- **Get the best network connections you can afford.** Ideally, each host should have a pair of Small Form-factor Pluggable (SFP) ports that you can run 10Gb fiber over. That way, your hosts can communicate with the core switches at top speed.
- **Provide redundancy in the host's subcomponents.** Most hosts support two processors, two memory banks, two network interfaces, and two power supplies. That provides for a maximum of uptime.

Understanding Windows Server 2019 Licensing

When planning your server architecture, you'll need to account for the fact that you must purchase sufficient licenses of Windows Server to cover all the servers you're running. Before virtualization, this was easy: Each server required its own license. With virtualization, things get tricky — and Microsoft doesn't make it easier by trying to simplify things.

Windows Server 2019 comes in three editions. These editions are as follows:

- **Standard Edition:** Ideal for customers who aren't virtualized or who are virtualized but have a relatively small number of servers (approximately 12 per host). Standard Edition costs $972 per license. Each license entitles you run two virtual machines, which seems like a heck of a deal. However, a major drawback is that the license is also limited to hosts that have a maximum of 8 cores per processor and two processors, for a total of 16 cores. If your host has more than 8 cores per processor or more than two processors, you'll need additional licenses.
- **Datacenter Edition:** Ideal for customers who are virtualized and have a large number of servers. Datacenter Edition costs $6,155 per license. Each license lets you run an unlimited number of virtual machines on a single host. Again, each license is limited to 8 cores per processor and two processors, for a total of 16 cores. If your host has more than 8 cores per processor or more than two processors, you'll need additional licenses.
- **Essentials Edition:** Designed for small businesses setting up their first server. It's limited to just 25 users. I won't consider this edition further here.

So, you've got to do some real math to figure out which licenses you'll need. Let's say you need to run a total of 16 servers on two hosts. Here are two licensing scenarios that would be permissible:

- Purchase eight Standard Edition licenses, for a total of $7,776. That works out to just $486 per server. These licenses will allow you to run 16 virtual machines (two per license), as long as your hosts have fewer than 8 processors per core and two processors per host. If your hosts have more than that — say, 12 processors per core, you'll need to purchase 16 Standard Edition licenses, for a total of $15,552, or $972 per server.
- Purchase two Datacenter Edition licenses, for a total of $12,310. That works out to about $769 per server. This is more than the Standard Edition licenses would cost, but it allows you to run an unlimited number of servers. In contrast, additional Standard Edition servers will require additional licenses. As with Standard Edition, if your hosts have more than 8 cores per processor or two processors per host, you'll have to purchase an additional license for each host, bringing the total to $24,620. That works out to about $1,539 per server.

It's obvious that Microsoft charges more to run Windows Server on more powerful hosts, which makes for an interesting pricing strategy. As it turns out, over the next few years, you'll be hard pressed to purchase hosts that fall below the single-license core limit of 8 cores per processor or two processors per host. That's because Intel's dual-socket Xeon processors are getting more and more cores with each successive generation. The current generation of Xeon processors sports up to 18 cores per processor. Intel still makes 4-, 6-, and 8-core versions of the Xeon processor, but who knows what the future will bring?

In any event, the per-core nature of Microsoft's licensing encourages you to purchase host processors with cores as close to but below 8 per processor increments. In other words, use 8- or 16-core processors in your hosts; avoid 10- or 18-core processors, because they nudge you just over the core limits for licensing.

Introducing Hyper-V

Virtualization is a complex subject, and mastering the ins and outs of working with a full-fledged virtualization system like VMware Infrastructure is a topic that's beyond the scope of this book. You can dip your toes into the shallow end of the virtualization pond, however, by downloading and experimenting with Microsoft's free virtualization product, called Hyper-V, which comes with all server versions of Windows since Windows Server 2008 and all desktop versions of Windows since Windows 8.

The version of Hyper-V that comes with desktop versions of Windows is called Client Hyper-V. The nice thing about starting with Client Hyper-V is that it's similar to the enterprise-grade version of Hyper-V that is included with Windows Server. Much of what you learn about Hyper-V on desktop Windows applies to the server version as well.

Understanding the Hyper-V hypervisor

Although Hyper-V is built into all modern versions of Windows, Hyper-V is *not* a type-2 hypervisor that runs as an application within Windows. Instead, Hyper-V is a true type-1 hypervisor that runs directly on the host computer hardware. This is true even for the Hyper-V versions that are included with desktop versions of Windows.

In Hyper-V, each virtual machine runs within an isolated space called a *partition*. Each partition has access to its own processor, RAM, disk, network, and other virtual resources.

There are two types of partitions in Hyper-V: a *parent partition* and one or more *child partitions.* The parent partition is a special partition that hosts the Windows operating system that Hyper-V is associated with. Child partitions host additional virtual machines that you create as needed.

When you activate the Hyper-V feature, the hypervisor is installed and the existing Windows operating system is moved into a virtual machine that runs in the parent partition. Then, whenever you start the host computer, the hypervisor is loaded, the parent partition is created, and Windows is started in a virtual machine within the parent partition.

Although it may appear that the hypervisor is running within Windows, actually the reverse is true: Windows is running within the hypervisor.

In addition to the Windows operating system, the parent partition runs software that enables the management of virtual machines on the hypervisor. This includes creating new virtual machines, starting and stopping virtual machines, changing the resources allocated to existing virtual machines (for example, adding more processors, RAM, or disk storage), and moving virtual machines from one host to another.

Understanding virtual disks

Every Hyper-V virtual machine must have at least one virtual disk associated with it. A *virtual disk* is nothing more than a disk file that resides in the file system of the host operating system. The file has one of two file extensions, depending on which of two data formats you choose for the virtual disk:

- `.vhd`: An older format that has a maximum virtual disk size of 2TB
- `.vhdx`: A newer format that can support virtual disks up to 64TB

For either of these virtual disk formats, Hyper-V lets you create two different types of virtual disks:

- **Fixed-size disk:** A virtual disk whose disk space is preallocated to the full size of the drive when you create the disk. For example, if you create a 100GB fixed-size disk using the `.vhdx` format, a `.vhdx` file of 100GB will be allocated to the drive. Even if the drive contains only 10GB of data, it will still consume 100GB of space on the host system's disk drive.

» **Dynamically expanding disk:** A virtual disk that has a maximum disk space, but that actually consumes only the amount of disk space that is required to hold the data on the disk. For example, if you create a dynamically expanding disk with a maximum of 100GB but only put 10GB of data on it, the .vhdx file for the disk will occupy just 10GB of the host system's disk drive.

TIP

Don't be confused by the names *fixed size* and *dynamically expanding*. Both types of disk can be expanded later if you run out of space. The main difference is whether the maximum amount of disk space allowed for the drive is allocated when the drive is first created or as needed when data is added to the drive. Allocating the space when the drive is created results in better performance for the drive, because Hyper-V doesn't have to grab more disk space every time data is added to the drive. Both types of drives can be expanded later if necessary.

Enabling Hyper-V

Hyper-V is not automatically enabled when you install Windows; you must first enable this feature before you can use Hyper-V.

To enable Hyper-V on a server version of Windows, call up the Server Manager and open the Add Roles and Features Wizard. Then enable the Hyper-V role. When you complete the wizard, Hyper-V will install the Type-1 hypervisor and move the existing Windows Server operating system into the parent partition. You can then start building virtual machines.

To enable Hyper-V on a desktop version of Windows, follow these steps:

1. **Open the Control Panel.**
2. **Choose Programs and Features.**

   The Programs and Features window appears.
3. **Click Turn Windows Features On or Off.**

   The Windows Features dialog box appears, as shown in Figure 10-1.
4. **Select the Hyper-V feature and click OK.**

   The Client Hyper-V hypervisor is installed as an application on the existing desktop Windows operating system, and you can begin using Hyper-V.
5. **When prompted, restart the computer.**

   The reboot is required to start the Hyper-V hypervisor. When your computer restarts, it's actually the Hyper-V hypervisor that starts, not Windows. The hypervisor then loads your desktop Windows into the parent partition.

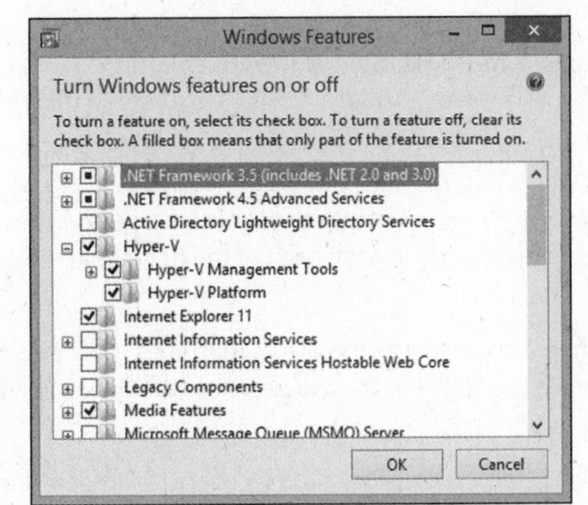

FIGURE 10-1: Enabling Hyper-V on a desktop version of Windows.

Getting Familiar with Hyper-V

To manage Hyper-V, you use the Hyper-V Manager, shown in Figure 10-2. To start this program, click the Start button, type **Hyper-V**, and then choose Hyper-V Manager.

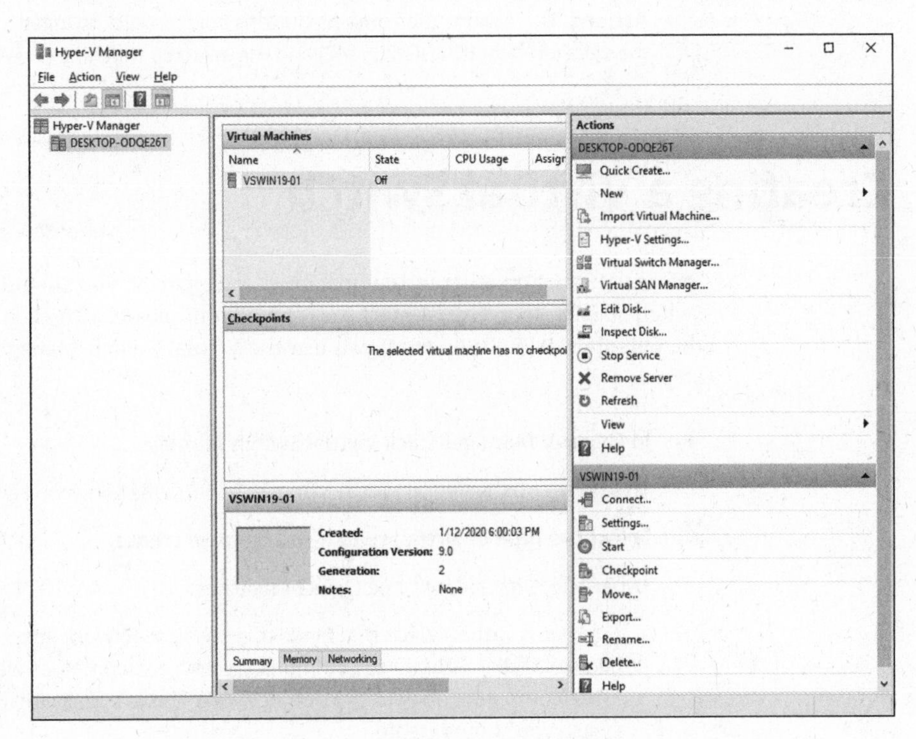

FIGURE 10-2: Hyper-V Manager.

The Hyper-V Manager window is divided into five panes:

- **Navigation:** On the left side of the window is a navigation pane that lists the Hyper-V hosts, which Hyper-V calls *virtualization servers.* In Figure 10-2, just one host is listed: my Windows computer. In an enterprise environment where you have more than one host, each of the hosts will be listed in this pane.
- **Virtual Machines:** This pane lists the virtual machines that are defined for the selected host. In Figure 10-2, I've created just one virtual machine so far, named VMWIN19-01.
- **Checkpoints:** In Hyper-V, a *checkpoint* is a recovery point for a virtual machine. You can create a checkpoint when you're going to make a modification to a virtual machine. Then, if something goes wrong, you can revert to the checkpoint.
- **Virtual Machine Summary pane:** Below the Checkpoints pane is a pane that provides summary information for the virtual machine selected in the Virtual Machines pane. In Figure 10-2, you can see the summary information for one of the Windows Server 2016 machines. This pane has three tabs: Summary, Memory, and Networking. In the figure, the Memory tab is selected so you can see the memory that has been allocated to the machine.
- **Actions:** The Actions tab contains buttons you can click to initiate actions for the selected host (DOUG-2014-I7) and the selected machine (VMWIN19-01).

Creating a Virtual Switch

Before you start creating virtual machines in Hyper-V, you should create a virtual switch so that your virtual machines can communicate with each other and with the outside world. To do that, you use the Virtual Switch Manager. Here are the steps:

1. **In Hyper-V Manager, click Virtual Switch Manager.**

   This brings up the Virtual Switch Manager window, as shown in Figure 10-3.

2. **Select the type of virtual switch you want to create.**

   Hyper-V lets you create three types of switches:

   - *External:* A virtual switch that binds to a physical network adapter, which allows virtual machines to communicate with each other, as well as with other computers on your physical network. This is usually the type of switch you should create.

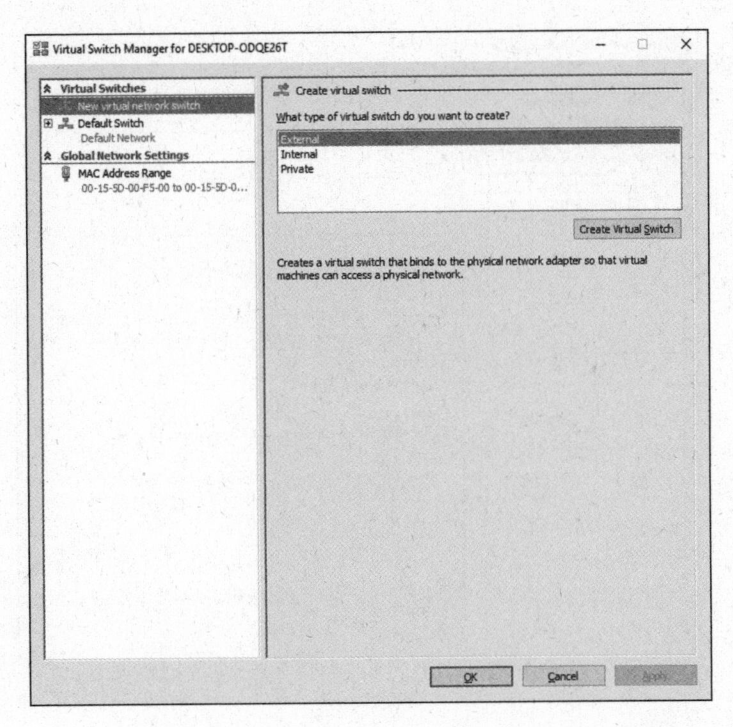

FIGURE 10-3: The Virtual Switch Manager window.

- *Internal:* A virtual switch that does not bind with a physical network adapter. This type of switch lets the virtual machines on this computer communicate with each other and with the host computer, but not with other computers on your physical network.
- *Private:* A virtual switch that lets virtual machines communicate with each other but not with the host computer or with any computers on your physical network.

3. **Click Create Virtual Switch.**

   The settings for the new virtual switch appear, as shown in Figure 10-4.

4. **Type a name for the new virtual switch in the Name field.**

   Use any name you want.

5. **Select the physical network adapter you want to bind the virtual switch to.**

   If your computer has more than one network adapter, select the one you want to use. Binding the virtual switch to a physical network adapter allows the virtual machines to communicate not only with each other but also with other computers connected via the adapter you select.

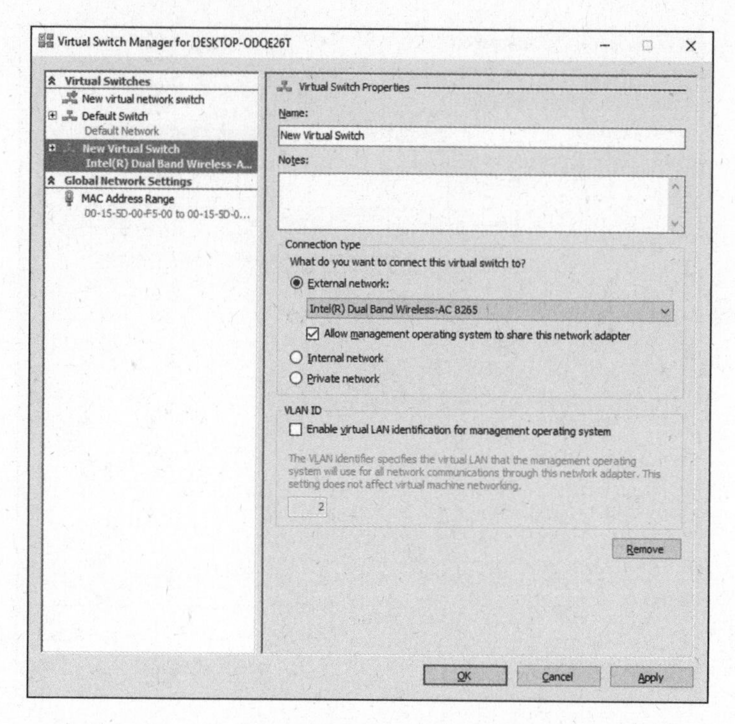

FIGURE 10-4: Creating a new virtual switch.

6. **If your network has multiple VLANs, click the Enable Virtual LAN Identification for Management Operating System check box and enter the VLAN ID for the VLAN you want this switch to connect to.**

   If your network doesn't have multiple VLANs, you can skip this step.

7. **Click OK.**

   The virtual switch is created. Your Hyper-V environment now has a virtual network in place, so you can start creating virtual machines.

Creating a Virtual Disk

Before you create a virtual machine, it's best to first create a virtual disk for the machine to use. Note that you can create a virtual disk at the same time that you create a virtual machine. However, creating the virtual disk first gives you more flexibility. So, I recommend you create virtual disks and virtual machines separately. Here are the steps to create a virtual disk:

1. **In Hyper-V Manager, click New and then choose Hard Disk.**

   This brings up the New Virtual Hard Disk Wizard, as shown in Figure 10-5.

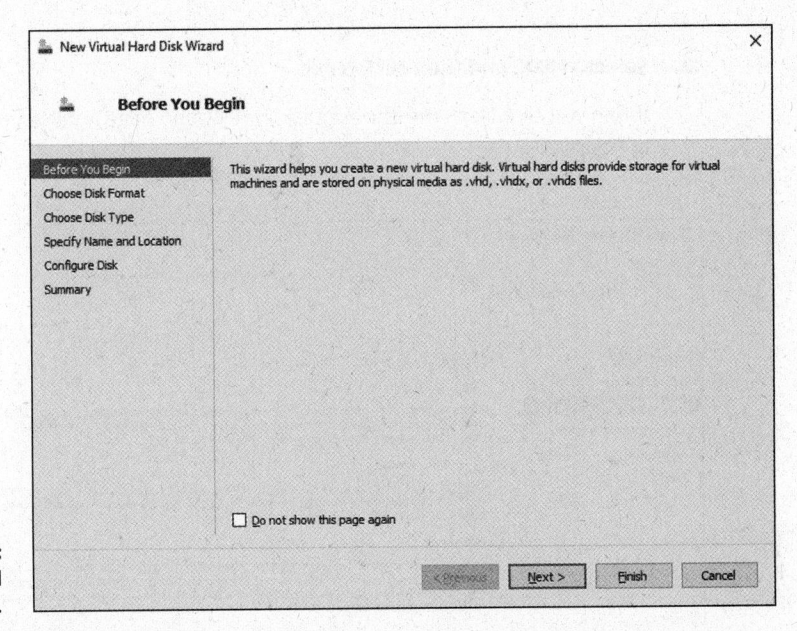

FIGURE 10-5: The New Virtual Hard Disk Wizard.

2. **Click Next.**

   You're asked which disk format to use, as shown in Figure 10-6. I recommend you always use the VHDX format, which can support drives larger than 2TB.

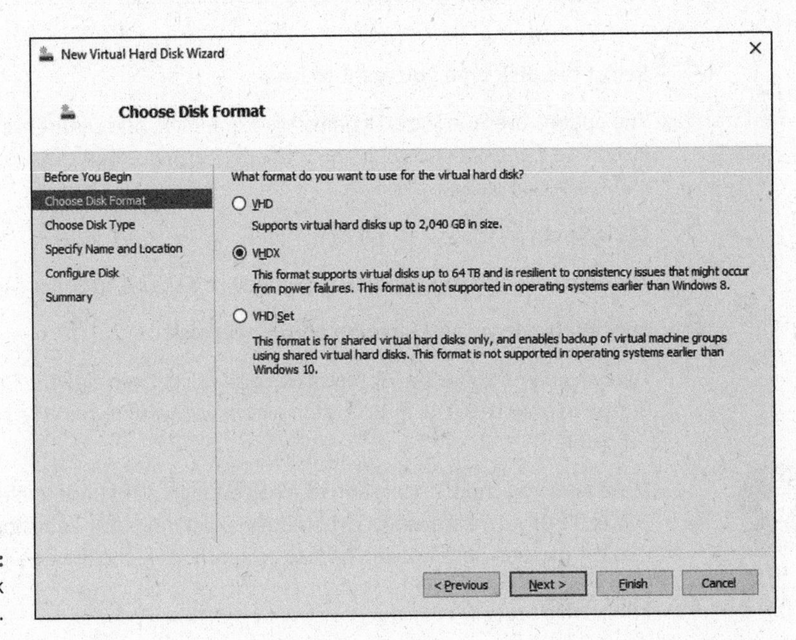

FIGURE 10-6: Choose your disk format.

3. **Select VHDX, and then click Next.**

   When you click Next, the Choose Disk Type option page is displayed, as shown in Figure 10-7.

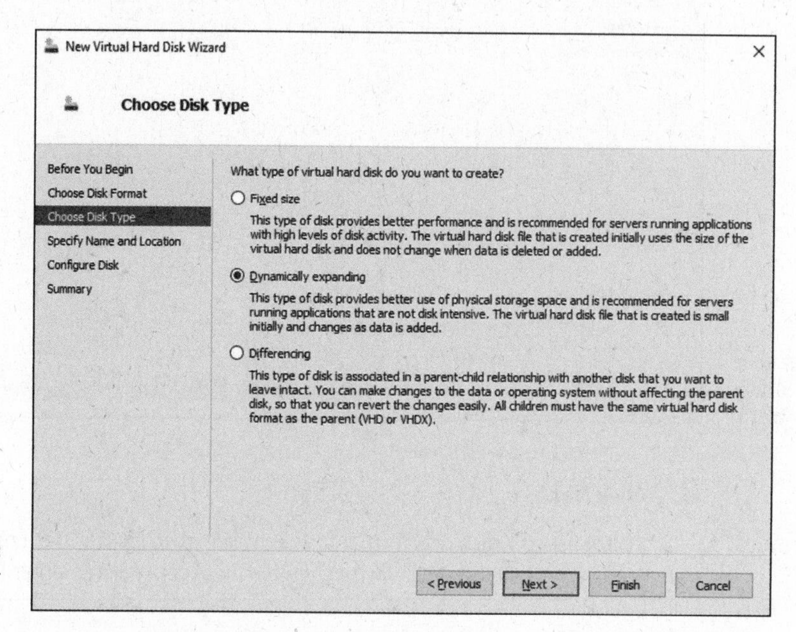

FIGURE 10-7: Choose your disk type.

4. **Select the disk type you want to use.**

   The options are Fixed Size, Dynamically Expanding, and Differencing. Choose Fixed Size if you're concerned about the performance of the disk; otherwise, choose Dynamically Expanding.

5. **Click Next.**

   The Specify Name and Location page, shown in Figure 10-8, appears.

6. **Specify the name and location of the new disk.**

   Type any name you want for the virtual disk drive. Then click the Browse button to browse to the disk location where you want Hyper-V to create the .vhdx file.

   Make sure you choose a location that has enough disk space to create the .vhdx file. If you're creating a dynamically expanding disk, you should ensure that the location has enough space to accommodate the drive as it grows.

7. **Click Next.**

   The Configure Disk page appears, as shown in Figure 10-9.

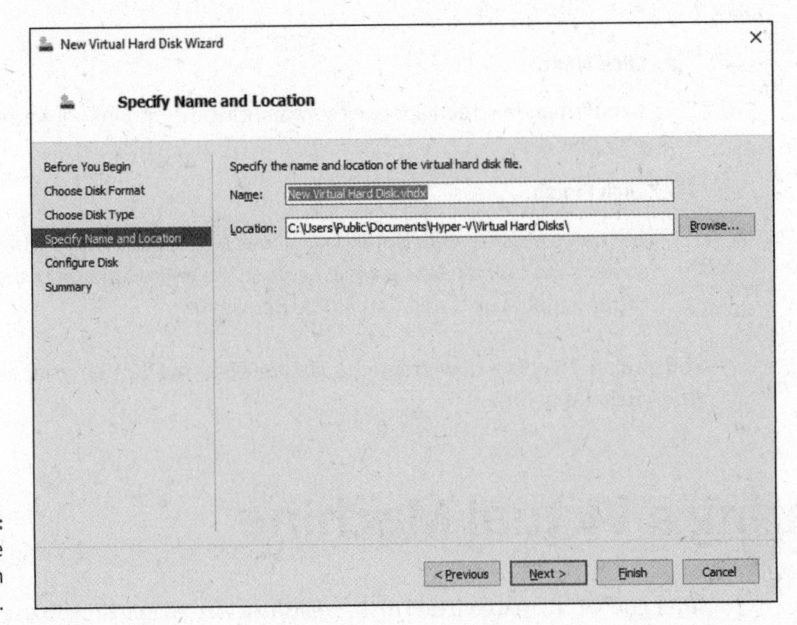

FIGURE 10-8: Specify the name and location of the disk.

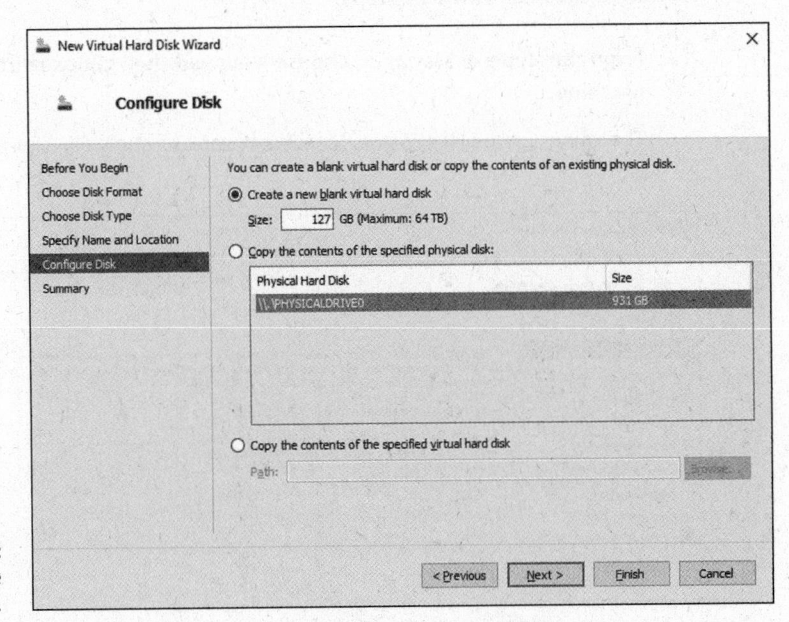

FIGURE 10-9: Specify the size of the disk.

8. Specify the maximum size for the disk drive.

TIP

This page also allows you to copy data either from an existing physical disk drive or from an existing virtual disk drive. Copying data from an existing physical drive is a quick way to convert a physical computer to a virtual computer; just copy the physical disk to a virtual disk, and then use the new virtual disk as the basis for a new virtual machine.

9. **Click Next.**

   A confirmation screen appears, summarizing the options you've selected for your new disk.

10. **Click Finish.**

   The new disk is created. Note that if you selected Fixed Disk as the disk type, creating the disk can take a while because the entire amount of disk storage you specified is allocated to the disk. Be patient.

You're done! You've now created a virtual disk that can be used as the basis for a new virtual machine.

Creating a Virtual Machine

After you've created a virtual disk, creating a virtual machine to use it is a straightforward affair. Follow these steps:

1. **From the Hyper-V Manager, choose New and then choose Virtual Machine.**

   This brings up the New Virtual Machine Wizard, as shown in Figure 10-10.

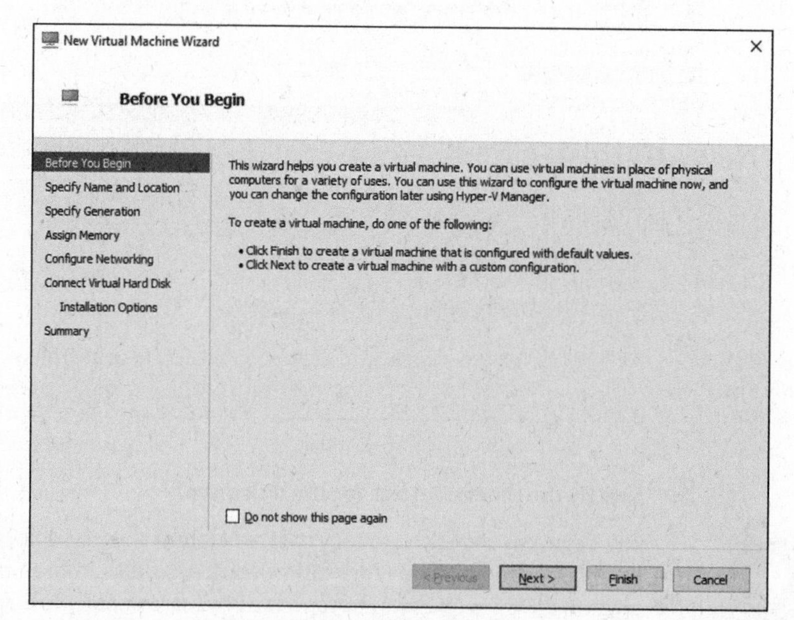

FIGURE 10-10: Say hello to the New Virtual Machine Wizard.

2. **Click Next.**

   The Specify Name and Location page appears, as shown in Figure 10-11.

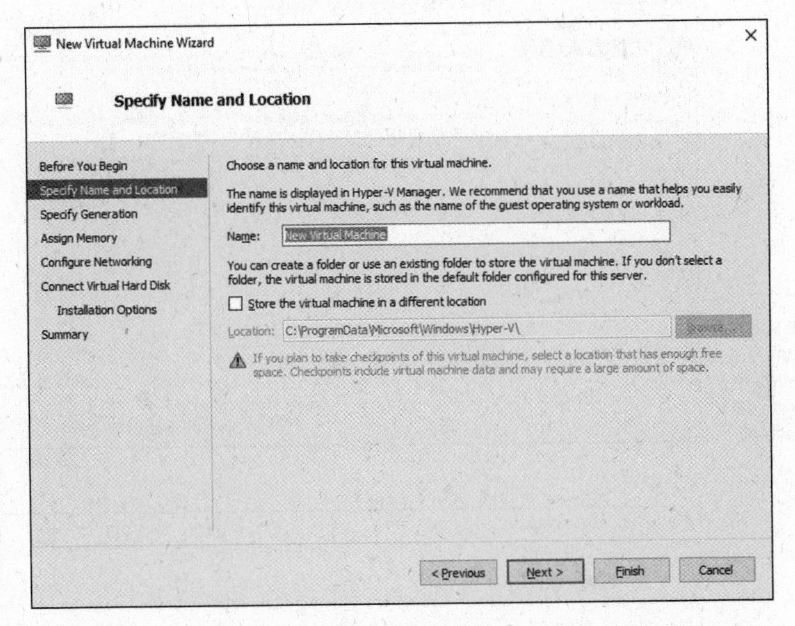

FIGURE 10-11: Specify the name and location of the virtual machine.

3. **Enter the name you want to use for your virtual machine.**

   You can choose any name you want here.

4. **Specify the location of the virtual machine's configuration file.**

   Every virtual machine has an XML file associated with it that defines the configuration of the virtual machine. You can allow this file to be stored in the default location, or you can override the default and specify a custom location.

5. **Click Next.**

   The Specify Generation page appears, as shown in Figure 10-12.

6. **Specify the generation you want to use for the new virtual machine.**

   In most cases, you should opt for Generation 2, which uses newer technology than Generation 1 machines. Use Generation 1 only if the guest operating system will be earlier than Windows Server 2012 or Windows 8.

7. **Click Next.**

   The Assign Memory page appears, as shown in Figure 10-13.

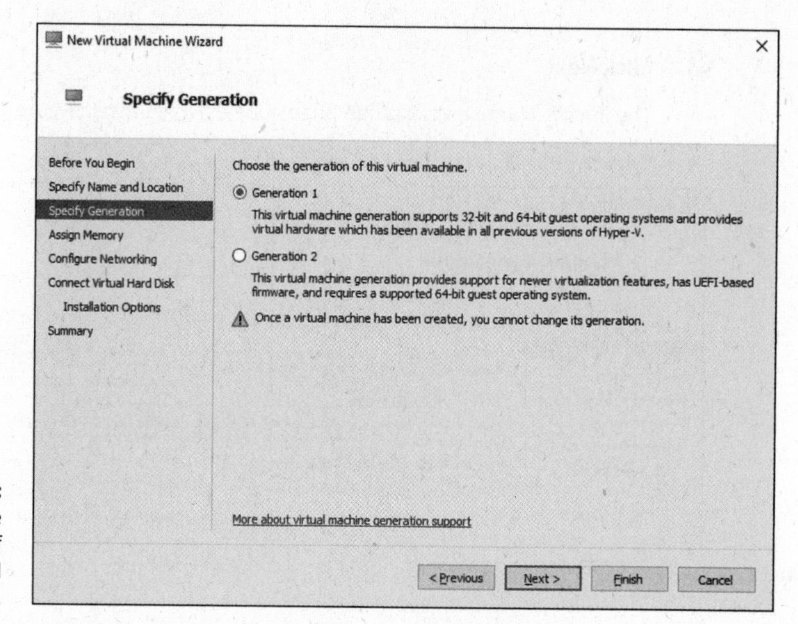

FIGURE 10-12: Specify the generation of the new virtual machine.

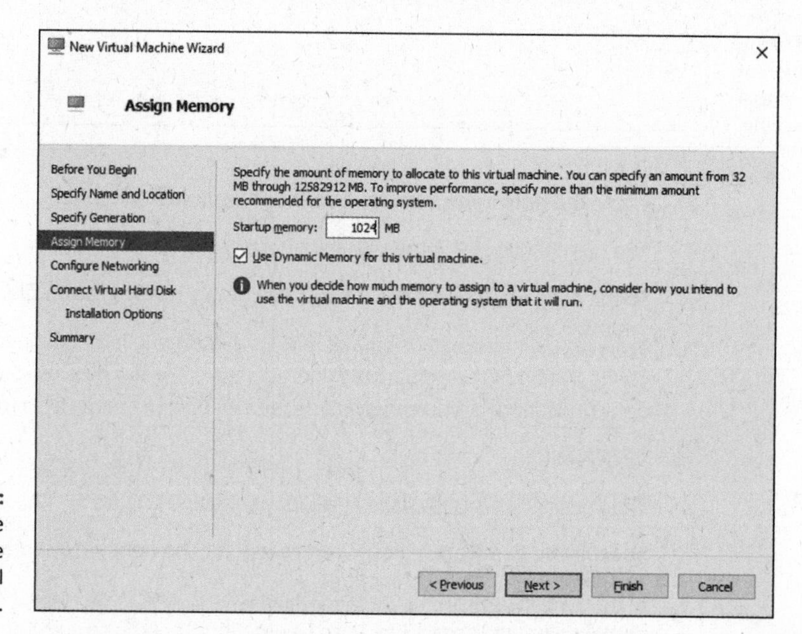

FIGURE 10-13: Specify the memory for the new virtual machine.

8. **Indicate the amount of RAM you want to allocate for the new machine.**

   The default is 2,048MB, but you'll almost certainly want to increase that.

   I also recommend that you select the Use Dynamic Memory for This Virtual Machine check box, which improves memory performance.

9. **Click Next.**

The Configure Networking page appears, as shown in Figure 10-14.

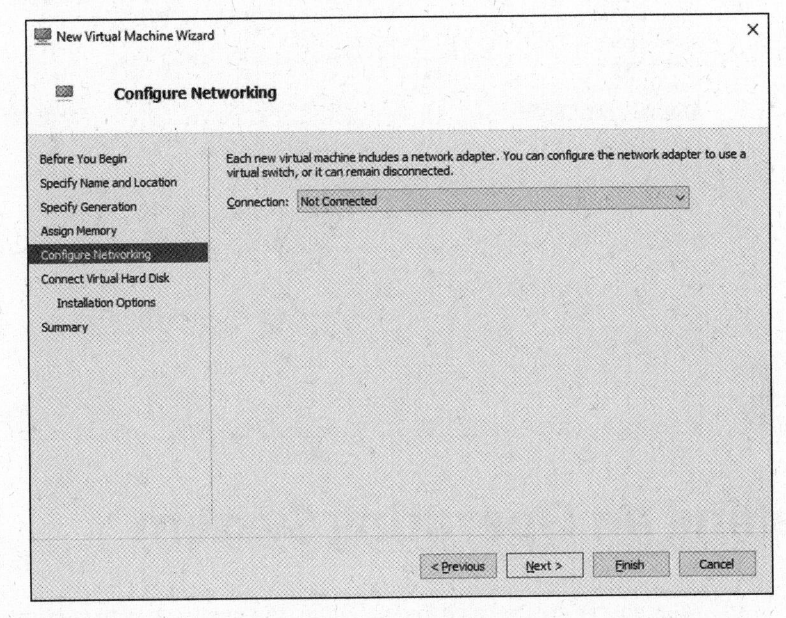

FIGURE 10-14: Configure the networking for the new virtual machine.

10. **Select the virtual switch you want to use for the virtual machine.**

This is the point where you realize why you needed to create a virtual switch before you started creating virtual machines. Use the Connection drop-down list to select the virtual switch you want to connect to this virtual machine.

11. **Click Next.**

The Connect Virtual Hard Disk page appears, as shown in Figure 10-15.

12. **Assuming you've already created a virtual disk for the virtual machine, choose the Use an Existing Virtual Hard Disk option, click Browse, and locate and select the virtual disk.**

If you haven't already created a virtual disk, you can use the Create a Virtual Hard Disk option and create one now.

13. **Click Next.**

A summary page is displayed indicating the selections you've made.

14. **Click Finish.**

The virtual machine is created.

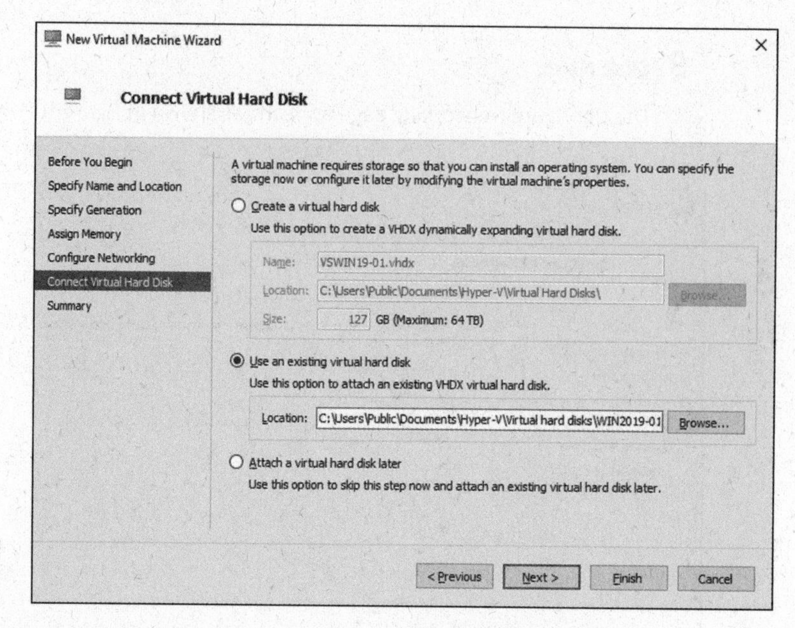

FIGURE 10-15: Connecting a virtual disk.

Installing an Operating System

After you've created a virtual machine, the next step is to configure it to install an operating system. First, you need to get the installation media in the form of an `.iso` file (an `.iso` file is a disk image of a CD or DVD drive). After you have the `.iso` file in place, follow these steps:

1. **From the Hyper-V Manager, choose the new virtual machine and click Settings.**

   The Settings dialog box appears, as shown in Figure 10-16.

2. **Click SCSI Controller in the Hardware list. Then select DVD Drive, and click Add.**

   The configuration page shown in Figure 10-17 appears.

3. **Click the Image File option, click Browse, and select the .iso file that contains the operating system's installation program.**
4. **Click OK.**

   You're returned to the Hyper-V Manager screen.

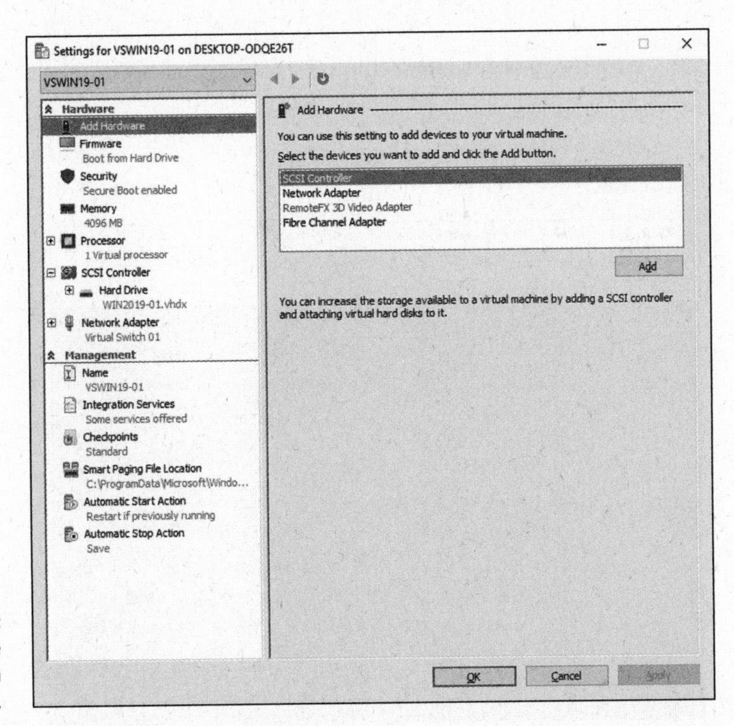

FIGURE 10-16: Editing the settings for a virtual machine.

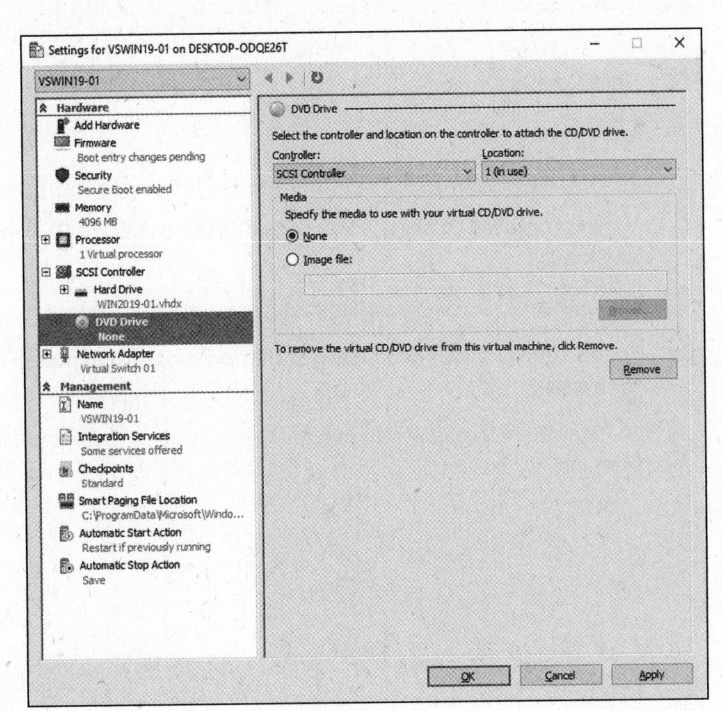

FIGURE 10-17: Configuring a DVD drive.

5. **With the new virtual machine still selected, click Connect.**

   A console window opens, showing that the virtual machine is currently turned off (see Figure 10-18).

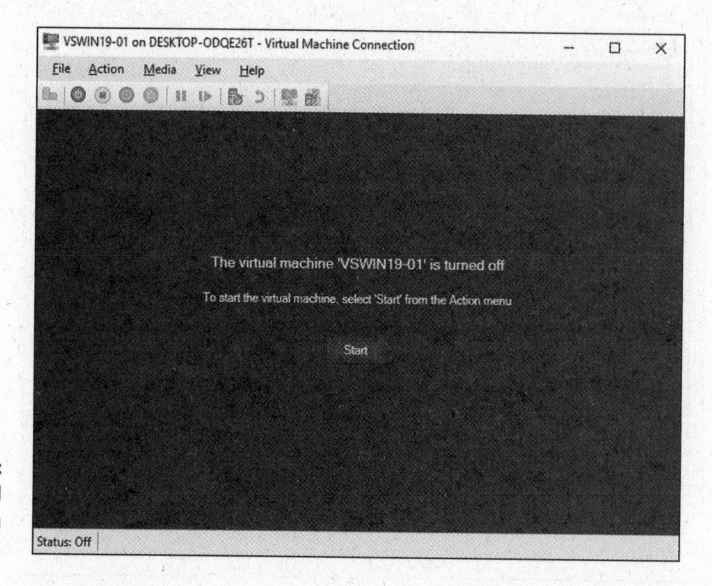

FIGURE 10-18: Viewing a virtual machine through a console.

6. **Click Connect.**
7. **Click Start.**

   The virtual machine powers up.

8. **When prompted to press a key to boot from the CD or DVD, press any key.**

   The operating system's installation program starts.

9. **Follow the instructions of the installation program to install the operating system.**

   That's all there is to it. You've now successfully created a Hyper-V virtual machine!

IN THIS CHAPTER

» Planning a Windows Server 2019 deployment

» Installing Windows Server 2019

» Configuring Active Directory and creating a new domain

Chapter 11

Setting Up a Windows Server

This chapter presents the procedures that you need to follow to install Windows Server — specifically, Windows Server 2019. Note that although the specific details provided are for Windows Server 2019, installing a previous version is very similar. So you won't have any trouble adapting these procedures if you're installing an older version.

In this chapter, I make a couple of important assumptions from the start. First, I assume that you're installing the server on a virtual machine. The procedures for installing a server directly onto physical hardware are similar, except that the installation media will be DVD disks instead of an ISO disk image.

Second, I assume that you're installing Windows Server 2019 into a brand new VM rather than trying to upgrade an existing virtual machine that already has a previous version of Windows Server 2019. Although it is possible to upgrade a previous Windows Server version to 2019, I don't generally recommend it because more than a few things can go wrong.

Planning a Windows Server Installation

Before you begin the Setup program to actually install a Windows Server operating system, you need to make several preliminary decisions, as the following sections describe.

Checking system requirements

Before you install a Windows Server operating system, you should make sure that the computer meets the minimum requirements. Table 11-1 lists the official minimum requirements for Windows Server 2019. (The minimums for Windows Server 2012 are the same.) Table 11-1 also lists what I consider to be more realistic minimums if you expect satisfactory performance from the server as a moderately used file server.

TABLE 11-1 Minimum Hardware Requirements for Windows Server 2019 (Standard Edition)

Item	Official Minimum	A More Realistic Minimum
CPU	1.4 GHz	3 GHz
RAM	2GB*	4GB
Free disk space	32GB	100GB

**Technically, the minimum RAM size is 512MB. But 2GB is required if you want to install the Desktop Experience option, which lets you interact with Windows Server using the same GUI interface as Windows 10. I recommend you install this option unless you have a lot of Windows Server experience under your belt and don't mind working in a command-line environment.*

Reading the release notes

Like all versions of Windows Server, Windows Server 2019 provides a set of release notes that you should read before you start Setup, just to check whether any of the specific procedures or warnings it contains applies to your situation.

You can get the release notes at `https://docs.microsoft.com/en-us/windows-server/get-started-19/rel-notes-19`.

Considering your licensing options

Two types of licenses are required to run a Windows Server operating system: a *server license*, which grants you permission to run a single instance of the server, and *Client Access Licenses* (CALs), which grant users or devices permission to connect to the server. When you purchase Windows Server, you ordinarily purchase a server license plus some number of CALs.

To complicate matters, there are two distinct types of CALs: per-user and per-device. *Per-user* CALs limit the number of users who can access a server simultaneously, regardless of the number of devices (such as client computers) in your organization. By contrast, *per-device* CALs limit the number of unique devices that can access the server, regardless of the number of users in your organization.

Note that you can download a six-month evaluation version of Windows Server 2019 from Microsoft's website at `www.microsoft.com/en-us/evalcenter/evaluate-windows-server-2019`.

TIP

If you want to follow along with the steps to install Windows Server onto a Hyper-V virtual machine as outlined later in this chapter, download the ISO version of the evaluation. Then you'll be able to mount the ISO to a DVD/CD drive in the virtual machine and install the server from the DVD.

Deciding your TCP/IP configuration

Before you install the operating system, you should have a plan for implementing TCP/IP on the network. Here are some of the things you need to decide or find out:

- » The IP subnet address and mask for your network.
- » The domain name for the network.
- » The host name for the server.
- » The static IP for the server. (All servers should have static IPs.)
- » Whether the server will be a DHCP server.
- » The Default Gateway for the server (that is, the IP address of the network's Internet router).
- » Whether the server will be a DNS server.

For more information about planning your TCP/IP configuration, see Chapter 5.

Choosing workgroups or domains

A *domain* is a method of placing user accounts and various network resources under the control of a single directory database. Domains ensure that security policies are applied consistently throughout a network and greatly simplify the task of managing user accounts on large networks.

A *workgroup* is a simple association of computers on a network that makes it easy to locate shared files and printers. Workgroups don't have sophisticated directory databases, so they can't enforce strict security.

Workgroups should be used only for tiny networks with just a few users. Truthfully, any network that is large enough to have a dedicated server running Windows Server 2019 is too large to use workgroups. So, if you're installing a Windows server, you should always opt for domains.

After you decide to use domains, you have to make two basic decisions:

- **What will the domain name be?** If you have a registered Internet domain name, such as `mydomain.com`, you may want to use it for your network's domain name. Otherwise, you can make up any name you want.
- **What computer or computers will be the domain controllers for the domain?** If this server is the first server in a domain, you must designate it as a domain controller and take steps to create the domain. (I explain how to do that later in this chapter.) If you already have a server acting as a domain controller, you can either add this computer as an additional domain controller or designate it a member server.

TIP

You can always change the role of a server from a domain controller to a member server, and vice versa, if the needs of your network change. If your network has more than one server, it's always a good idea to create at least two domain controllers. That way, if one fails, the other one can take over.

Running Setup

Now that you've planned your installation and prepared the computer, you're ready to run the Setup program. The following procedure describes the steps that you must follow to install Windows Server 2019 on a virtual machine using an ISO image of the installation DVD. Before you begin this procedure, you'll need to download the ISO file from Microsoft's website. And you should set up a virtual machine and mount the ISO file to the VM's DVD/CD drive. (To find out how to set up a VM using Microsoft's Hyper-V, refer to Chapter 10.)

1. **Configure the new virtual machine with the specifications you want to use, mount the installation ISO file on the virtual DVD drive, and start the VM.**

   After a few moments, the Windows Setup Wizard fires up and asks for your language, time and currency format, and keyboard layout.

2. **Click Next.**

   The Welcome screen appears, as shown in Figure 11-1.

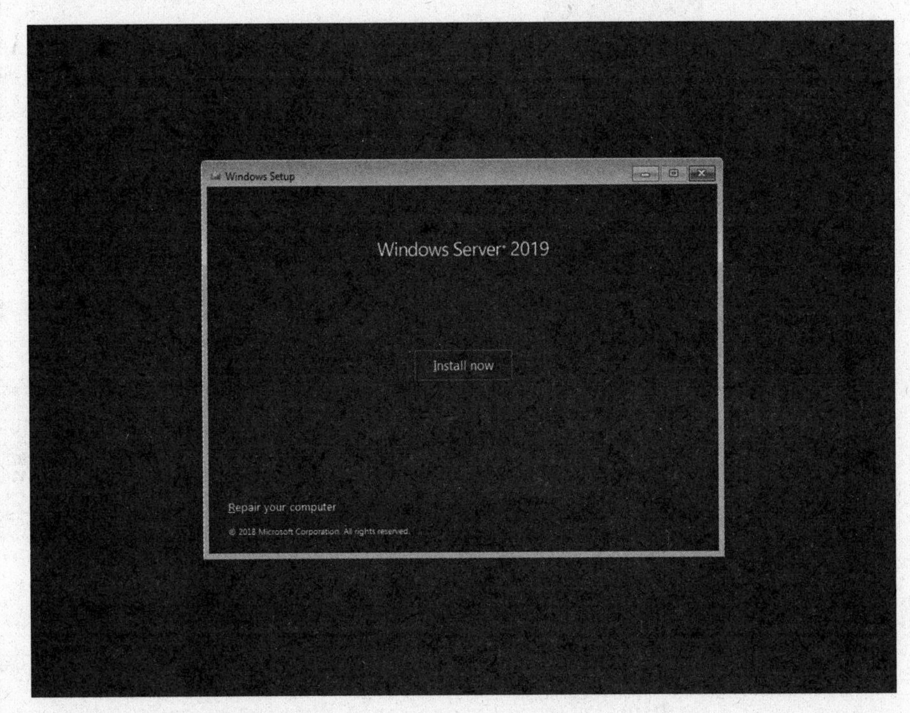

FIGURE 11-1: Welcome to Windows Setup!

3. **Click Install Now.**

   Windows asks which edition of the operating system you want to install, as shown in Figure 11-2.

   Four editions are available:

   - *Windows Server 2019 Standard:* A streamlined version of the operating system that does not have a graphical user interface (GUI). You must manage this edition of the server remotely using PowerShell or Windows Management Console.
   - *Windows Server 2019 Standard (Desktop Experience):* This is the version you want to pick so that you'll have a friendly GUI modeled after Windows 10.
   - *Windows Server 2019 Datacenter:* A high-powered version of Windows Server designed for large data centers with a lot of servers.
   - *Windows Server 2019 Datacenter (Desktop Experience):* The GUI version of the Datacenter edition.

   For this book, I use the Standard Desktop Experience version to demonstrate how to manage various aspects of Windows Server 2019.

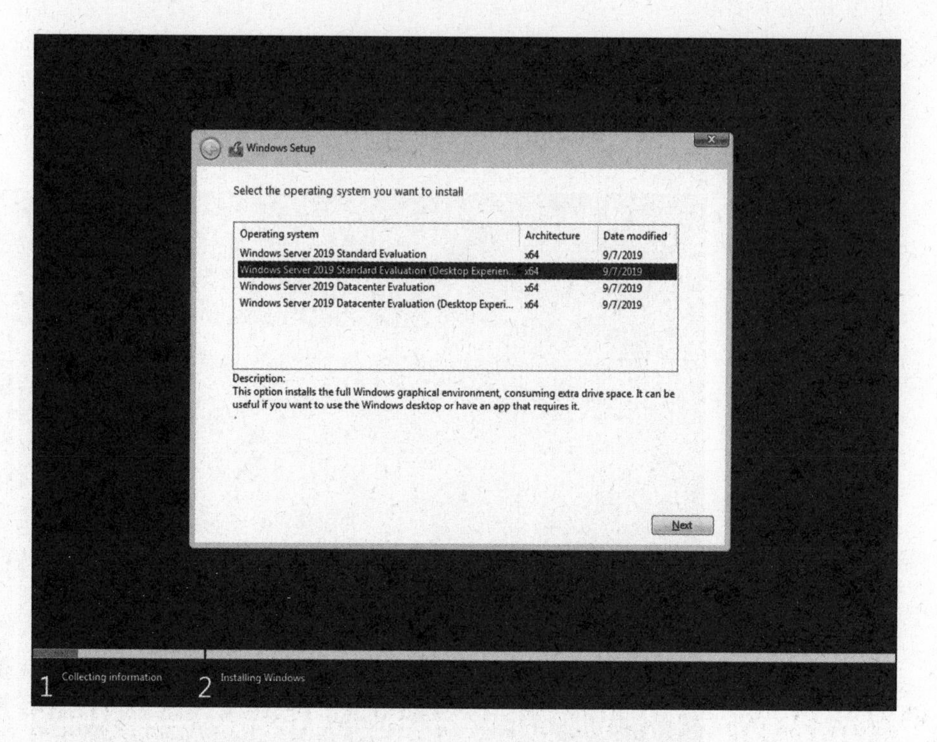

FIGURE 11-2: Selecting the OS edition to install.

4. **Select the edition you want to install and then click Next.**

   The Setup Wizard displays the License Agreement information. Read it if you enjoy legalese.

5. **Click I Accept the License Terms and then click Next.**

   The Setup Wizard asks whether you want to perform an upgrade installation or a full installation.

6. **Click the installation option you want to use.**

   Setup continues by displaying the disk drives available on the computer so that you can choose where to install Windows, as shown in Figure 11-3. In this example, there is just one drive, so there isn't really a choice.

7. **Select the disk on which you want to install Windows and then click Next.**

   Setup formats the drive and then copies files to the newly formatted drive. This step usually takes a while. I suggest you bring along your favorite book. Start reading at Chapter 1.

   After all the files have been copied, Setup reboots your computer. Then Setup examines all the devices on the computer and installs any necessary device drivers. You can read Chapter 2 of your book during this time.

   When Setup finishes installing drivers, it asks for the password you want to use for the computer's Administrator account, as shown in Figure 11-4.

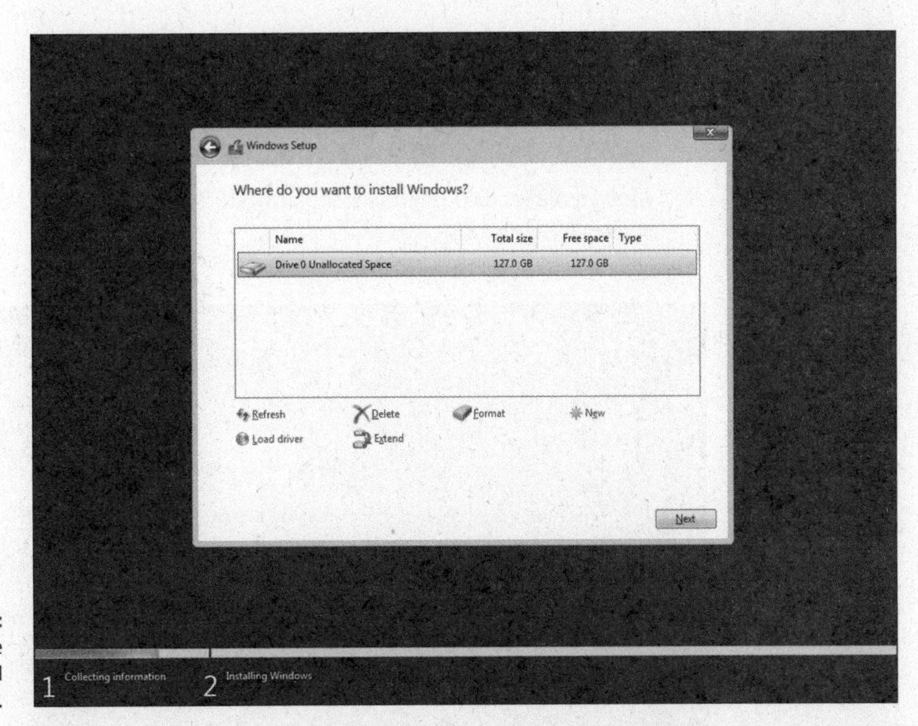

FIGURE 11-3: Deciding where to install Windows.

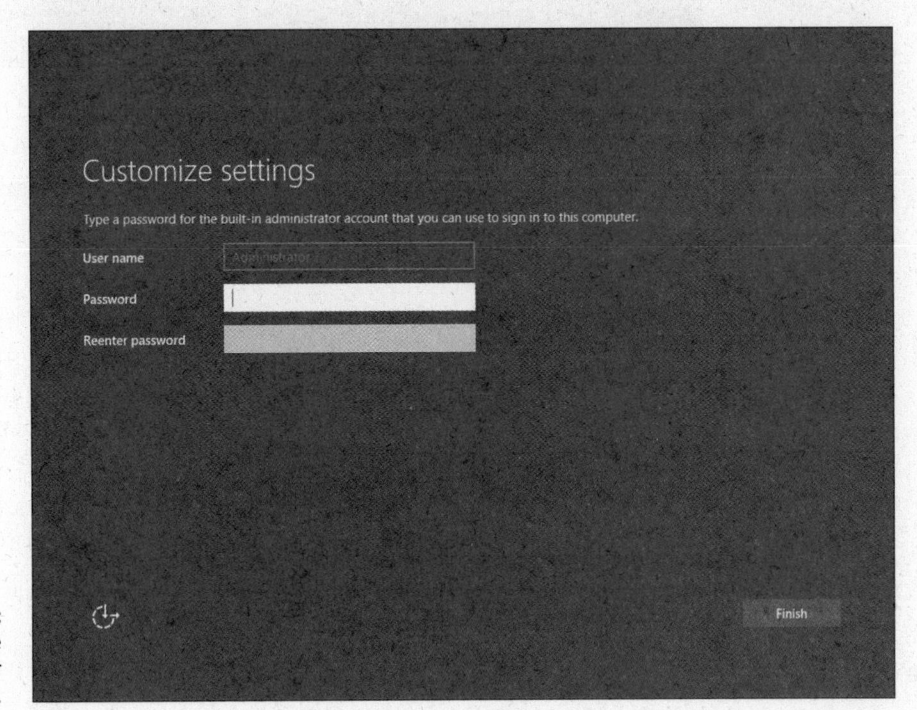

FIGURE 11-4: Setting the Administrator password.

8. **Enter the Administrator password twice and then click Finish.**

REMEMBER

Be sure to write this password down somewhere and keep it in a secure place. If you lose this password, you won't be able to access your server.

After you've set the Administrator password, Windows displays the login screen shown in Figure 11-5.

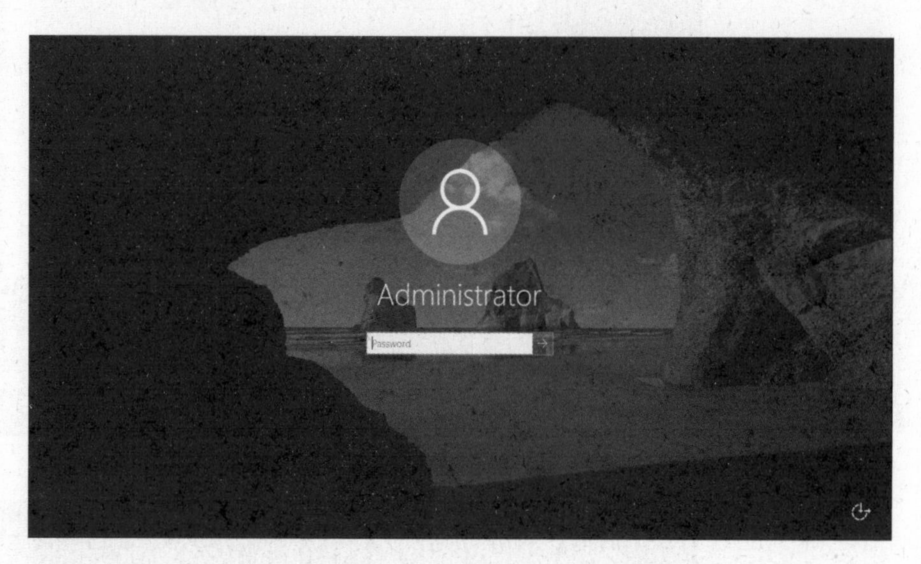

FIGURE 11-5: Log in to your new server.

9. **Log in using the Administrator account.**

You have to enter the password you created in Step 7 to gain access.

When you're logged in, Windows displays the Server Manager Dashboard, as shown in Figure 11-6.

The Server Manager Dashboard provides links that let you complete the configuration of your server. Specifically, you can

- Click Configure This Local Server to configure server settings such as the computer's name and the domain it belongs to, network settings such as the static IP address, and so on.
- Click Add Roles and Features to add server roles and features. (For more information, see "Adding Server Roles and Features," later in this chapter.)
- Click Add Other Servers to Manage to manage other servers in your network.
- Click Create a Server Group to create a customized group of servers.
- Click Connect This Server to Cloud Services if you integrate cloud services with your server.

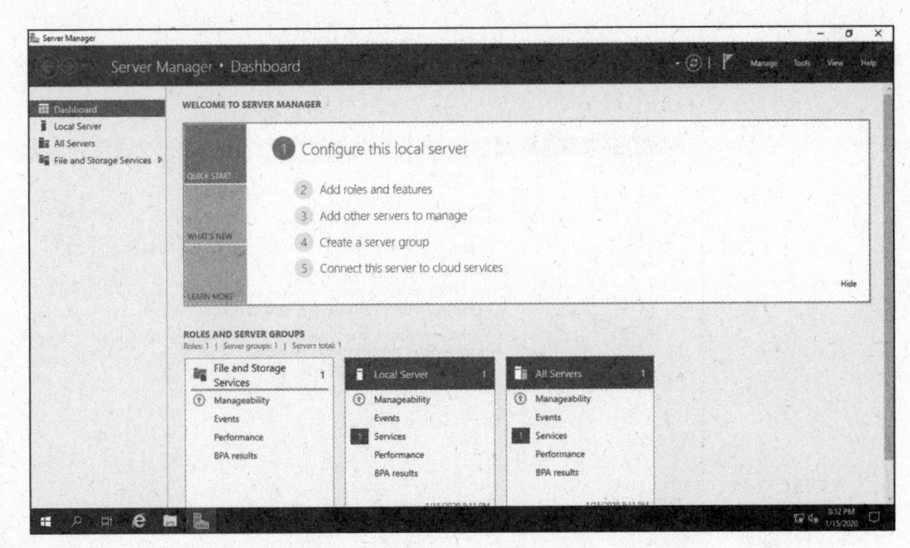

FIGURE 11-6: The Server Manager Dashboard.

Adding Server Roles and Features

Server roles are the roles that your server can play on your network — roles such as file server, web server, or DHCP or DNS server. *Features* are additional capabilities of the Windows operating system itself, such as the .NET Framework or Windows Backup. Truthfully, the distinctions between roles and features are a bit arbitrary. The web server is considered to be a role, for example, but the Telnet server is a feature. Go figure.

The following procedure describes how to install server roles. The procedure for installing server features is similar. For this example, we'll install the Active Directory Domain Services and DNS Server roles. These roles are required to create a domain and designate this server as a domain controller.

1. **Click Add Roles and Features on the Server Manager Dashboard.**

   The Add Roles and Features Wizard appears, as shown in Figure 11-7.

2. **Click Next.**

   The wizard asks which of two installation types you want to perform. In most cases, you want to leave the default choice (Role-Based or Feature-Based Installation) selected. Select the alternative (Remote Desktop Services Installation) only if you're configuring a remote virtual server.

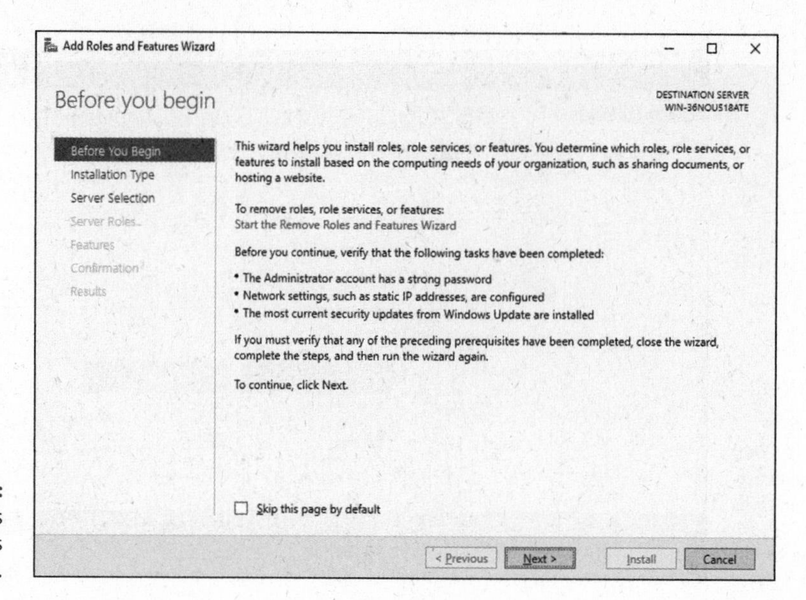

FIGURE 11-7: The Add Roles and Features Wizard.

3. **Click Next.**

   The wizard lets you select the server you want to install roles or features for, as shown in Figure 11-8. In this example, only one server is listed. If you'd chosen Add Other Servers to Manage in the Server Manager Dashboard to add other servers, those servers would appear on this page as well.

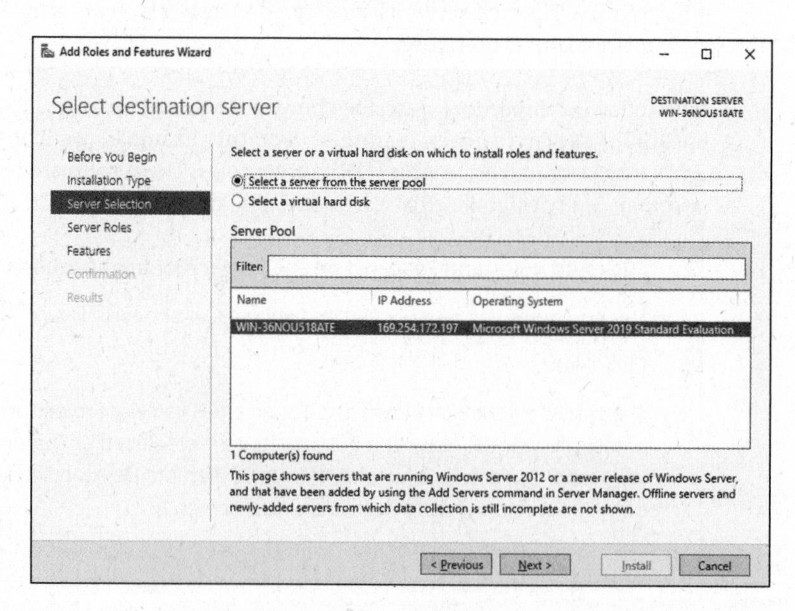

FIGURE 11-8: Selecting the server to manage.

4. **Click Next.**

The Select Server Roles page, shown in Figure 11-9, appears. This page lets you select one or more roles to add to your server.

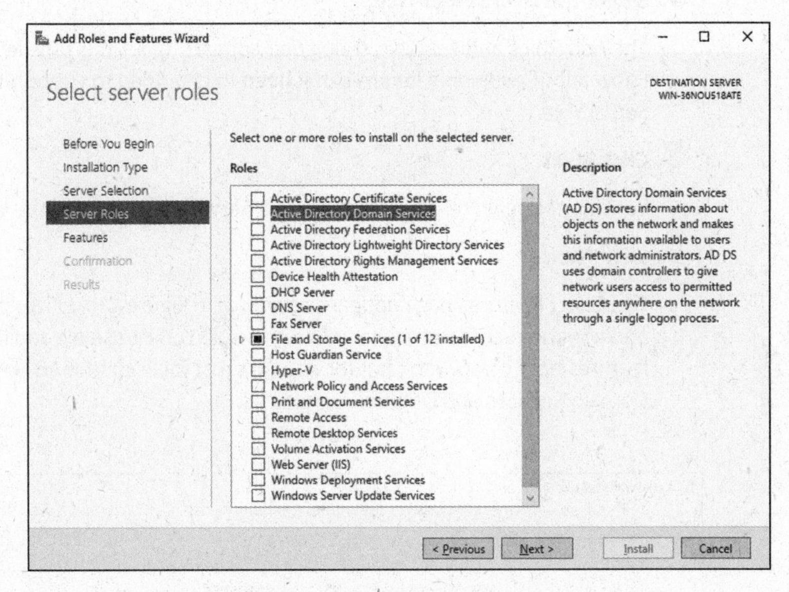

FIGURE 11-9: The Select Server Roles page.

5. **Select the Active Directory Domain Services role.**

The Add Roles and Features dialog box appears, as shown in Figure 11-10. Here, Windows is telling you that in order to install the Active Directory Domain Services role, you'll also have to install several other features, such as Group Policy Management and Remote Server Administration tools.

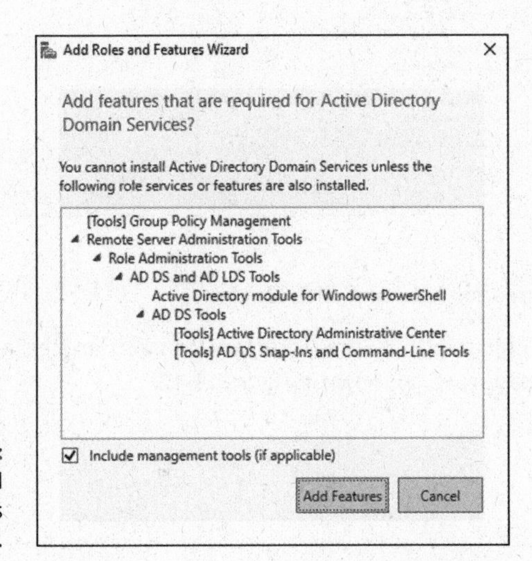

FIGURE 11-10: Adding additional required roles and services.

6. **Click Add Features.**

   The Add Roles and Features dialog box disappears and you're returned to the Select Server Roles page of the Wizard.

7. **Select the DNS Server role.**

   The Add Roles and Features dialog box appears again, this time letting you know about additional features that need to be added to support the DNS Server role.

8. **Click Next.**

   You're once again returned to the Select Server Roles page of the Wizard.

9. **Click Next.**

   The Select Features page appears, as shown in Figure 11-11. This page lists additional server features that you can install. You can select additional features here if you want, but for now let's just focus on the Active Directory Domain Services and DNS Server roles.

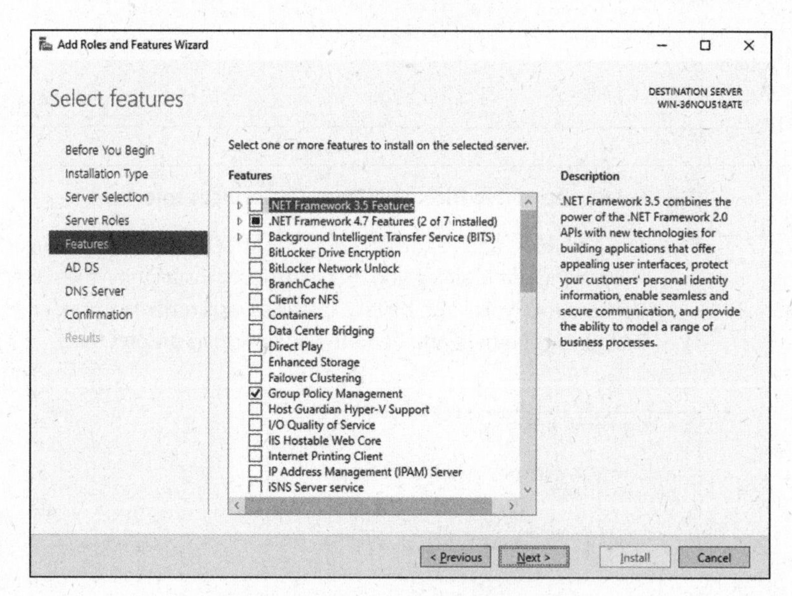

FIGURE 11-11: The Select Features page.

10. **Skip these features for now and click Next.**

    Because you've opted to install Active Directory Domain Services, a helpful information page appears, as shown in Figure 11-12.

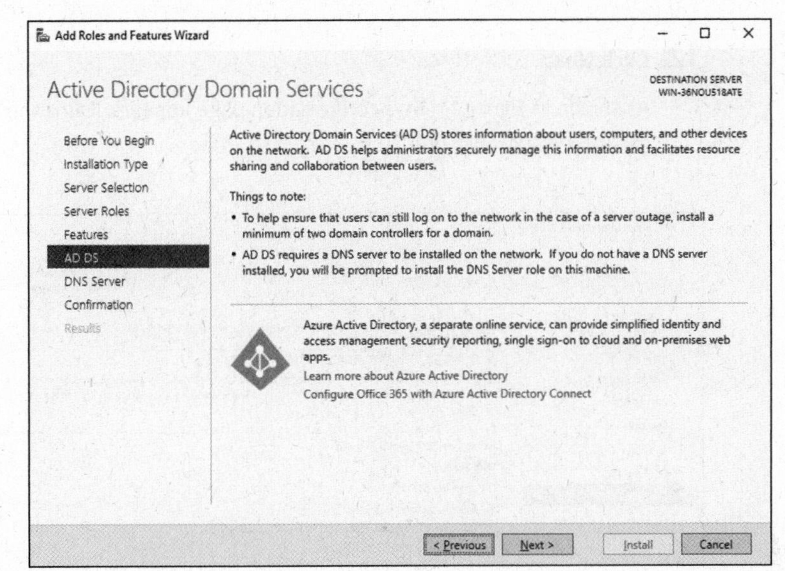

FIGURE 11-12: Stuff you need to know about Active Directory Domain Services.

11. Click Next.

Now, because you also selected the DNS Server role, the Wizard displays another page of helpful information about the DNS Server role, as shown in Figure 11-13.

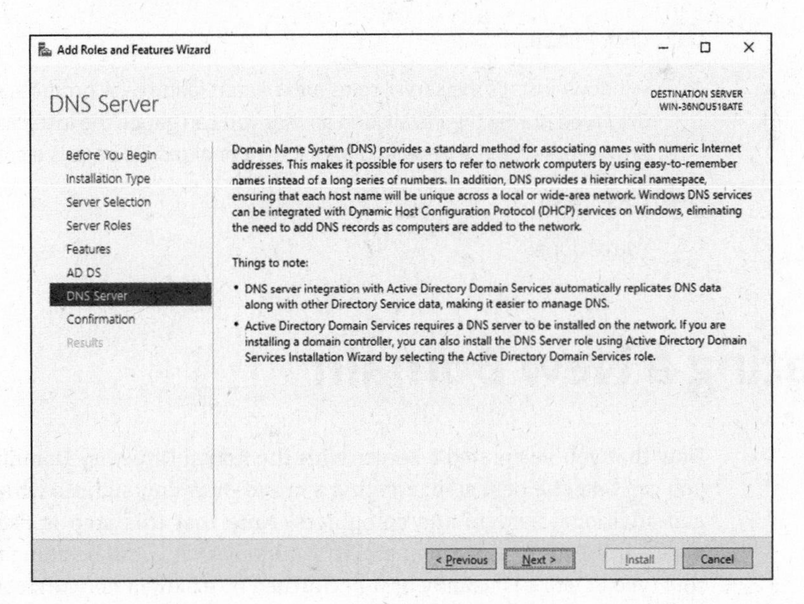

FIGURE 11-13: Stuff you need to know about DNS Server role.

12. Click Next.

As shown in Figure 11-14, a confirmation page appears, listing the roles and features you've selected.

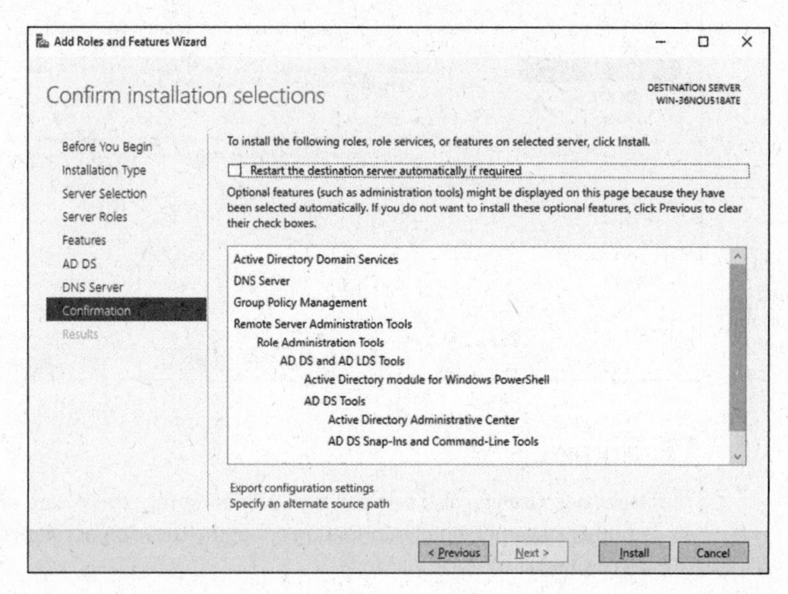

FIGURE 11-14: Are you ready to install the new server roles?

13. Click Install.

Windows installs the server roles and related features. A progress screen is displayed during the installation so that you can gauge the installation's progress. When the installation finishes, a final results page is displayed.

14. Click Close.

You're done!

Creating a New Domain

Now that you've created a server with the Active Directory Domain Services role, you can take the next step: creating a brand-new domain into which you can later add additional servers and computers. Note that this step is usually done only once within an organization, and it's unlikely that you'll be doing this yourself — this type of work is usually best performed by a skilled network consultant. However, seeing how it's done will help you better understand how Active Directory works and how it's configured in your organization.

To get the process started, notice that after you complete the Add Roles and Features Wizard, you're returned to the Server Manager screen, with an alert message indicating that you need to configure Active Directory Domain Services. The alert message is displayed in yellow and includes a More link, which you can click to get the configuration started. Here are the steps:

1. **Click More in the Configuration Required alert.**

   The All Servers Task Details and Notifications screen is displayed, as shown in Figure 11-15.

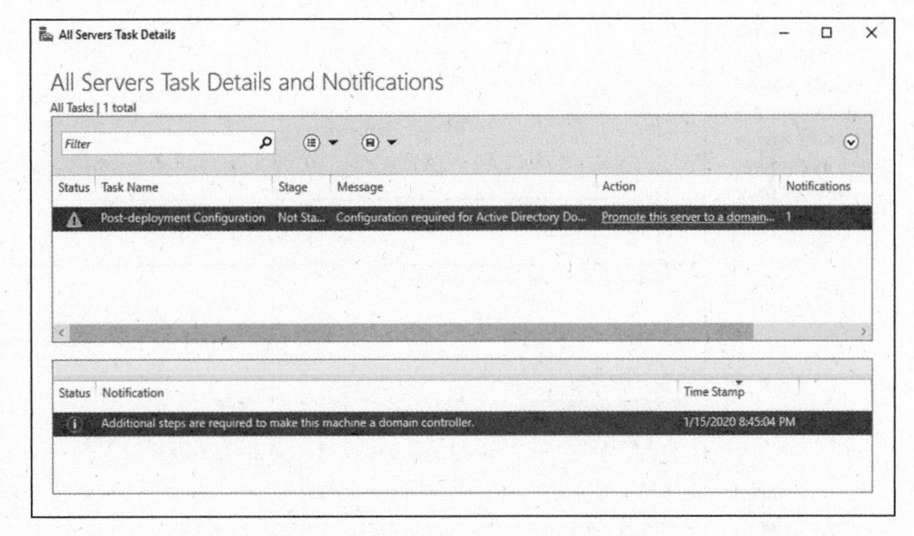

FIGURE 11-15: Additional configuration needs to be done for the Active Directory Domain Services role.

2. **Click the Promote This Server to a Domain Controller link.**

   The Active Directory Domain Services Configuration Wizard comes to life, as shown in Figure 11-16. For this exercise, we're going to start a new domain forest whose root domain name is `lowewriter.pri`.

3. **Select Add a New Forest, and then enter a name for the new forest's root domain.**

   Here's, I'm entering `lowewriter.pri` as the root domain name.

4. **Click Next.**

   The next page of the Wizard, shown in Figure 11-17, lets you set various configuration options for the domain controller. You can leave these options at their default settings, but you'll need to enter a password that can be used in the event that you need to fix a damaged Active Directory controller. You'll probably never need this password, but write it down and store it in a secure location just in case.

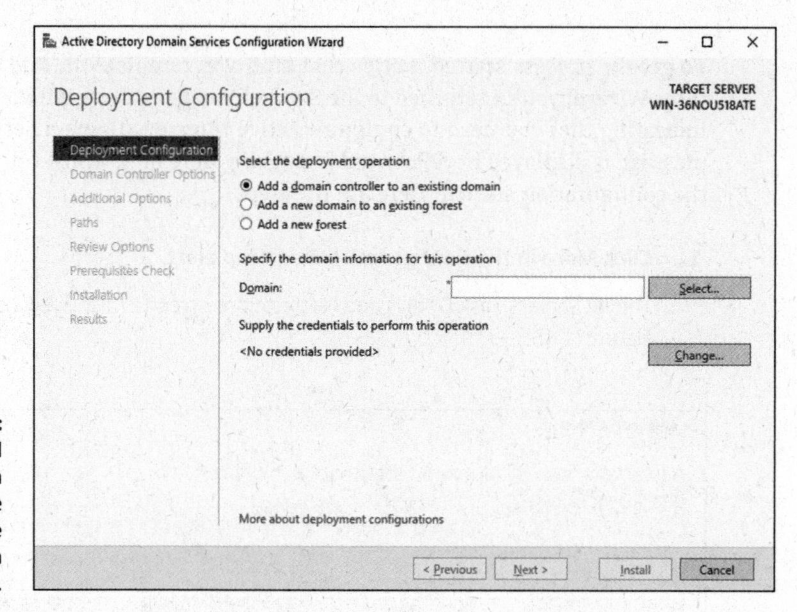

FIGURE 11-16: Additional configuration needs to be done for the Active Directory Domain Services role.

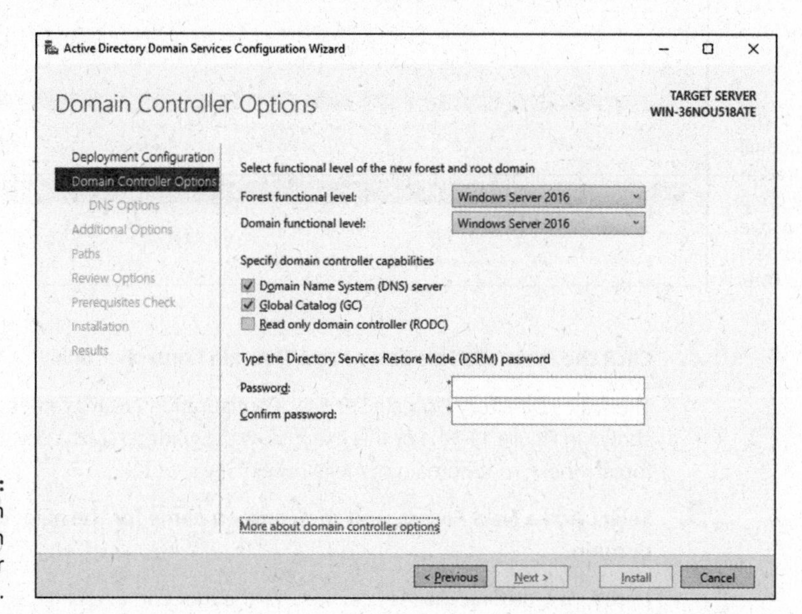

FIGURE 11-17: Here you can set domain controller options.

5. **Enter your recovery password twice, and then click Next.**

   You see a DNS Options screen that indicates a potential problem regarding DNS delegation, which, for the purposes of our little exercise, you can safely ignore.

6. **Click Next.**

   The next screen shows that the wizard has proposed a NETBIOS name based on the root domain name you entered. NETBIOS is an ancient (circa 1983) networking standard that, believe it or not, still exists deep under the hood of modern networks. As such, a NETBIOS name is a necessity. The name proposed by the wizard is almost always appropriate, so you can move on — there's nothing to see here.

7. **Click Next again.**

   The next screen of the wizard lets you set the location for various files created and used by Active Directory. You should almost always leave these at their defaults.

8. **Click Next again.**

   Now you get to a review screen that summarizes the options you've selected. Look over the details to make sure you've chosen wisely, and then move on.

9. **Click Next again.**

   In this step, the wizard checks to ensure that all required prerequisites are in place so that the new domain can be created, as shown in Figure 11-18. If any red errors are shown, you'll need to correct the issues before you can continue.

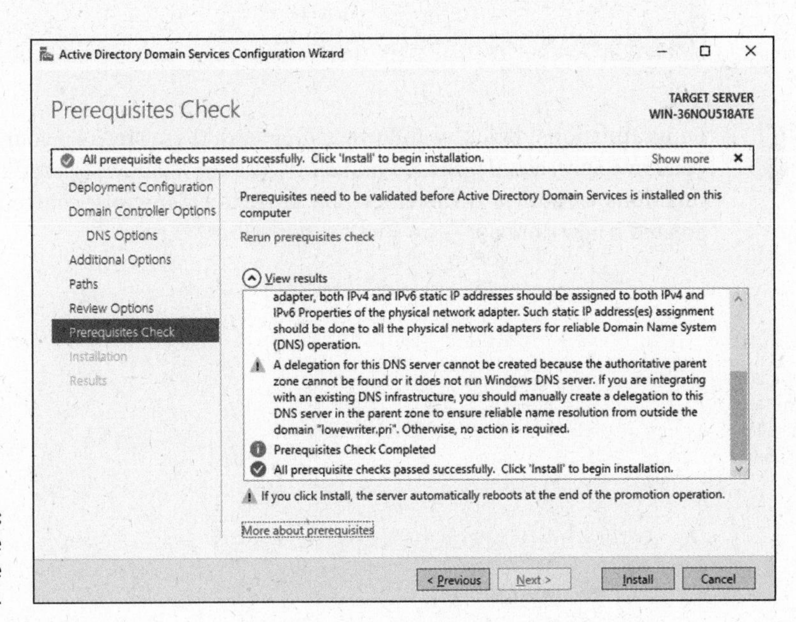

FIGURE 11-18: Hooray! All the prerequisites are good to go.

10. Click Install.

The wizard grinds and whirs for a bit while it configures Active Directory. During this process, the server will reboot. When it comes up again, the new domain will be ready to go, as evidenced by the login screen shown in Figure 11-19. Notice that the login screen now indicates that you'll be logging in to the `LOWEWRITER` domain.

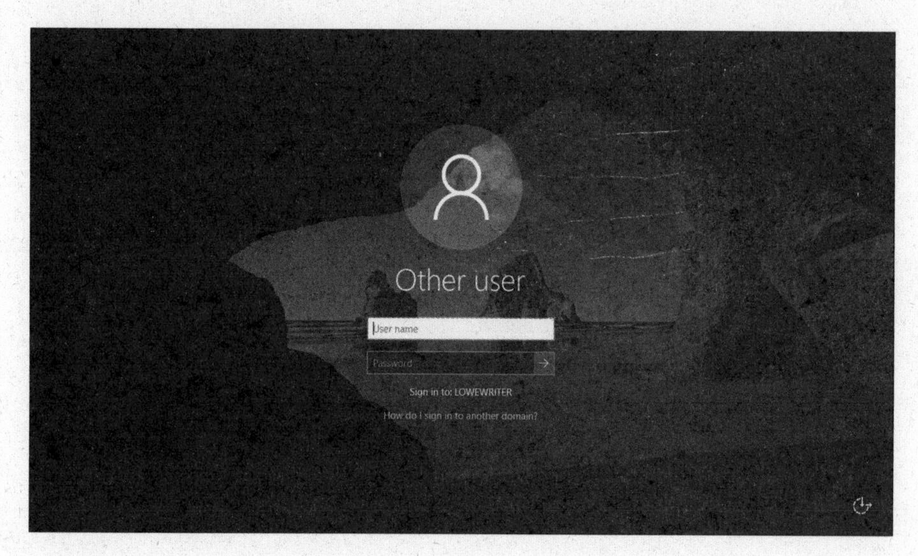

FIGURE 11-19: Logging in to the new domain.

Congratulations! If you've followed along with these steps on your own computer, you have successfully created and configured a Windows Server 2019 server that functions as both a DNS server and an Active Directory controller, and you've created a new domain — no small task!

» Examining user accounts

» Creating user accounts

» Setting account options

» Working with groups

» Creating a login script

Chapter 12
Managing Windows User Accounts

Every user who accesses a network must have a user account. User accounts allow you — as network administrator — to control who can access the network and who can't. In addition, user accounts let you specify what network resources each user can use. Without user accounts, all your resources would be open to anyone who casually dropped by your network.

User accounts are managed by Active Directory, the directory service used by Windows to manage all sorts of entities that exist on a network — not just users, but computers, printers, groups, and other types of objects as well.

This chapter gives you an overview of how Active Directory is organized. Then it dives into the details of working with Active Directory user accounts.

Understanding How Active Directory Is Organized

Active Directory is essentially a database management system. The Active Directory database is where the individual objects tracked by the directory are stored. Active Directory uses a *hierarchical* database model, which groups items in a treelike structure.

The terms *object, organizational unit, domain, tree,* and *forest* are used to describe the way Active Directory organizes its data. The following sections explain the meaning of these important Active Directory terms.

Objects

The basic unit of data in Active Directory is called an *object.* Active Directory can store information about many kinds of objects. The objects you work with most are users, groups, computers, and printers.

Figure 12-1 shows the Active Directory Manager displaying a list of built-in objects that come preconfigured with Windows Server 2019. To get to this management tool, choose Start⇨Administrative Tools⇨Active Directory Users and Computers. Then click the Built-in node to show the built-in objects.

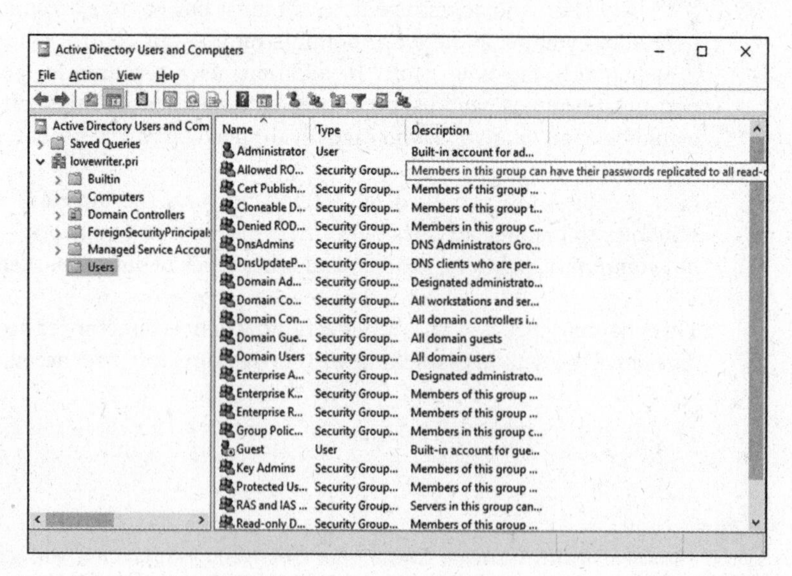

FIGURE 12-1: Objects displayed by the Active Directory Manager console.

Objects have descriptive characteristics called *properties* or *attributes*. You can call up the properties of an object by double-clicking the object in the management console.

Domains

A *domain* is the basic unit for grouping related objects in Active Directory. Typically, domains correspond to departments in a company. A company with separate Accounting, Manufacturing, and Sales departments might have domains named (you guessed it) `Accounting`, `Manufacturing`, and `Sales`. Or the domains may correspond to geographical locations. A company with offices in Detroit, Dallas, and Denver might have domains named `det`, `dal`, and `den`.

Note that because Active Directory domains use DNS naming conventions, you can create subdomains that are considered to be child domains. You should always create the top-level domain for your entire network before you create any other domain. If your company is named Nimbus Brooms, and you've registered `nimbusbroom.com` as your domain name, you should create a top-level domain named `nimbusbroom.com` before you create any other domains. Then you can create subdomains such as `accounting.nimbusbroom.com`, `manufacturing.nimbusbroom.com`, and `sales.nimbusbroom.com`.

Every domain must have at least one *domain controller*, which is a server that's responsible for the domain. Although one domain controller is the minimum, it's best to provide at least two domain controllers within your domain. That way, if one fails, Active Directory will continue working for the domain. Active Directory uses a feature called *replication* to keep all the domain controllers in sync.

Organizational units

Many domains have too many objects to manage together in a single group. Fortunately, Active Directory lets you create one or more *organizational units*, also known as OUs. OUs let you organize objects within a domain, without the extra work and inefficiency of creating additional domains. (In this respect, an OU is similar to a folder in the file system.)

One reason to create OUs within a domain is to assign administrative rights to each OU of different users. Then these users can perform routine administrative tasks such as creating new user accounts or resetting passwords.

Suppose that the domain for the Denver office, named `den`, houses the Accounting and Legal departments. Rather than create separate domains for these departments, you could create organizational units for the departments.

Trees

A *tree* is a set of Active Directory names that share a namespace. The domains `nimbusbroom.com`, `accounting.nimbusbroom.com`, `manufacturing.nimbusbroom.com`, and `sales.nimbusbroom.com` make up a tree that's derived from a common root domain, `nimbusbroom.com`.

The domains that make up a tree are related to one another through *transitive trusts*. In a transitive trust, if DomainA trusts DomainB and DomainB trusts DomainC, DomainA automatically trusts DomainC.

TIP

Note that a single domain all by itself is still considered to be a tree.

Forests

As its name suggests, a *forest* is a collection of trees. In other words, a forest is a collection of one or more domain trees that do *not* share a common parent domain. Every domain must belong to a forest, so even if your organization has just one domain, you'll also have one forest.

But suppose your company (Nimbus Brooms) acquires Tracorum Technical Enterprises, which already has its own root domain named `tracorumtech.com`, with several subdomains of its own. You can create a forest from these two domain trees so that the domains can trust each other.

Understanding Windows User Accounts

Now that we've reviewed the basic organizational structure of Active Directory, let's look at how users accounts are created and managed. As you might guess, user accounts are among the basic tools for managing a Windows server. As a network administrator, you'll spend a large percentage of your time dealing with user accounts — creating new ones, deleting expired ones, resetting passwords for forgetful users, granting new access rights, and so on.

The following sections describe some of the pertinent characteristics of user accounts.

Local accounts versus domain accounts

A *local account* is a user account stored on a particular computer, applicable to that computer only. Typically, each computer on your network has a local account for each person who uses that computer.

By contrast, a *domain account* is a user account that's stored by Active Directory (AD) and can be accessed from any computer that's a part of the domain. Domain accounts are centrally managed. This chapter deals primarily with setting up and maintaining domain accounts.

User account properties

Every user account has several important account properties that specify the characteristics of the account. The three most important account properties are

- **Username:** A unique name that identifies the account. The user must enter the username when logging on to the network. The username is public information. In other words, other network users can (and often should) find out your username.
- **Password:** A secret word that must be entered to gain access to the account. You can set up Windows so that it enforces password policies, such as the minimum length of the password, whether the password must contain a mixture of letters and numerals, and how long the password remains current before the user must change it.
- **Group membership:** The group(s) to which the user account belongs. Group memberships are the key to granting access rights to users so that they can access various network resources (such as file shares or printers) or perform certain network tasks (such as creating new user accounts or backing up the server).

Many other account properties record information about the user, such as the user's contact information, whether the user is allowed to access the system only at certain times or from certain computers, and so on.

Creating a New User

That's enough theory; let's get to the practical stuff. Here's the procedure for creating an Active Directory user account:

1. **Choose Start ➪ Windows Administrative Tools ➪ Active Directory Users and Computers.**

   This command fires up the Active Directory Users and Computers management console (refer to Figure 12-1).

2. **Right-click the Users Organizational Unit for the domain that you want to add the user to and then choose New ➪ User from the contextual menu.**

   Selecting this command launches the New Object – User Wizard, shown in Figure 12-2, to create a new user object in the OU you selected.

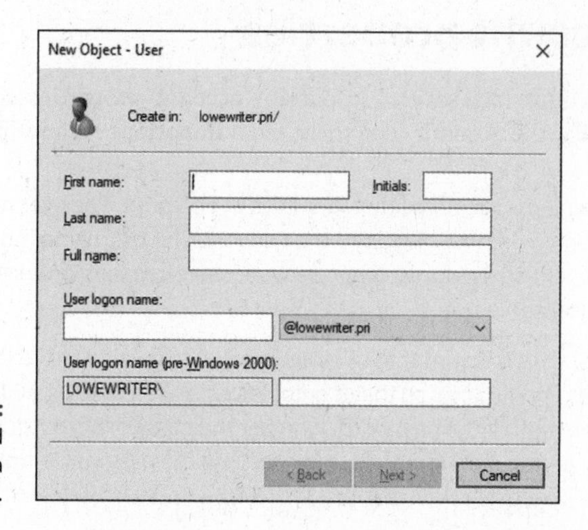

FIGURE 12-2: Use the wizard to create a new user.

3. **Enter the user's first name, middle initial, and last name.**

   As you fill in these fields, the New Object Wizard automatically fills in the Full Name field.

4. **Change the Full Name field if you want it to appear different from what the wizard proposes.**

   You may want to reverse the first and last names so the last name appears first, for example.

5. **Enter the user logon name.**

   This name must be unique within the domain. (Don't worry, if you try to use a name that isn't unique, you'll get an error message.)

TIP

   Pick a naming scheme to follow when creating user logon names. You can use the first letter of the first name followed by the complete last name, the complete first name followed by the first letter of the last name, or any other scheme that suits your fancy.

6. **Click Next.**

   The second page of the New Object – User Wizard appears, as shown in Figure 12-3.

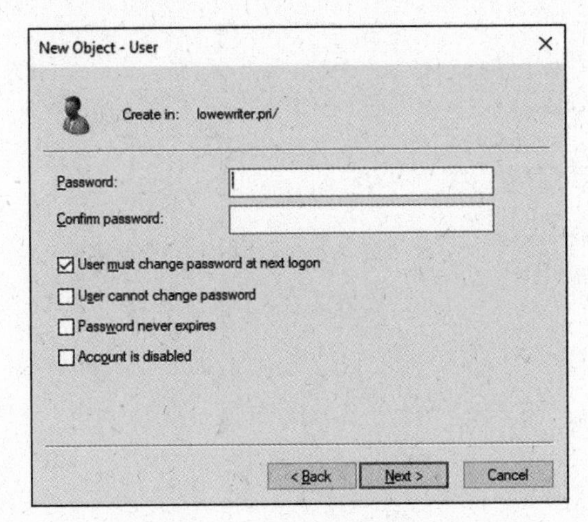

FIGURE 12-3: Set the user's password.

7. **Enter the password twice.**

   You're asked to enter the password and then confirm it, so type it correctly. If you don't enter it identically in both boxes, you're asked to correct your mistake.

8. **Specify the password options that you want to apply.**

   The following password options are available:

   - User Must Change Password at Next Logon
   - User Cannot Change Password
   - Password Never Expires
   - Account Is Disabled

   For more information about these options, see the section "Setting account options," later in this chapter.

9. **Click Next.**

   You're taken to the final page of the New Object – User Wizard, as shown in Figure 12-4.

10. **Verify that the information is correct and then click Finish to create the account.**

    If the account information isn't correct, click the Back button, and correct the error.

You're done! Now you can customize the user's account settings. At minimum, you'll probably want to add the user to one or more groups. You may also want to add contact information for the user or set up other account options.

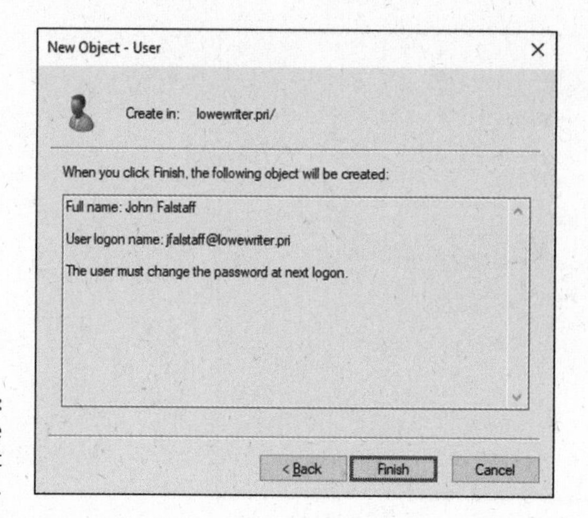

FIGURE 12-4: Verifying the user account information.

TIP

An alternative way to create a new user is simply to copy an existing user. When you copy an existing user, you provide a new username and password and Windows copies all the other property settings from the existing user to the new user.

Setting User Properties

After you create a user account, you can set additional properties for the user by right-clicking the new user and choosing Properties from the contextual menu. This command brings up the User Properties dialog box, which has about a million tabs that you can use to set various properties for the user. Figure 12-5 shows the General tab, which lists basic information about the user, such as the user's name, office location, and phone number.

The following sections describe some of the administrative tasks that you can perform via the various tabs of the User Properties dialog box.

Changing the user's contact information

Several tabs of the User Properties dialog box contain contact information for the user, such as

- **Address:** Change the user's street address, post office box, city, state, zip code, and so on.
- **Telephones:** Specify the user's phone numbers.
- **Organization:** Record the user's job title and the name of his boss.

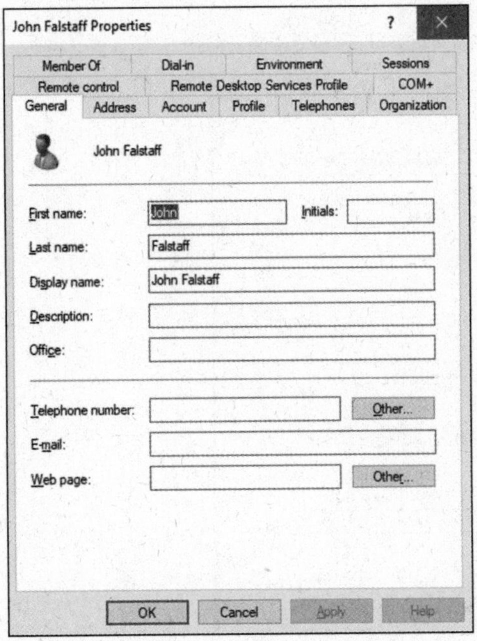

FIGURE 12-5: The General tab.

Setting account options

The Account tab of the User Properties dialog box, shown in Figure 12-6, features a variety of interesting options that you can set for the user. You can change the user's logon name, change the password options that you set when you created the account, and set an expiration date for the account.

The following account options are available in the Account Options list box:

- **User Must Change Password at Next Logon:** This default option allows you to create a one-time-only password that can get the user started with the network. The first time the user logs on to the network, he is asked to change the password.
- **User Cannot Change Password:** Use this option if you don't want to allow users to change their passwords. (Obviously, you can't use this option and the preceding one at the same time.)
- **Password Never Expires:** Use this option to bypass the password-expiration policy for this user so that the user will never have to change her password.
- **Store Password Using Reversible Encryption:** This option stores passwords by using an encryption scheme that hackers can easily break, so you should avoid it like the plague.

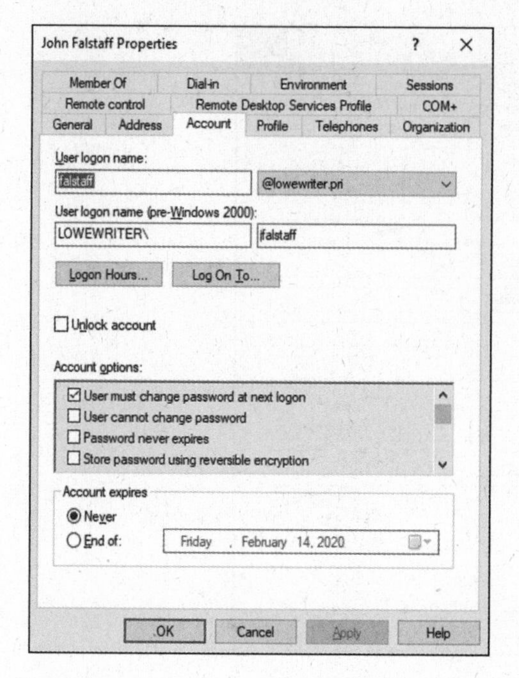

FIGURE 12-6: Set user account info here.

- **Account Is Disabled:** This option allows you to create an account that you don't yet need. As long as the account remains disabled, the user won't be able to log on. See the upcoming section, "Disabling and Enabling User Accounts," to find out how to enable a disabled account.
- **Smart Card Is Required for Interactive Logon:** If the user's computer has a smart card reader to read security cards automatically, select this option to require the user to use it.
- **Account Is Trusted for Delegation:** This option indicates that the account is trustworthy and can set up delegations. This advanced feature usually is reserved for Administrator accounts.
- **Account Is Sensitive and Cannot Be Delegated:** This option prevents other users from impersonating this account.
- **Use DES Encryption Types for This Account:** This option beefs up the encryption for applications that require extra security.
- **Do Not Require Kerberos Preauthentication:** *Kerberos* refers to a common security protocol used to authenticate users. Select this option only if you are using a different type of security.

Specifying logon hours

You can restrict the hours during which the user is allowed to log on to the system. Click the Logon Hours button on the Account tab of the User Properties dialog box to open the Logon Hours for [User] dialog box, as shown in Figure 12-7.

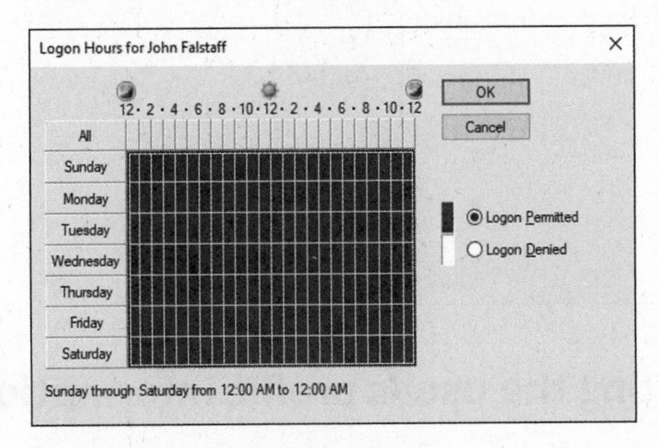

FIGURE 12-7: Restrict a user's logon hours.

Initially, the Logon Hours dialog box is set to allow the user to log on at any time of day or night. To change the hours that you want the user to have access, click a day and time or a range of days and times, select Logon Permitted or Logon Denied, and then click OK.

Restricting access to certain computers

Typically, a user can use his user account to log on to any computer that's part of the user's domain. You can restrict a user to certain computers, however, by clicking the Log On To button on the Account tab of the User Properties dialog box. This button brings up the Logon Workstations dialog box, as shown in Figure 12-8.

To restrict the user to certain computers, select the Following Computers radio button. Then, for each computer you want to allow the user to log on from, enter the computer's name in the text box and click Add.

TIP

If you make a mistake, you can select the incorrect computer name and then click Edit to change the name or click Remove to delete the name.

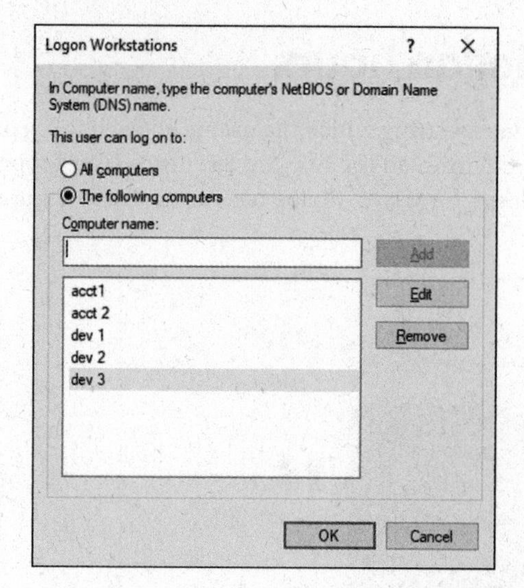

FIGURE 12-8: Restricting the user to certain computers.

Setting the user's profile information

From the Profile tab, as shown in Figure 12-9, you can configure three bits of information about the user's profile information:

- **Profile Path:** This field specifies the location of the user's roaming profile.
- **Logon Script:** This field is the name of the user's logon script. A *logon script* is a batch file that's run whenever the user logs on. The main purpose of the logon script is to map the network shares that the user requires access to. Logon scripts are carryovers from early versions of Windows NT Server. In Windows Server 2012, profiles are the preferred way to configure the user's computer when the user logs on, including setting up network shares. Many administrators still like the simplicity of logon scripts, however. For more information, see the section "Creating a Logon Script," later in this chapter.
- **Home Folder:** This section is where you specify the default storage location for the user.

TIP

From the Profile tab, you can specify the location of an existing profile for the user, but it doesn't actually let you set up the profile.

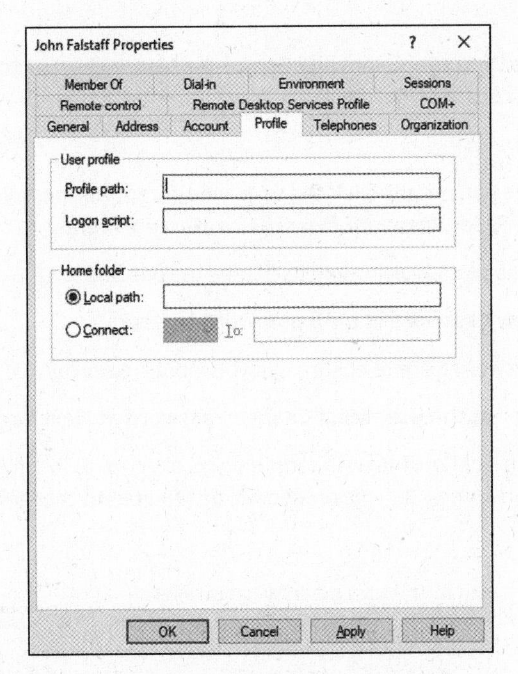

FIGURE 12-9: The Profile tab.

Resetting User Passwords

By some estimates, the single most time-consuming task of most network administrators is resetting user passwords. Lest you assume that all users are forgetful idiots, put yourself in their shoes, being made to set their passwords to something incomprehensible (94kD82leL384K) that they have change a week later to something more unmemorable (dJUQ63DWd8331) that they don't write down. Then network admins get mad when they forget their passwords.

So, when a user calls and says that he forgot his password, the least you can do is (appear to) be cheerful when you reset it. After all, the user probably spent 15 minutes trying to remember it before finally giving up and admitting failure.

Here's the procedure to reset the password for a user domain account:

1. **Log on as an administrator.**

   REMEMBER

   You must have administrator privileges to perform this procedure.

2. **Choose Start ⇨ Administrative Tools ⇨ Active Directory Users and Computers.**

   The Active Directory Users and Computers management console appears.

3. **In the Active Directory Users and Computers management console, click Users in the console tree.**

   Refer to Figure 12-1.

4. **In the Details pane, right-click the user who forgot her password and then choose Reset Password from the contextual menu.**

   A dialog box appears allowing you to change the password.

5. **Enter the new password in both password boxes.**

   Enter the password twice to ensure that you input it correctly.

TIP

6. **(Optional) Select the User Must Change Password at Next Logon option.**

   If you select this option, the password that you assign will work for only one logon. As soon as the user logs on, she will be required to change the password.

7. **Click OK.**

   That's all there is to it! The user's password is reset.

Disabling and Enabling User Accounts

To temporarily prevent a user from accessing the network, you can disable his account. You can always enable the account later, when you're ready to restore the user to full access. Here's the procedure:

1. **Log on as an administrator.**

   You must have administrator privileges to perform this procedure.

2. **From Server Manager, choose Tools ➪ Active Directory Users and Computers.**
3. **In the Active Directory Users and Computers management console that appears, click Users in the console tree.**
4. **In the Details pane, right-click the user that you want to enable or disable; then choose either Enable Account or Disable Account from the contextual menu to enable or disable the user, respectively.**

Deleting a User

People come, and people go. And when they go, so should their user account. Deleting a user account is surprisingly easy. Just follow these steps:

1. **Log on as an administrator.**

   You must have administrator privileges to perform this procedure.

2. **Choose Start ⇨ Administrative Tools ⇨ Active Directory Users and Computers.**
3. **In the Active Directory Users and Computers management console that appears, click Users in the console tree.**
4. **In the Details pane, right-click the user that you want to delete and then choose Delete from the contextual menu.**

   Windows asks whether you really want to delete the user, just in case you're kidding.

5. **Click Yes.**

   Poof! The user account is deleted.

WARNING

Deleting a user account is a permanent, nonreversible action. Do it only if you're absolutely sure that you never ever want to restore the user's account. If there's any possibility of restoring the account later, disable the account instead of deleting it. (See the preceding section.)

Working with Groups

A *group* is a special type of account that represents a set of users who have common network access needs. Groups can dramatically simplify the task of assigning network access rights to users. Rather than assign access rights to each user individually, you can assign rights to the group itself. Then those rights automatically extend to any user you add to the group.

The following sections describe some of the key concepts that you need to understand to use groups, along with some of the most common procedures you'll employ when setting up groups for your server.

Creating a group

Here's how to create a group:

1. **Log on as an administrator.**

   You must have administrator privileges to perform this procedure.

2. **From Server Manager, choose Tools ⇨ Active Directory Users and Computers.**

   The Active Directory Users and Computers management console appears.

3. **Right-click the domain to which you want to add the group and then choose New ⇨ Group from the contextual menu.**
4. **In the New Object – Group dialog box that appears, as shown in Figure 12-10, enter the name for the new group.**

   Enter the name in both text boxes.
5. **Choose the group scope.**

   The choices are

   - *Domain Local:* For groups that will be granted access rights to network resources
   - *Global:* For groups to which you'll add users and Domain Local groups
   - *Universal:* If you have a large network with multiple domains
6. **Choose the group type.**

   The choices are Security and Distribution. In most cases, choose Security.
7. **Click OK.**

   The group is created. However, at this point, it has no members. To remedy that, keep reading.

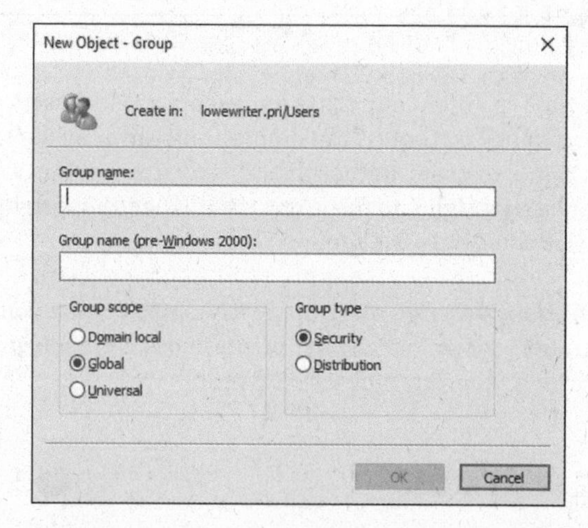

FIGURE 12-10: Create a new group.

Adding a member to a group

Groups are collections of objects called *members.* The members of a group can be user accounts or other groups. A newly created group (see the preceding section) has no members. As you can see, a group isn't useful until you add at least one member.

Follow these steps to add a member to a group:

1. **Log on as an administrator.**

   You must have administrator privileges to perform this procedure.

2. **Choose Start ➪ Administrative Tools ➪ Active Directory Users and Computers.**

   The Active Directory Users and Computers management console appears.

3. **Open the folder that contains the group to which you want to add members and then double-click the group.**

   The Group Properties dialog box appears.

4. **Click the Members tab.**

   The members of the group are displayed, as shown in Figure 12-11.

5. **Click Add, type the name of a user or other group that you want to add to this group, and then click OK.**

   The member is added to the list.

6. **Repeat Step 5 for each user that you want to add.**

   Keep going until you add everyone!

7. **Click OK.**

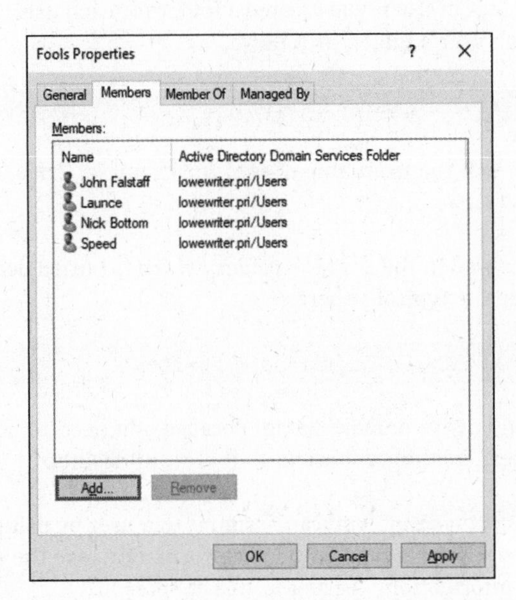

FIGURE 12-11: Adding members to a group.

That's all there is to it.

TIP

On the Member Of tab of the Group Properties dialog box, you can see a list of each group that the current group is a member of.

Creating a Logon Script

A *logon script* is a batch file that's run automatically whenever a user logs on. The most common reason for using a logon script is to map the network shares that the user needs access to. Here's a simple logon script that maps three network shares:

```
echo off
net use m: \\server1\shares\admin
net use n: \\server1\shares\mktg
net use o: \\server2\archives
```

Here, two shares on server1 are mapped to drives `M:` and `N:`, and a share on server2 is mapped as drive `O:`.

If you want, you can use the special variable `%username%` to get the user's user-name. This variable is useful if you created a folder for each user, and you want to map a drive to each user's folder, as follows:

```
net use u: \\server1\users\%username%
```

If a user logs on with the username `dlowe`, for example, drive `U:` is mapped to `\\server1\users\dlowe`.

TIP

Scripts should be saved in the `Scripts` folder, which is buried deep in the bowels of the `SYSVOL` folder — typically, here:

```
c:\Windows\SYSVOL\Sysvol\domainname\Scripts
```

where *domainname* is your domain name. Because you need to access this folder frequently, I suggest creating a shortcut to it on your desktop.

After you create a logon script, you can assign it to a user by using the Profile tab of the User Properties dialog box. For more information, see the section "Setting the user's profile information," earlier in this chapter.

» Looking at file servers

» Creating shares

» Setting permissions

Chapter 13
Managing Network Storage

One key purpose of most computer networks is to provide shared access to disk storage. In this chapter, you find out about several ways that a network can provide shared disk storage. Then you discover how to configure Windows Server 2019 to operate as a file server.

Understanding Disk Storage

We're all familiar with using disk drives to store data: Every Windows computer system in use today has some sort of disk storage on which data is stored. The following sections explain some of the underlying concepts of disk storage you need to understand as you plan how disk storage can be shared on a network.

Hard disk drives

A traditional disk drive stores data on rotating disks that are covered with magnetic material. A disk of this type is often called a *hard disk drive* (HDD), though sometimes the term *spinning disk* is used.

The modern HDD was invented by IBM in 1956, just a few years before I was born. It had a capacity of a whopping 5MB and had to be moved with a forklift. The price to lease one was $650 per month, equivalent to more than $6,000 today.

We've come a long way in 64 years: Today, you can buy a disk drive that has more than 10,000 times the capacity for under $200.

Other than being physically much smaller, having thousands of times the storage capacity, and costing much less, today's spinning disks are not really all that different from the original 1956 version. Like the original version, data is stored in concentric rings on the surface of the spinning disk platter. This rings are called *tracks*. The data itself is written to each track as a sequence of sectors; a sector is the smallest unit of data that the disk drive can read or write. On most drives, each sector contains 512 bytes of data. But on some more expensive drives, the sector size is a whopping 4K.

You'll sometimes here the term *block* used in lieu of *sector*. A block is the smallest unit of data that an operating system recognizes on a formatted disk drive. Typically, blocks are 2K in size, meaning that each block requires four sectors on the disk.

Hard disk drives have many advantages — principal among them being low cost and high capacity. The biggest disadvantage of hard disk drives are their speed. The speed of a disk drive is limited by the fact that it relies on mechanical motion to access data.

To understand this limitation, consider what has to happen for a disk drive to read a specific sector of data:

1. Because each disk platter has thousands of tracks, the *read/write head* (the mechanism that actually reads or writes data to the tracks) must be moved so that it is positioned over the track that contains the sector to be read.
2. When the read/write head is in place, the disk has to wait for the data to rotate around the platter until it's under the read/write head. On average, the data will be exactly halfway around the platter from the read/write head; sometimes it will be closer, but the data is just as likely to be farther. Either way, the disk drive has to wait for the data to spin around the circle.

The exact amount of time that the disk drive has to wait for the data depends on how fast the disk drive spins. A typical disk drive spins at 7,200 revolutions per minute (RPM). That sounds really fast, but in computer terms it's interminably slow. 7,200 RPM is the equivalent of 120 rotations per second. That's 0.00833 second per revolution. Cut that in half and you can see that the disk drive has to wait an average of 0.004167 second for the data to arrive at the read/write head.

Let's put that into perspective by comparing it to the typical clock speed of a modern CPU. CPUs typically purr along with clock speeds ranging from 2.0 GHz to 3.5 GHz; we'll pick 2.0 GHz to keep the math simple. That's two billion clock ticks *per second.* In the 0.004167 second it takes for the disk drive to wait for data to arrive, the CPU clock will have ticked more than 8 million times. Thus, the CPU can do 8 million other things while it waits for data to spin around the disk.

Here's an even better way to get a perspective on how slow disk drives are: Imagine that you could slow both the CPU clock and the disk drive down to where the CPU clock ticked once per second. Snap your fingers and count *one one thousand, two one thousand* to get this mental exercise going. At that pace, you would have to wait more than three months for the data to spin around to the point where it can be read by the read/write heads.

For the fun of it, let's go back a moment to Step 1 in the process, in which the read/write head moved to the track that contains the sector to be accessed. In most disk drives, this move can take anywhere from a single revolution to three full revolutions to accomplish. In our one-click-per-second illustration, we had to wait on average six months before we could even start to wait for the data. So now we've waited a total of nine months, on average, just to be in position to read the data.

3. And we're still not done: When the data has arrived at the read/write head, the head is activated to either read or write the data. And then, the read/write head can only read the data at the speed that it passes under the read/write head. Figuring out the amount of time it takes for the data to fly under the head is a much more complicated task than accounting for moving the read/write heads and waiting for the data to arrive. Let's just figure that it will take a few additional days or weeks to actually read the data.

That was fun, wasn't it? The whole point of this detailed discussion is to illustrate why disk storage is one of the most troublesome bottlenecks when it comes to the overall performance of your network.

To be fair, disk drive manufacturers go to great lengths to minimize the impact of the inherent slowness of disk technology. For example, all disk drives include a relatively large amount of internal storage that is used as a *cache*, which stores the most recently read data in fast memory so that it can quickly be read again if it's needed. Even so, cache memory can only marginally improve disk performance. What's needed is an inherently faster storage technology, which is where solid state drives come in.

Solid state drives to the rescue!

A *solid state drive* (SSD) uses integrated circuits to store data in electronic memory rather than on magnetic spinning disks.

Some people use the term *solid state disk*, which I'll excuse because SSDs function as replacements for spinning disks. But because an SSD does not contain a disk, those folks are technically incorrect. Pointing out their inaccuracy is a sure way to lose them as friends, so I suggest you smile quietly to yourself and let it be.

SSDs are based on a type of memory technology called *flash memory*, which has been around since the 1980s but has only recently become affordable as an alternative to spinning disks. An SSD is dramatically faster than a spinning disk because there are no mechanical components involved when accessing data: There is no read/write head to move and no spinning platter to wait for. An SSD can read and write data at electronic speeds, not at mechanical speeds. (Because of the lack of moving parts, SSDs are also more reliable than spinning disks.)

This performance benefit comes at a price, though. SSD storage typically costs from two to five times as much as HDD storage, depending on the type of drive you need.

An important point to realize about SSDs is that they're designed to mimic traditional spinning disk storage. So, SSDs use the same electronic connections as spinning disks, which means you can use SSDs interchangeably with HDDs. Whenever I mention *disks* or *disk storage* in this book, you can assume that I mean either SSD or HDD storage.

It's a RAID!

I mention in Chapter 10 that individual disk drives are usually combined in groups using a technique called *RAID*, which stands for *redundant array of inexpensive disks*. In fact, RAID arrays are a universal requirement for disk storage for two reasons:

- RAID allows you to combine several disk drives into a single unit that has a total capacity that exceeds the capacity of the individual disks that make up the array.
- RAID provides an important safety precaution that enables your data to survive the loss of one of the disks in the array. (In some cases, a RAID array can survive the loss of two or even more disks.)

There are several basic *levels* of RAID you need to be aware of:

- **RAID 0**, which uses a technique called *striping* to combine two or more disk drives to create a single image whose total capacity is the sum of the capacities of the individual disks. RAID 0 is almost never used because although it allows for greater capacity, it doesn't provide any protection against the failure of one of the drives in the array. In fact, if one of the drives in a RAID 0 array fails, you'll lose all the data stored on all the drives in the array.
- **RAID 1,** in which a pair of disk drives are *mirrored* so that if one of them fails, the data on the other drive protects the data on the array.
- **RAID 10,** in which an even number of disk drives are both striped and mirrored. In other words, each pair of drives in the array is mirrored, and the resulting mirror pairs are striped to increase the array's capacity. Thus, the total capacity of the array is equal to half of the capacity of the drives combined; for example, if a RAID 10 array is created from eight 2TB drives, the total capacity of the array is 8TB (one half of the total 16TB raw capacity of the eight drives). RAID 10 is one of the safest types of RAID arrays because the array can survive the loss of any single drive within the array, and it can survive the loss of more than one drive provided that neither of the two lost drives is in the same mirror pair.

  RAID 10 is also one of the fastest forms of RAID. That's because there is very little overhead involved in the mirroring and striping required for RAID 10. Because of its speed and safety, RAID 10 arrays are often used for an organization's most critical data.
- **RAID 5,** in which disks are striped but not mirrored. Instead, redundancy is used across the striped drives so that if any one of the drives is lost, the data that was on that drive can be recreated using the data stored on the surviving drives.

  RAID 5 achieves its redundancy using the principal of *parity.* The simplest way to understand parity is to consider an example: Suppose that you write three numbers — say, 5, 10, and 15 — on three separate pieces of paper. Unfortunately, if you lose one of the pieces of paper, you'll lose the number that was written on it. To safeguard against this, you decide to calculate a second number such that the sum of all four numbers will be zero. In this case, the fourth number would be –30 (because 5 + 10 + 15 equals 30, so adding –30 to 30 gives you zero). Now, if you lose one of the pieces of paper — say, the one containing the number 10 — you just reverse the mathematical process you used to determine the fourth number. In other words, you *subtract* the three numbers you do know. After scratching your head to remember how to subtract negative numbers, you'll arrive at the answer — 10 — which happens to be the missing number.

In a RAID 5 array, parity is calculated for every bit of data written to the array, and the resulting parity values are distributed evenly across all the disks in the array such that if any one disk is lost, the missing data can be reconstructed from the data on the surviving drives.

The effective capacity of a RAID 5 array is equal to the total capacity of all the drives less one. So, if a RAID 5 array is built from five 2TB drives, the effective capacity is 8TB (4 × 2TB).

- **RAID 6** is almost the same as RAID 5 except that two parity values are calculated and distributed across the disks. This allows you to lose two drives in the array and still be able to recover your data. The effective capacity of a RAID 6 array is equal to the total capacity of all the drives less two. So, a RAID 6 array built from five 2TB arrays will have an effective capacity of 6TB (3 × 2TB).

Three ways to attach disks to your servers

I also mention in Chapter 10 that there are three primary ways to connect disks to a server computer. This is true whether your servers are virtual or physical servers. To recap:

- **Local disk storage:** All server computers have at least one disk drive that is directly attached to the server, mounted in the same case as the server. Most servers can accommodate additional disk drives, which can be consolidated into RAID arrays to create disk storage that is available to that server and that server only.
- **Storage Area Network (SAN):** A SAN is essentially a network of disk controllers that manage arrays of disk drives. The controllers themselves are connected to server computers and to each other via a special network designed specifically for this purpose. The most common network technology used in SANs is called *Fibre Channel;* it runs at speeds considerably higher than most Ethernet networks — at the time I wrote this book, the fastest available Fibre Channel speed was 256 Gbps.

  A SAN allows you to separate your disk storage from your server computers so that you can manage storage and servers independently. If you need more servers, you can add more and connect them to the SAN. If you need more disk space, you can add additional disks to the SAN and your servers will be able to use it.

» **Network attached storage (NAS):** NAS is considerably simpler than SAN; NAS storage is simply disk storage that is connected to your Ethernet network. To use NAS, you purchase one or more NAS appliances, which contain disk storage and a controller that presents the disk storage to the network. Because NAS is accessed via your Ethernet network, it's considerably slower than SAN storage. But it's also considerably less expensive.

Focusing on File Servers

Now that we've looked at the basics of how disk storage works and how it can be attached, let's turn to the main way in which you can create and manage storage on your network: by setting up one or more *file servers*. A *file server* is simply a network server whose primary role is to share its disk storage. Using a file server is the most common way to provide shared network storage.

A file server can be anything from a simple desktop computer that has been pressed into service as a file server to an expensive ($25,000 or more) server with redundant components so that the server can continue to run when a component fails.

TIP

The rest of this chapter is devoted to showing you how to configure Windows Server 2019 to run as a file server.

Understanding permissions

Before I get into the details of setting up a file server, you need to have a solid understanding of the concept of permissions. *Permissions* allow users to access shared resources on a network. Simply sharing a resource, such as a disk folder or a printer, doesn't guarantee that a given user is able to access that resource. Windows makes this decision based on the permissions that have been assigned to various groups for the resource and group memberships of the user. For example, if the user belongs to a group that has been granted permission to access the resource, the access is allowed. If not, access is denied.

In theory, permissions sound pretty simple. In practice, however, they can get pretty complicated. The following paragraphs explain some of the nuances of how access control and permissions work:

» Every object — that is, every file and folder — on an NTFS volume has a set of permissions — the Access Control List (ACL) associated with it.

» The ACL identifies which users and groups can access the object and specifies what level of access each user or group has. A folder's ACL may specify that one group of users can read files in the folder, whereas another group can read and write files in the folder, and a third group is denied access to the folder.

» Container objects — files and volumes — allow their ACLs to be inherited by the objects that they contain. As a result, if you specify permissions for a folder, those permissions extend to the files and child folders that appear within it.

Table 13-1 lists the six permissions that can be applied to files and folders on an NTFS volume.

TABLE 13-1 File and Folder Permissions

Permission	Description
Full Control	The user has unrestricted access to the file or folder.
Modify	The user can change the file or folder's contents, delete the file or folder, read the file or folder, or change the attributes of the file or folder. For a folder, this permission allows you to create new files or subfolders within the folder.
Read & Execute	For a file, this permission grants the right to read or execute the file. For a folder, this permission grants the right to list the contents of the folder or to read or execute any of the files in the folder.
List Folder Contents	This permission applies only to folders; it grants the right to list the contents of the folder.
Read	This permission grants the right to read the contents of a file or folder.
Write	This permission grants the right to change the contents of a file or its attributes. For a folder, this permission grants the right to create new files and subfolders within the folder.

Actually, the six file and folder permissions comprise various combinations of special permissions that grant more detailed access to files or folders. Table 13-2 lists the special permissions that apply to each of the six file and folder permissions.

TIP

Assign permissions to groups rather than to individual users. that way, if a particular user needs access to a particular resource, add that user to a group that has permission to use the resource.

TABLE 13-2 Special Permissions

Special Permission	Full Control	Modify	Read & Execute	List Folder Contents	Read	Write
Traverse Folder/ Execute File	*	*	*	*		
List Folder/Read Data	*	*	*	*	*	
Read Extended Attributes	*	*	*	*	*	
Create Files/Write Data	*	*				*
Create Folders/ Append Data	*	*				*
Write Attributes	*	*				*
Write Extended Attributes	*	*				*
Delete Subfolders and Files	*					
Delete	*	*				
Read Permissions	*	*	*	*	*	*
Change Permissions	*					
Take Ownership	*					
Synchronize	*	*	*	*	*	*

Understanding shares

A *share* is simply a folder that is made available to other users via the network. Each share has the following elements:

- **Share name:** The name by which the share is known over the network
- **Path:** The path to the folder on the local computer that's being shared, such as `C:\Accounting`
- **Description:** A one-line description of the share
- **Permissions:** A list of users or groups who have been granted access to the share

When you install Windows and configure various server roles, special shared resources are created to support those roles. You shouldn't disturb these special shares unless you know what you're doing. Table 13-3 lists some of the most common special shares.

TABLE 13-3 Special Shares

Share Name	Description
`drive$`	The root directory of a drive.
`ADMIN$`	Used for remote administration of a computer. This share points to the OS folder (usually, C:\Windows).
`IPC$`	Used by named pipes, a programming feature that lets processes communicate with one another.
`NETLOGON`	Required for domain controllers to function.
`SYSVOL`	Another required domain controller share.
`PRINT$`	Used for remote administration of printers.
`FAX$`	Used by fax clients.

Notice that some of the special shares end with a dollar sign ($). These shares are hidden shares, not visible to users. You can still access them, however, by typing the complete share name (including the dollar sign) when the share is needed. The special share `C$`, for example, is created to allow you to connect to the root directory of the `C:` drive from a network client. You wouldn't want your users to see this share, would you? (Shares such as `C$` are also protected by permissions, of course, so if an ordinary user finds out that `C$` is the root directory of the server's `C:` drive, he still can't access it.)

Managing Your File Server

To manage shares on a Windows Server 2019 system, open the Server Manager, and select File and Storage Services in the task pane on the left side of the window. Then click Shares to reveal the management console shown in Figure 13-1.

The following sections describe some of the most common procedures that you'll use when managing your file server.

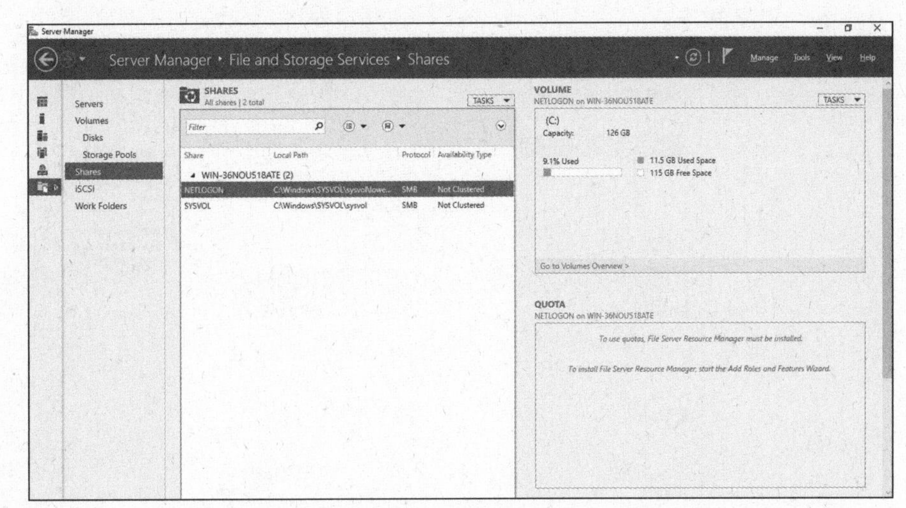

FIGURE 13-1: Managing shares in Windows Server 2019.

Using the New Share Wizard

To be useful, a file server should offer one or more *shares* — folders that have been designated as publicly accessible via the network. To create a new share, use the New Share Wizard:

1. **In Server Manager, select File and Storage Services, click Shares, and then choose New Share from the Tasks drop-down menu.**

   The opening screen of the New Share Wizard appears, as shown in Figure 13-2. Here, the wizard presents several options for how you'd like to proceed. The most common option is SMB Share, which is the standard file sharing protocol for Windows networks.

2. **Select SMB Share – Quick in the list of profiles and then click Next.**

   The New Share Wizard asks for the location of the share, as shown in Figure 13-3.

3. **Select the server where you want the share to reside.**

4. **Select the location of the share by choosing one of these two options:**

   - *Select by Volume:* This option selects the volume on which the shared folder will reside while letting the New Share Wizard create a folder for you. If you select this option, the wizard will create the shared folder on the designated volume. Use this option if the folder doesn't yet exist and you don't mind Windows placing it in the default location, which is inside a folder called `Shares` on the volume you specify.
   - *Type a Custom Path:* Use this option if the folder exists or if you want to create one in a location other than the `Shares` folder.

   For this example, I chose the Select by Volume example to allow the wizard to create the share in the `Shares` folder on the `C:` drive.

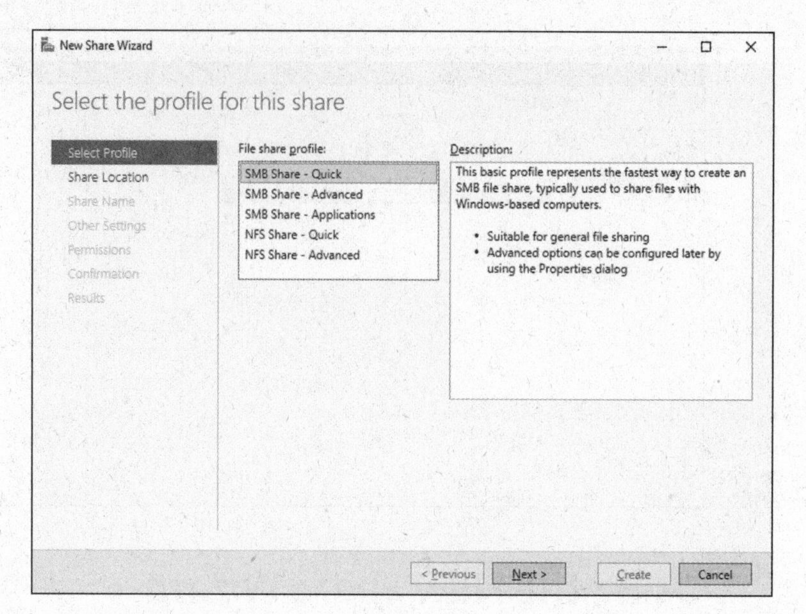

FIGURE 13-2: The New Share Wizard comes to life.

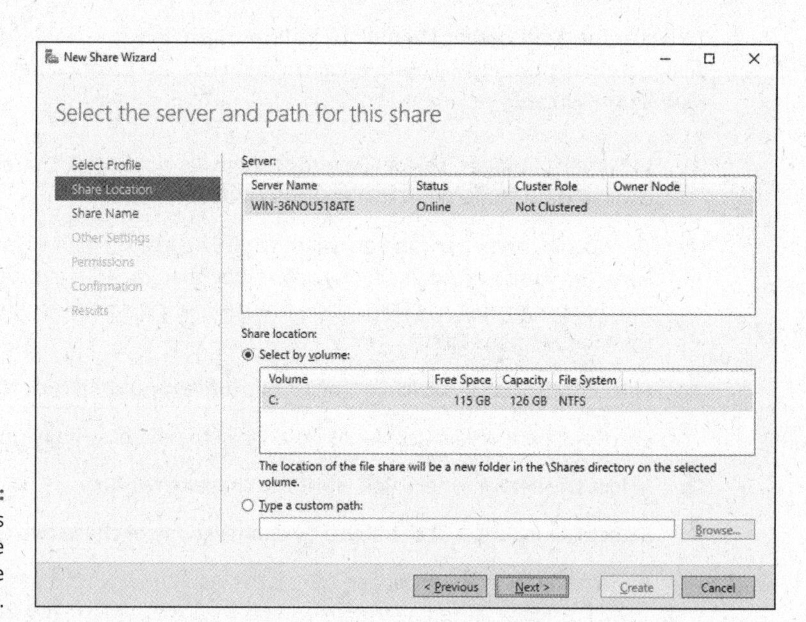

FIGURE 13-3: The wizard asks where you'd like to locate the share.

5. **Click Next.**

   The dialog box shown in Figure 13-4 appears.

6. **Enter the name that you want to use for the share in the Share Name field.**

   The default name is the name of the folder being shared. If the folder name is long, you can use a more succinct name here.

   For this example, I entered the share name `Data`.

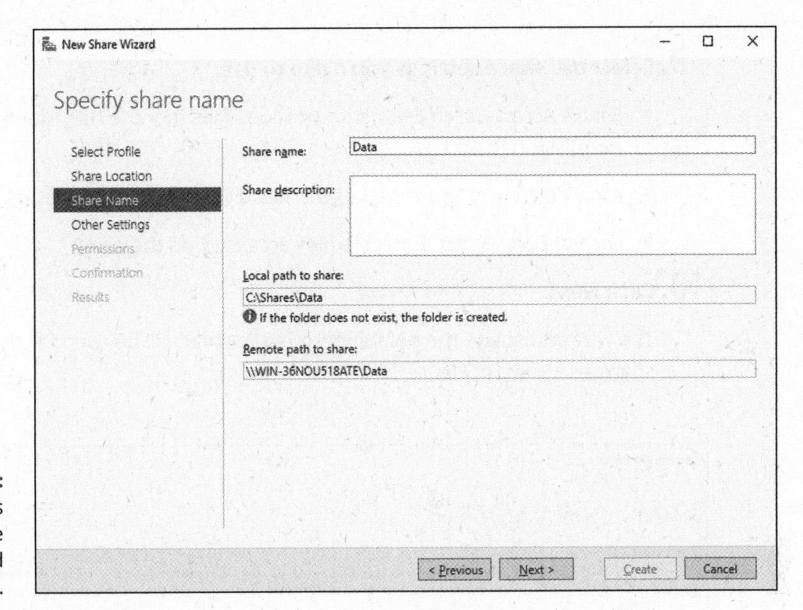

FIGURE 13-4: The wizard asks for the share name and description.

7. **Enter a description for the share.**
8. **Click Next.**

   The dialog box shown in Figure 13-5 appears.

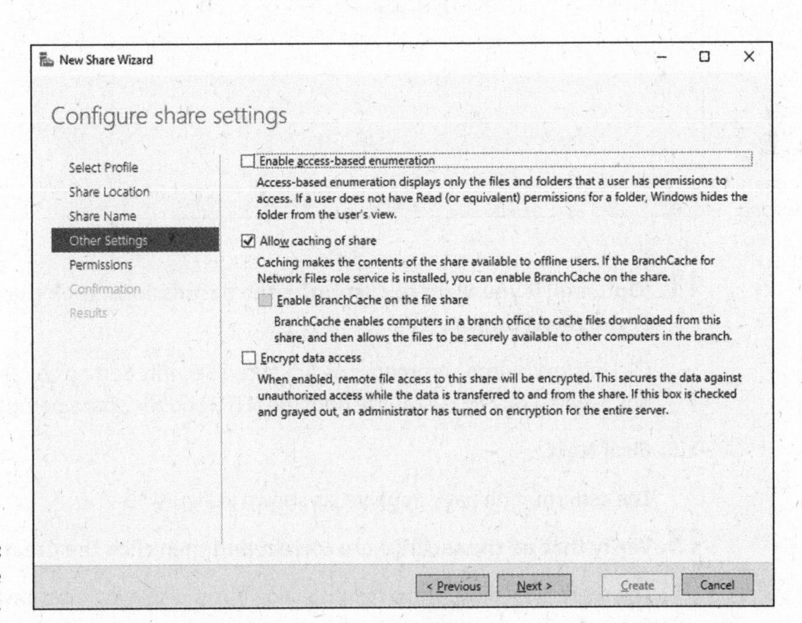

FIGURE 13-5: Specify the share settings.

9. **Select the share settings you'd like to use:**

   - *Enable Access-Based Enumeration:* Hides files that the user does not have permission to access
   - *Allow Caching of Share:* Makes the files available to offline users
   - *Encrypt Data Access:* Encrypts files accessed via the share

10. **Click Next.**

    The wizard displays the default permissions that will be used for the new share, as shown in Figure 13-6.

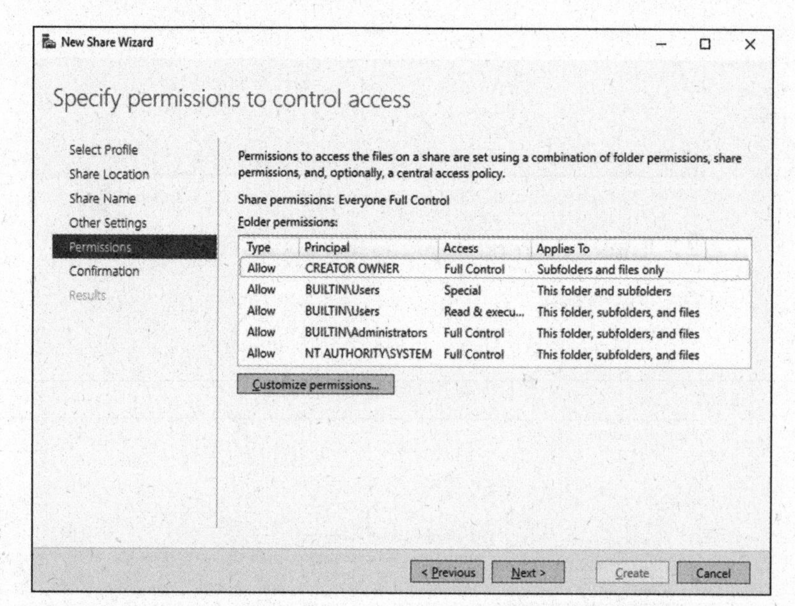

FIGURE 13-6: Setting the share permissions.

11. **(Optional) If you want to customize the permissions, click the Customize Permissions button.**

    Clicking this button summons the Advanced Security Settings for Data dialog box, where you can customize both the NTFS and the share permissions.

12. **Click Next.**

    The confirmation page appears, as shown in Figure 13-7.

13. **Verify that all the settings are correct and then click the Create button.**

    The share is created, and a results dialog box is displayed, as shown in Figure 13-8.

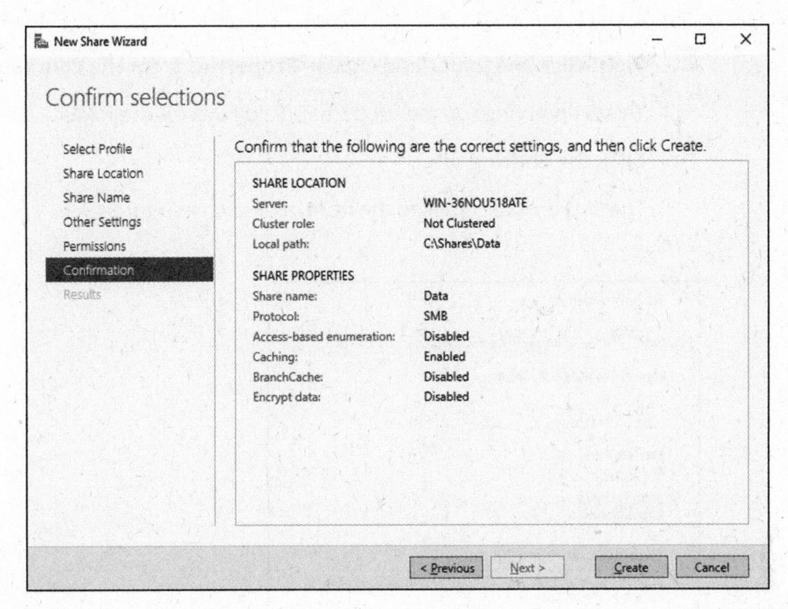

FIGURE 13-7: Confirming your share settings.

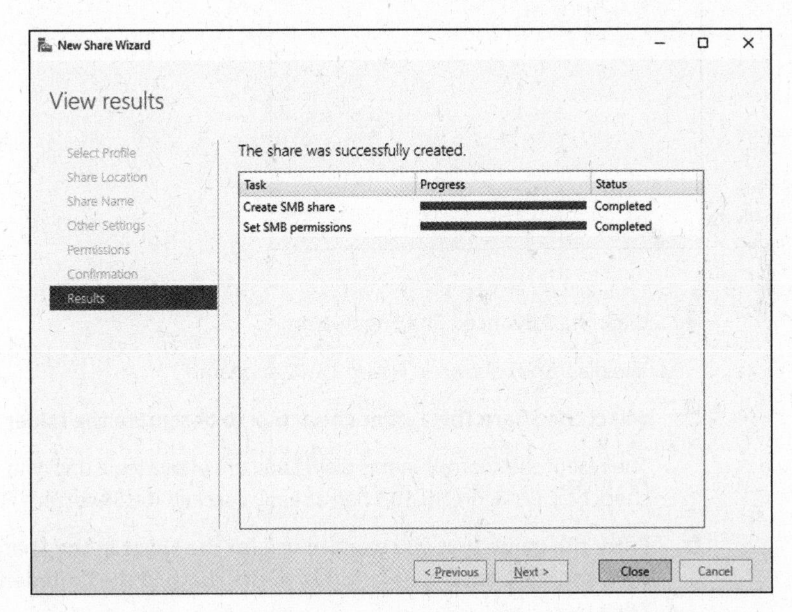

FIGURE 13-8: You're done!

Sharing a folder without the wizard

If you think wizards should be confined to *Harry Potter* movies, you can set up a share without bothering with the wizard. Just follow these steps:

1. **Press the Windows key, click Computer, and navigate to the folder that you want to share.**

2. **Right-click the folder and choose Properties from the contextual menu.**

   This action brings up the Properties dialog box for the folder.

3. **Click the Sharing tab.**

   The Sharing tab comes to the front, as shown in Figure 13-9.

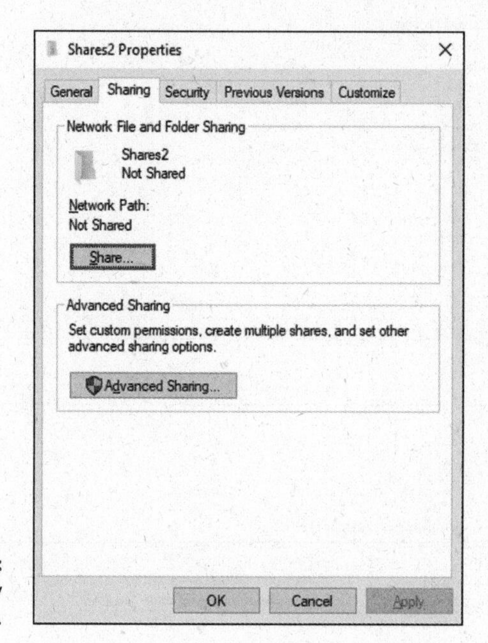

FIGURE 13-9: Manually share a folder.

4. **Click the Advanced Sharing button.**

   The dialog box shown in Figure 13-10 appears.

5. **Select the Share This Folder check box to designate the folder as shared.**

   The rest of the controls in this dialog box are unavailable until you select this check box. (In Figure 13-10, I had already checked the Share This Folder box.)

6. **Enter the name that you want to use for the share in the Share Name field and then enter a description of the share in the Comments field.**

   The default name is the name of the folder being shared. If the folder name is long, you can use a more succinct name here.

   The description is strictly optional but sometimes helps users determine the intended contents of the folder.

7. **Click the Permissions button and then set the permissions you want to apply to the share.**

   For more information, see the next section.

8. **Click OK.**

   The folder is now shared.

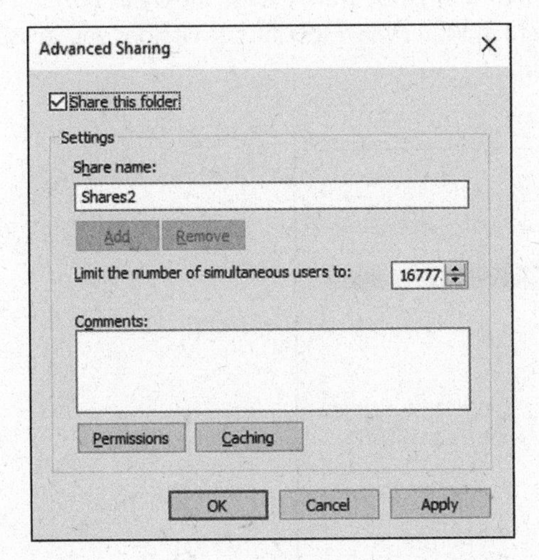

FIGURE 13-10: Set the share name.

Granting permissions

When you first create a file share, all users are granted read-only access to the share. If you want to allow users to modify files in the share or allow them to create new files, you need to add permissions. Here's how to do this using Windows Explorer:

1. **Open Windows Explorer by pressing the Windows key and clicking Computer; then browse to the folder whose permissions you want to manage.**

2. **Right-click the folder you want to manage and then choose Properties from the contextual menu.**

   The Properties dialog box for the folder appears.

3. **Click the Sharing tab; then click Advanced Sharing.**

   The Advanced Sharing dialog box appears.

4. **Click Permissions.**

The dialog box shown in Figure 13-11 appears. This dialog box lists all the users and groups to whom you've granted permission for the folder. Initially, read permissions are granted to a group called Everyone, which means that anyone can view files in the share but no one can create, modify, or delete files in the share.

When you select a user or group from the list, the check boxes at the bottom of the list change to indicate which specific permissions you've assigned to each user or group.

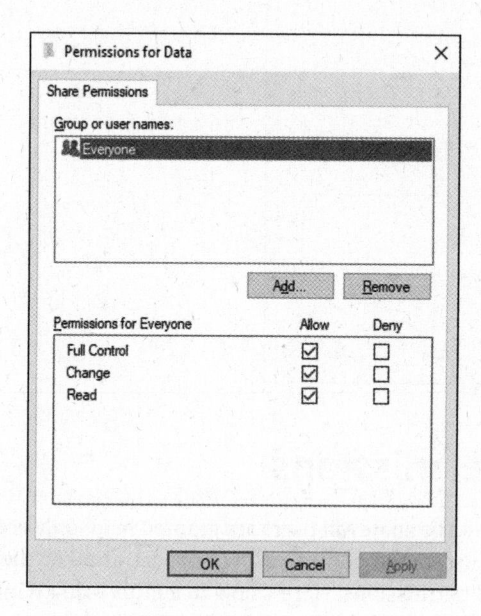

FIGURE 13-11: Set the share permissions.

5. **Click the Add button.**

The dialog box shown in Figure 13-12 appears.

6. **Enter the name of the user or group to whom you want to grant permission and then click OK.**

TIP

If you're not sure of the name, click the Advanced button. This action brings up a dialog box from which you can search for existing users.

When you click OK, you return to the Share Permissions tab (refer to Figure 13-11), with the new user or group added.

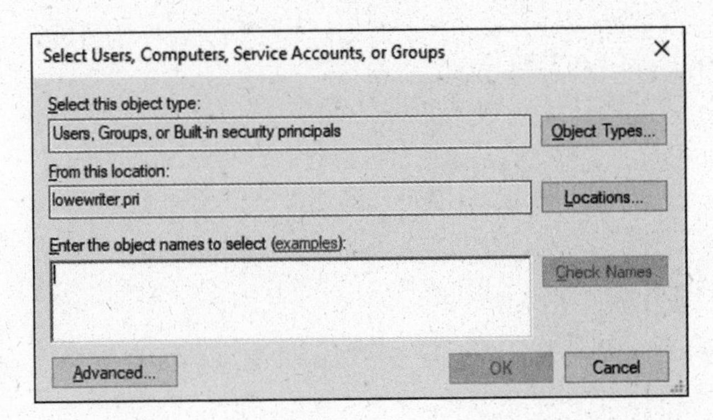

FIGURE 13-12: Adding permissions.

7. **Select the appropriate Allow and Deny check boxes to specify which permissions to allow for the user or group.**
8. **Repeat Steps 5–7 for any other permissions that you want to add.**
9. **When you're done, click OK.**

Here are a few other thoughts to ponder concerning adding permissions:

- If you want to grant full access to everyone for this folder, don't bother adding another permission. Instead, select the Everyone group and then select the Allow check box for each permission type.
- You can remove a permission by selecting the permission and then clicking the Remove button.
- If you'd rather not fuss with the Share and Storage Management console, you can set the permissions from My Computer. Right-click the shared folder, choose Sharing and Security from the contextual menu, and then click Permissions. Then you can follow the preceding procedure, picking up at Step 5.

REMEMBER

- The permissions assigned in this procedure apply only to the share itself. The underlying folder can also have permissions assigned to it. If that's the case, whichever of the restrictions is most restrictive always applies. If the share permissions grant a user Full Control permission but the folder permission grants the user only Read permission, for example, the user has only Read permission for the folder.

4 Managing Your Network

IN THIS PART . . .

Learn what network management is all about.

Take care of your users.

Work with Group Policy.

Manage all the software on your network.

Work with mobile devices such as laptops and smartphones.

IN THIS CHAPTER

» Exploring the many jobs of the network administrator

» Documenting the network

» Dusting, vacuuming, and mopping

» Managing network users

» Choosing the right tools

» Building your library

» Getting certified

Chapter 14 Welcome to Network Management

Help wanted. Network administrator to help small business get control of a network run amok. Must have sound organizational and management skills. Only moderate computer experience required. Part-time only.

Does this sound like an ad that your company should run? Every network needs a network administrator, whether the network has 2 computers or 2,000. Of course, managing a 2,000-computer network is a full-time job, whereas managing a 2-computer network isn't. At least, it shouldn't be.

This chapter introduces you to the boring and tedious job of network administration. Oops — you're probably reading this chapter because you've been elected to be the network manager, so I'd better rephrase that:

This chapter introduces you to the wonderful, exciting world of network management! Oh, boy! This is going to be fun!

What a Network Administrator Does

A network administrator "administers" a network: installing, configuring, expanding, protecting, upgrading, tuning, and repairing the network.

A network administrator takes care of the network hardware (such as cables, hubs, switches, routers, servers, and clients) and the network software (such as network operating systems, email servers, backup software, database servers, and application software). Most important, the administrator takes care of network users by answering their questions, listening to their troubles, and solving their problems.

On a big network, these responsibilities constitute a full-time job — or a staff of full-timers. Large networks tend to be volatile: Users come and go, equipment fails, software chokes, and life in general seems to be one crisis after another.

Smaller networks are much more stable. After you get your network up and running, you probably won't have to spend much time managing its hardware and software. An occasional problem may pop up, but with only a few computers on the network, problems should be few and far between.

Regardless of the network's size, the administrator attends to common chores:

- **Get involved in every decision to purchase new computers, printers, or other equipment.**
- **Put on the pocket protector whenever a new computer is added to the network.** The network administrator's job includes considering changes in the cabling configuration, assigning a computer name to the new computer, integrating the new user into the security system, and granting user rights.
- **Whenever a software vendor releases a new version of its software, read about the new version and decide whether its new features warrant an upgrade.** In most cases, the hardest part of upgrading to new software is determining the *migration path* — that is, upgrading your entire network to the new version while disrupting the network and its users as little as possible. This statement is especially true if the software in question happens to be your network operating system because any change to the network operating system can potentially impact the entire network.

  REMEMBER

  Between upgrades, software vendors periodically release patches and service packs that fix minor problems. For more information, see Chapter 19.

- **Perform routine chores, such as backing up the servers, archiving old data, and freeing up server disk space.** Much of the task of network administration involves making sure that things keep working by finding and correcting problems before users notice that something is wrong. In this sense, network administration can be a thankless job.
- **Gather, organize, and track the entire network's software inventory.** You never know when something will go haywire on the ancient Windows Vista computer that Joe in Marketing uses, and you have to reinstall that old copy of Lotus Approach. Do you have any idea where the installation disks are?

Choosing the Part-Time Administrator

The larger the network, the more technical support it needs. Most small networks — with just a dozen or so computers — can get by with a part-time network administrator. Ideally, this person should be a closet computer geek: someone who has a secret interest in computers but doesn't like to admit it. Someone who will take books home and read them over the weekend. Someone who enjoys solving computer problems just for the sake of solving them.

The job of managing a network definitely requires computer skills, but it isn't entirely a technical job. Much of the work that the network administrator does is routine housework. Basically, the network administrator dusts, vacuums, and mops the network periodically to keep it from becoming a mess.

Here are some additional ideas on picking a part-time network administrator:

- The network administrator needs to be an organized person. Conduct a surprise office inspection and place the person with the neatest desk in charge of the network. (Don't warn them in advance, or everyone may mess up their desks intentionally the night before the inspection.)
- Allow enough time for network administration. For a small network (say, no more than 20 or so computers), an hour or two each week is enough. More time is needed upfront as the network administrator settles into the job and discovers the ins and outs of the network. After an initial settling-in period, though, network administration for a small office network doesn't take more than an hour or two per week. (Of course, larger networks take more time to manage.)

- Make sure that everyone knows who the network administrator is and that the network administrator has the authority to make decisions about the network, such as what access rights each user has, what files can and can't be stored on the server, how often backups are done, and so on.
- Pick someone who is assertive and willing to irritate people. A good network administrator should make sure that backups are working *before* a hard drive fails and make sure that antivirus protection is in place *before* a virus wipes out the entire network. This policing will irritate people, but it's for their own good.
- In most cases, the person who installs the network is also the network administrator. This is appropriate because no one understands the network better than the person who designs and installs it.
- The network administrator needs an understudy — someone who knows almost as much about the network, is eager to make a mark, and smiles when the worst network jobs are delegated.
- The network administrator has some sort of official title, such as Network Boss, Network Czar, Vice President in Charge of Network Operations, or Dr. Network. A badge, a personalized pocket protector, or a set of Spock ears helps, too.

The Three "Ups" of Network Management

Much of the network manager's job is routine stuff — the equivalent of vacuuming, dusting, and mopping, or changing your car's oil and rotating the tires.

Three of the most important routine tasks that a network administrator must do vigilantly are what I call the "Three Ups of Network Management." They are

- **Back up:** The network manager must ensure that the network is properly backed up. If something goes wrong and the network isn't backed up, guess who gets the blame? On the other hand, if disaster strikes yet you're able to recover everything from yesterday's backup with only a small amount of work lost, who gets the pat on the back, the fat bonus, and the vacation in the Bahamas? Chapter 22 describes the options for network backups. Read it *soon.*
- **Lock up:** Another major task for a network administrator is sheltering the network from the evils of the outside world. These evils come in many forms, including hackers trying to break into your network and virus programs arriving through email or untrustworthy websites. Chapter 19 describes this task in more detail.

- **Clean up:** Users think that the network server is like the attic: They want to throw files up there and leave them forever. No matter how much disk storage your network has, your users will fill it up sooner than you think, so the network manager gets the fun job of cleaning up the attic once in a while. The best advice I can offer is to continually complain about how messy it is up there and warn your users that spring cleaning is on the to-do list.

Managing Network Users

Managing network technology is the easiest part of network management. Computer technology can be confusing at first, but computers aren't as confusing as people. The real challenge of managing a network is managing the network's users.

The difference between managing technology and managing users is obvious: You can figure out computers, but who can ever really figure out people? The people who use the network are much less predictable than the network itself. Here are some tips for dealing with users:

- **Make user training a key part of the network manager's job.** Make sure that everyone who uses the network understands how it works and how to use it. If the network users don't understand how the network works, they may unintentionally do all kinds of weird things to it.
- **Treat network users respectfully.** If users don't understand how to use the network, it's not their fault. Explain it to them. Offer a class. Buy each one a copy of this book, and tell them to read it during the lunch hour. Hold their hands. Just don't treat them like idiots.
- **Create a network cheat sheet.** It should contain everything users need to know about using the network — on one page. Everyone needs a copy.
- **Be as responsive as possible.** If you don't quickly fix a network user's problem, he may try to fix it. You don't want that to happen.

TIP

The better you understand the psychology of network users, the more prepared you are for the strangeness they often serve up. Toward that end, I recommend that you read the *Diagnostic and Statistical Manual of Mental Disorders* (also known as *DSM-5*) from cover to cover.

Acquiring Software Tools for Network Administrators

Network managers need certain tools to get their jobs done. Managers of big, complicated, expensive networks need big, complicated, expensive tools. Managers of small networks need small tools.

Some tools that a manager needs are hardware tools, such as screwdrivers, cable crimpers, and hammers. The tools I'm talking about, however, are software tools. I mention a couple of them earlier: Visio (to help you draw network diagrams) and a network-discovery tool to help you map your network. Here are a few others:

- **Built-in TCP/IP commands:** Many of the software tools that you need in order to manage a network come with the network itself. As the network manager, you should read through the manuals that come with your network software to see which management tools are available. For example, Windows includes a `net diag` command that you can use to make sure that all the computers on a network can communicate with each other. (You can run `net diag` from an MS-DOS prompt.) For TCP/IP networks, you can use the TCP/IP diagnostic commands that I summarize in Table 14-1.
- **System Information:** This program, which comes with Windows, is a useful utility for network managers.
- **Hotfix Checker:** This handy PowerShell tool from Microsoft scans your computers to see which patches need to be applied. You can download it for free from the Microsoft website. Just go to `www.microsoft.com` and search for **get-hotfix**.
- **Protocol analyzer:** A *protocol analyzer* (or *packet sniffer*) can monitor and log the individual packets that travel along your network. You can configure the protocol analyzer to filter specific types of packets, watch for specific types of problems, and provide statistical analysis of the captured packets.

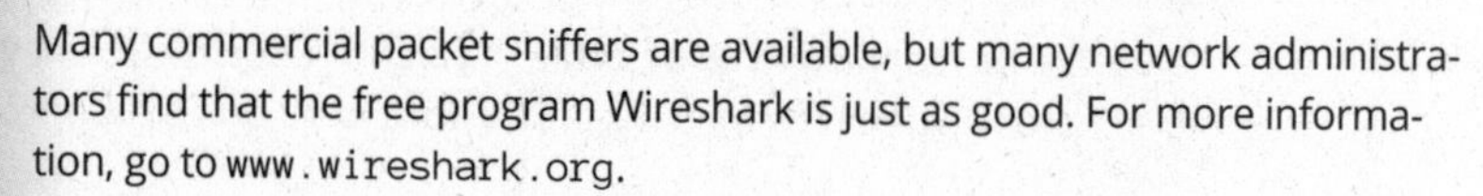

Many commercial packet sniffers are available, but many network administrators find that the free program Wireshark is just as good. For more information, go to `www.wireshark.org`.

- **Network Monitor:** All current versions of Windows include a program called Network Monitor, which provides basic protocol analysis and can often help solve pesky network problems.

TABLE 14-1 TCP/IP Diagnostic Commands

Command	What It Displays
`arp`	Address resolution information used by the Address Resolution Protocol (ARP)
`hostname`	Your computer's host name
`ipconfig`	Current TCP/IP settings
`nbtstat`	The status of NetBIOS over TCP/IP connections
`netstat`	Statistics for TCP/IP
`nslookup`	DNS information
`ping`	Verification that a specified computer can be reached
`route`	The PC's routing tables
`tracert`	The route from your computer to a specified host

Building a Library

Scotty delivered one of his best lines in the original *Star Trek* series when he refused to take shore leave so that he could get caught up on his technical journals. "Don't you ever relax?" asked Kirk. "I am relaxing!" Scotty replied.

To be a good network administrator, you need to read computer books — lots of them. And you need to enjoy doing it. If you're the type who takes computer books with you to the beach, you'll make a great network administrator.

Read books on a variety of topics. I don't recommend specific titles, but I do recommend that you get a good, comprehensive book on each of these topics:

- Network security and hacking
- Wireless networking
- Network cabling and hardware
- Ethernet
- Windows Server 2008, 2012, and 2016
- Windows 7, 8, 8.1, and 10
- Linux
- TCP/IP

- DNS
- Sendmail or Microsoft Exchange Server, depending on which email server you use

In addition to books, you may also want to subscribe to some magazines to keep up with what's happening in the networking industry. Here are a few you should probably consider, along with their web addresses:

- *InformationWeek:* `www.informationweek.com`
- *InfoWorld:* `www.infoworld.com`
- *Network Computing:* `www.networkcomputing.com`
- *Network World:* `www.networkworld.com`
- *2600 The Hacker Quarterly* (a great magazine on computer hacking and security): `www.2600.com`

TIP

The Internet is one of the best sources of technical information for network administrators. You'll want to stock your browser's Favorites menu with plenty of websites that contain useful networking information. In addition, you may want to subscribe to one of the many online newsletters that deliver fresh information on a regular basis via email.

Pursuing Certification

Remember the scene near the end of *The Wizard of Oz* when the Wizard grants the Scarecrow a diploma, the Cowardly Lion a medal, and the Tin Man a testimonial?

Network certifications are kind of like that. I can picture the scene now:

> *The Wizard:* "And as for you, my network-burdened friend, any geek with thick glasses can administer a network. Back where I come from, there are people who do nothing but configure Cisco routers all day long. And they don't have any more brains than you do. But they do have one thing you don't have: certification. And so, by the authority vested in me by the Universita Committeeatum E Pluribus Unum, I hereby confer upon you the coveted certification of CND."
>
> *You:* "CND?"

> *The Wizard:* "Yes, that's, uh, *Certified Network Dummy.*"
>
> *You:* "The Seven Layers of the OSI Reference Model are equal to the Sum of the Layers on the Opposite Side. Oh, joy, rapture! I feel like a network administrator already!"

My point is that certification in and of itself doesn't guarantee that you really know how to administer a network. That ability comes from real-world experience — not exam crams.

Nevertheless, certification is becoming increasingly important in today's competitive job market. So you may want to pursue certification, not just to improve your skills, but also to improve your resume. Certification is an expensive proposition. Each test can cost several hundred dollars, and depending on your technical skills, you may need to buy books to study or enroll in training courses before you take the tests.

You can pursue two basic types of certification: vendor-specific certification and vendor-neutral certification. The major software vendors such as Microsoft and Cisco provide certification programs for their own equipment and software. CompTIA, a nonprofit industry trade association, provides the best-known vendor-neutral certification.

Helpful Bluffs and Excuses

As network administrator, you just won't be able to solve a problem sometimes, at least not immediately. You can do two things in this situation. The first is to explain that the problem is particularly difficult and that you'll have a solution as soon as possible. The second solution is to look the user in the eyes and, with a straight face, try one of these phony explanations:

- Blame it on the version of whatever software you're using. "Oh, they fixed that with version 39."
- Blame it on cheap, imported memory chips.
- Blame it on Democrats. Or Republicans. Doesn't matter.
- Blame it on oil company executives.
- Blame it on the drought.
- Blame it on the rain.

- Blame it on climate change.
- Hope that the problem wasn't caused by stray static electricity. Those types of problems are very difficult to track down. Tell your users that not properly discharging themselves before using their computers can cause all kinds of problems.
- You need more memory.
- You need a bigger disk.
- You need a faster processor.
- Blame it on Jar Jar Binks.
- You can't do that in Windows 10.
- You can only do that in Windows 10.
- Could be a virus.
- Or sunspots.
- No beer and no TV make Homer something something something.

- Looking at support tickets
- Seeing how users communicate with the help desk
- Helping your users remotely
- Building a knowledge base
- Building a self-service portal
- Surveying your users' satisfaction
- Monitoring how your help desk is performing
- Using a commercial help-desk solution

Chapter 15
Supporting Your Users

A few years ago, I helped purchase and configure a cloud-based gadget that displays company information on a large-screen TV located in the office's main lobby. It took a bit of fiddling to get it figured out, but when we did, it turned out to be a great solution for the problem at hand.

We recently needed to expand to a second large-screen TV that would display the exact same information on a second screen. So, I purchased another one of the same gadgets and spent most of an entire morning trying to figure out how to get it integrated into our account. Finally, in exasperation, I called the vendor's help desk for support. They politely asked if I had read the instructions, which I hadn't, and then gently guided me through the super-simple two-step process. The second gadget was up and running within about 5 minutes.

The moral of the story is that sometimes even people who should know better (a.k.a. *me*) don't follow the instructions and end up needing help. I needed the help desk.

One of the most important functions of an IT administrator is supporting your users. The best way to do that is to set up a help desk.

If you work at a small company where you're pretty much the entire IT department, you probably *are* the help desk. In a small organization (say, less than 40 people), it may be possible for one person to manage the entire IT infrastructure, including setting up and maintaining server hardware and software, as well as network equipment and application software, and still manage to field occasional calls from users who need help. But as more employees are brought on, you'll eventually need to set up a formal help desk with staff dedicated to solving users' problems.

In this chapter, I give you a brief overview of what to consider when setting up and managing an effective help desk.

Establishing the Help Desk's Charter

Whether you're starting from scratch or evaluating the effectiveness of an existing help desk, the best place to start is to create a charter for the help desk. Without a clear charter, you'll never be able to measure your help desk's effectiveness.

The charter should spell out the core mission of the help desk, which certainly includes solving IT problems encountered by users of your organization's computer systems. The charter should also address considerations such as the following:

- How does the mission of the help desk fit in with the company's core values?
- Where does the help desk fit within the company's overall organization? Who does it report to?
- Does the help desk seek to solve *all* IT problems, or does it have a more limited scope? In other words, if a user doesn't understand how to create and use a Microsoft Excel PivotTable, is it within the scope of the help desk to provide that training? Or is the help desk's mission limited to helping users when Excel is broken, not when the user doesn't know how to use it?

» Should the help desk prioritize the needs of some users over others when deciding how to allocate resources? For example, is one department, such as factory production or payroll, a higher priority than other, less time-critical departments? Should executive management get higher-priority response than other users?

» Are there specific performance metrics you can identify to measure the effectiveness of the help desk? Is it the total number of calls handled, the percentage of problems solved, on the first call? Or is it a positive score on user satisfaction surveys?

» How responsive do you expect your help desk to be? Is it important that a live human being always answers the help desk phone? Does that person need to be a technician who can actually solve problems, or is it acceptable to have a nontechnical person field the calls and assign the calls to the appropriate technician?

The help desk charter should be developed by a team representing all the stakeholders. This should include not just IT staff and executive management, but also representatives from the various departments that will be supported by the help desk.

The charter should be reviewed periodically to assess whether the help desk is meeting the needs of its users. This assessment should also include representation from all the stakeholders.

Tracking Support Tickets

The most important tool that a help desk uses to manage support requests is the *support ticket*. A support ticket tracks the status of a support request, from inception to completion, and should record every pertinent detail related to the request, including the following:

» **Basic identification information for the person requesting the ticket:** This may include name and contact information, a short title (which can be created by the user requesting the support or by help desk staff), the date and time the ticket was created, a detailed description of the problem, and identification of the support technician initially assigned to the ticket.

» **The ticket category:** You'll need to devise a list of categories that is appropriate for your company, but you'll likely include categories similar to the following:
  - Login and password
  - New user onboarding
  - Employee separation
  - Hardware
  - Email
  - Printing
  - Intranet
  - Microsoft Office
  - Virus/malware
  - Phones
  - Other
» **An indication of the ticket's status:** Here's an example of status options:
  - Received
  - Assigned
  - In Progress
  - Escalated
  - Resolved
  - Closed
» **All correspondence, including emails, messages, and detailed notes of all phone conversations:** These records should include the date and time of the correspondence.
» **Screen shots, event logs, and any other pertinent files.**
» **Detailed descriptions of everything that has been done to resolve the issue.**
» **Details of how the issue was ultimately resolved.**

TIP

Ideally, the user who initiated a support request should have access to the ticket and the opportunity to contribute notes to it. Keeping users in the loop will help with overall satisfaction of the trouble resolution process and can help them appreciate that their problem has not been forgotten but is indeed being worked on.

The entire ticketing database should also be available to help desk staff so that they can review previous issues. A support technician should be able to search the database for tickets created by the same user or other users who have reported similar problems. That way, the technician can avoid reinventing the wheel.

You may be tempted to keep track of your help desk tickets in a home-grown database, perhaps using a database program such as Microsoft Access or a more general program such as Microsoft OneNote. However, you'll soon outgrow the home-grown solution when you discover that it isn't as flexible as you need it to be, isn't very searchable, and doesn't have the features that can make your help desk shine, such as automatic reminders or built-in workflows for delegating or escalating issues.

Fortunately, there is plenty of good software available to help you manage your help desk's trouble tickets. Just do a web search for the keywords *help desk software*, and you'll find dozens of options to choose from.

Deciding How to Communicate with Users

Users must have a reliable and responsive method of communicating with the help desk. Here are some of the best options to consider:

- **Phone:** The most obvious way to communicate is via phone. If your help desk is small, with just a few support technicians, you might simply publish the phone numbers or extensions of each of the technicians. For a larger help desk, you'll want to provide a single number for incoming calls and *not* publish the numbers for individual support technicians.

  The advantage of the phone is that the contact is immediate (assuming someone actually answers the phone) and human (assuming the phone is answered by an actual person). The disadvantage is that it's difficult to triage incoming calls. Ideally, a single person should be responsible for answering incoming calls and transferring them to the appropriate technician.

  You may want to consider setting up a phone tree, with options such as "Press 1 for help logging on," "Press 2 for help with accounting software," and so on. This is fine, as long as the tree is simple and users are quickly able to get to a real person who can help. If users have to make a bunch of touch-tone selections that ultimately drop them into a voicemail, they'll be frustrated.

A disadvantage of the phone is that there is no detailed record of the conversation. After each phone call, the technician should record the details of the call in the support ticket. However, it isn't easy to note every detail accurately, so errors or misunderstandings are bound to creep in.

- **Email:** This is a common and useful means of communicating with the help desk. Just set up a support mailbox using an easy-to-remember name like "Support" or "Help," and connect your technician's Outlook profiles to the support mailbox. Then support technicians can monitor the support mailbox and respond when new help requests are received.

  One of the major advantages of email is that it creates an accurate record of correspondence. Emails can be attached to tickets, so technicians can quickly review previous activity on a ticket.

- **Online chat:** An increasingly popular way for users to communicate with support technicians is through online chat. Many users prefer chat to email because it's more responsive, and prefer chat to phone because chat is less intrusive. Support staff often prefers chat as well because the lag time between messages allows them to work on more than one issue simultaneously. Win-win!

  As an added bonus, a transcript of the chat can be copied into the ticket, preserving a record of the entire conversation.

  Many chat services are available that lend themselves to use by support staff. One I'm fond of is Slack (`www.slack.com`). In addition to simple chats, it offers video chats, screen sharing, file transfer, and many other features that can be useful to support teams.

Using Remote Assistance

One of the most annoying aspects of providing technical support for network users is that you often have to go to the user's desk to see what's going on with his or her computer. That's annoying enough if the other user's desk is across the room or down the hall, but it's almost unworkable if the user you need to support is across town or in a different city or state altogether.

Fortunately, Windows includes a handy feature called Remote Assistance, which is designed to let you provide technical support to an end user without going to the user's location. With Remote Assistance, you can see the user's screen in a window on your own screen, so you can watch what the user is doing. You can even take control when necessary to perform troubleshooting or corrective actions to help solve the user's problems.

Note that there are commercial alternatives to Windows Remote Assistance that do a much better job at this task. This chapter shows you how to use Remote Assistance because it's free, and all Windows computers since Windows XP have it.

Enabling Remote Assistance

Before you can lend assistance to a remote computer, Remote Assistance must be enabled on that computer. You should enable Remote Assistance before you need it, so that when the time comes, you can easily access your users' computers. But the procedure is simple enough that you can probably walk a user through the steps over the phone so that you can then gain access.

Here are the steps:

1. **Click the Start button, type** Remote Assistance**, and click Enable Remote Assistance Invitations to Be Sent from This Computer.**

   This brings up the System Properties dialog box, shown in Figure 15-1.

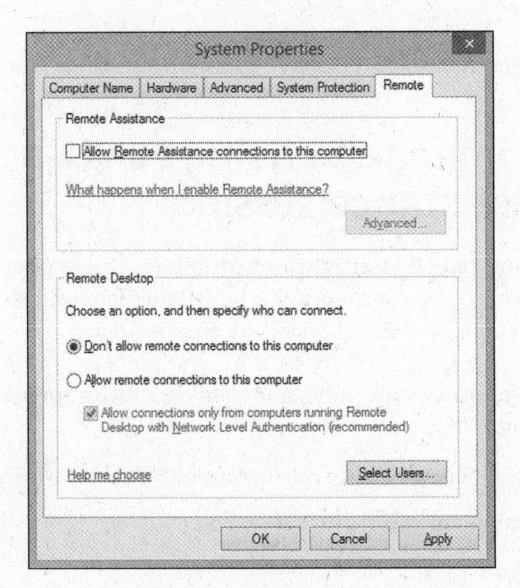

FIGURE 15-1: Enabling Remote Assistance.

2. **Select the Allow Remote Assistance Connections to This Computer check box.**
3. **Click the Advanced button.**

   The Remote Assistance Settings dialog box appears, as shown in Figure 15-2.

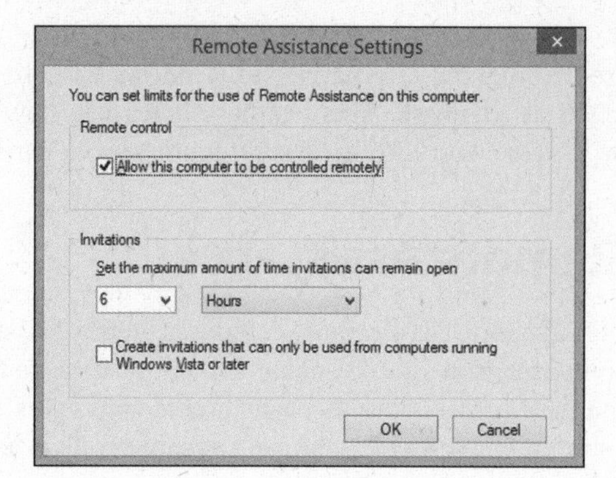

FIGURE 15-2: Setting the advanced Remote Assistance options.

4. **Select the Allow This Computer to Be Controlled Remotely check box.**

   This option will allow you to later take control of this computer remotely.

5. **Click OK.**

   You're returned to the System Properties dialog box.

6. **Click OK.**

You're done. You can now initiate Remote Assistance sessions from this computer.

Inviting someone to help you via a Remote Assistance session

The user requesting remote assistance must initiate the request before you can connect to the user's computer to provide help. You may need to guide your user through this procedure over the phone. Here are the steps:

1. **Click the Start button, type** Invite**, and then click Invite Someone to Connect to Your PC.**

   The Windows Remote Assistance window appears, as shown in Figure 15-3.

2. **Click Invite Someone You Trust to Help You.**

   The screen shown in Figure 15-4 appears.

3. **Click Save This Invitation as a File, and save the invitation file to a convenient disk location.**

   This option creates a special file that you can save to disk. A Save As dialog box appears, allowing you to save the invitation file in any disk location you want. You can then email the invitation to your helper, who can use the invitation to connect to your PC.

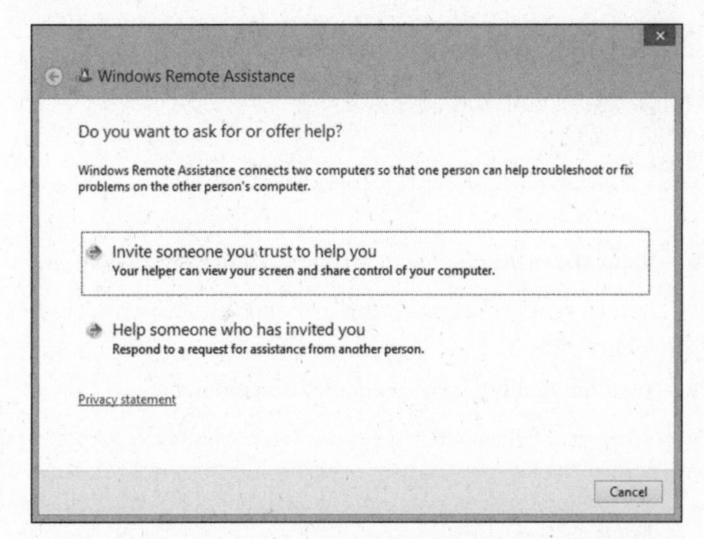

FIGURE 15-3: The Windows Remote Assistance window.

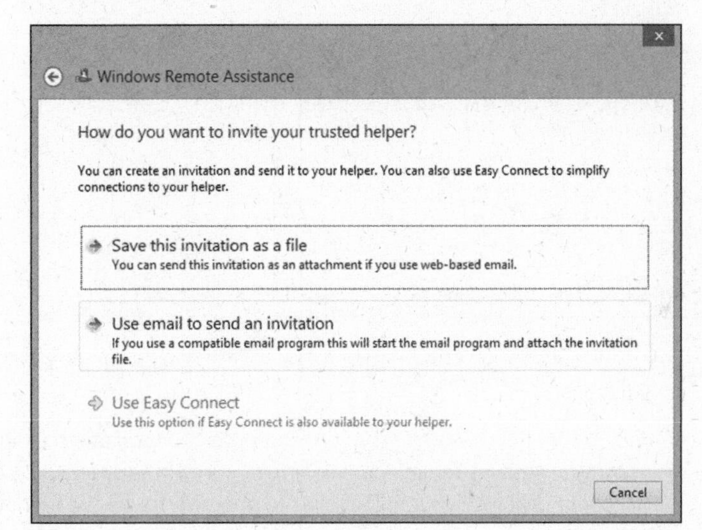

FIGURE 15-4: Inviting someone to help you.

4. **If you use Microsoft Outlook, you can choose Use Email to Send an Invitation.**

   This option fires up Outlook and creates an email with the invitation file already attached.

   Either way, a password will be displayed, as shown in Figure 15-5. You'll need to provide this password to your helper when requested. (Typically, you'll do that over the phone.)

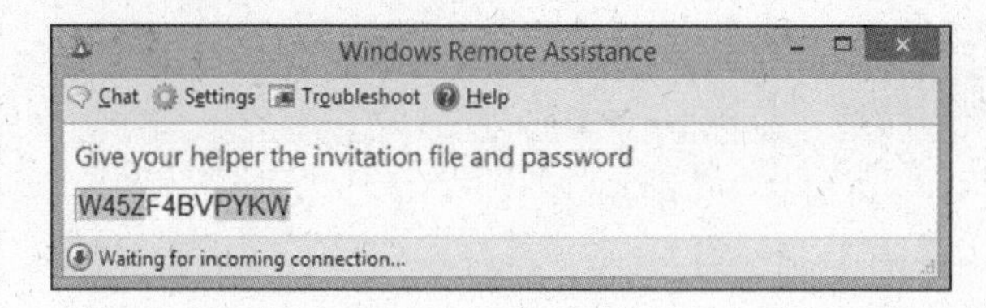

FIGURE 15-5: You'll need to tell your helper the password.

5. **Email the invitation file to the person you want to help you.**

   Use your preferred email program to send the invitation file as an attachment.

6. **Wait for your helper to request the password.**

   When your helper enters the password into his or her Remote Assistance screen, the Remote Assistance session is established. You'll be prompted for permission to allow your helper to take control of your computer, as shown in Figure 15-6.

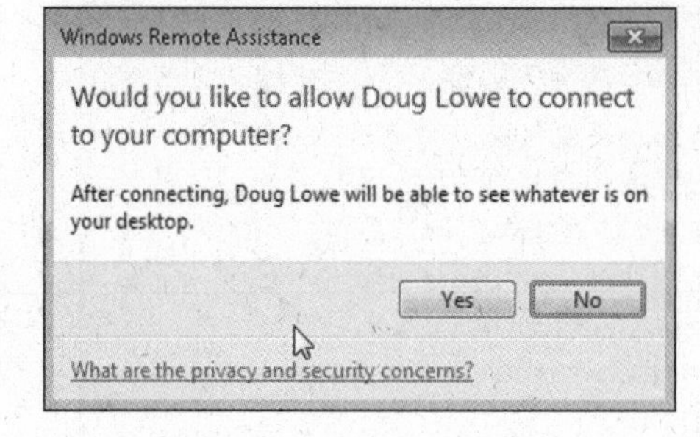

FIGURE 15-6: Granting your helper permission to take control.

7. **Click Yes.**

   Your helper now has access to your computer. To facilitate the Remote Assistance session, the toolbar shown in Figure 15-7 appears. You can use this toolbar to chat with your helper or to temporarily pause the Remote Assistance session.

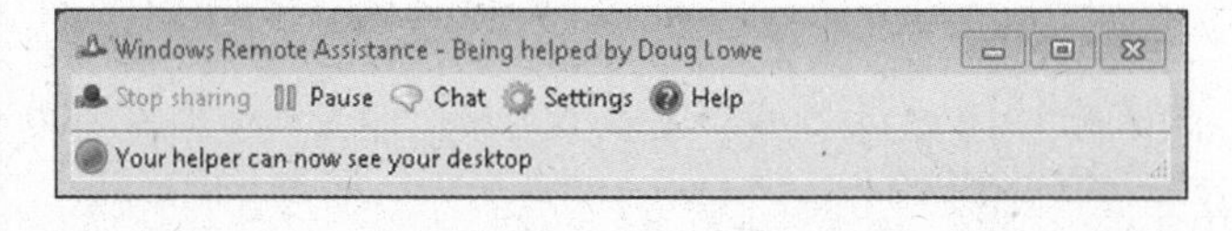

FIGURE 15-7: The Windows Remote Assistance toolbar.

Responding to a Remote Assistance invitation

If you've received an invitation to a Remote Assistance session, you can establish the session by following these steps:

1. **Click the Start button, type** Invite**, and then click Invite Someone to Connect to Your PC.**

   This brings up the Windows Remote Assistance window (refer to Figure 15-3).

2. **Click Help Someone Who Has Invited You.**
3. **Click Use an Invitation File.**

   An Open dialog box appears.

4. **Locate the invitation file you were sent, select it, and click Open.**

   You're prompted to enter the Remote Assistance password, as shown in Figure 15-8.

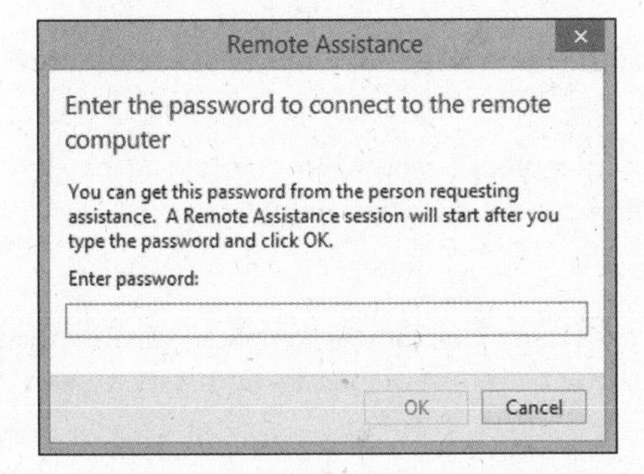

FIGURE 15-8: Enter the Remote Assistance password.

As an alternative to Steps 1 through 3, you can simply double-click the invitation file you received. Windows Remote Assistance launches and prompts you for the password.

5. **Enter the password given to you by the user requesting help, and then click OK.**

   The remote user is prompted to grant you permission to start the Remote Assistance session (this is where the remote user sees the screen shown in Figure 15-6). When the user grants permission, the Remote Assistance session is established. You can now see the user's screen in the Remote Assistance window, as shown in Figure 15-9.

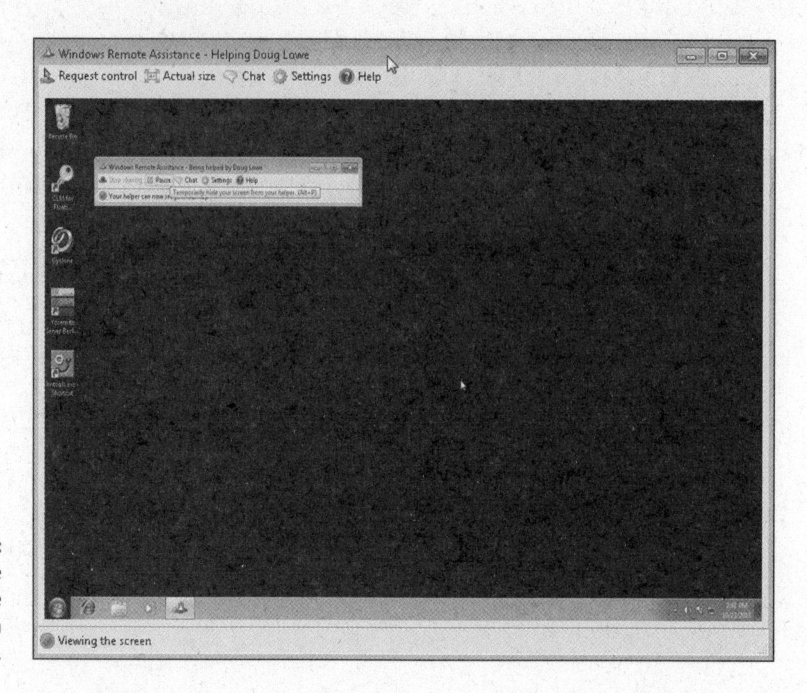

FIGURE 15-9: A Remote Assistance session in progress.

6. **To take control of the remote user's computer, click Request Control.**

   The remote user is prompted to allow control. Assuming that permission is granted, you can now control the other computer.

7. **Do your thing.**

   Now that you're connected to the remote computer, you can perform whatever troubleshooting or corrective actions are necessary to solve the user's problems.

8. **If necessary, use the Chat window to communicate with the user.**

   You can summon the Chat window by clicking the Chat button in the toolbar. Figure 15-10 shows a chat in progress.

9. **To conclude the Remote Assistance session, simply close the Remote Assistance window.**

   The remote user is notified that the Remote Assistance session has ended.

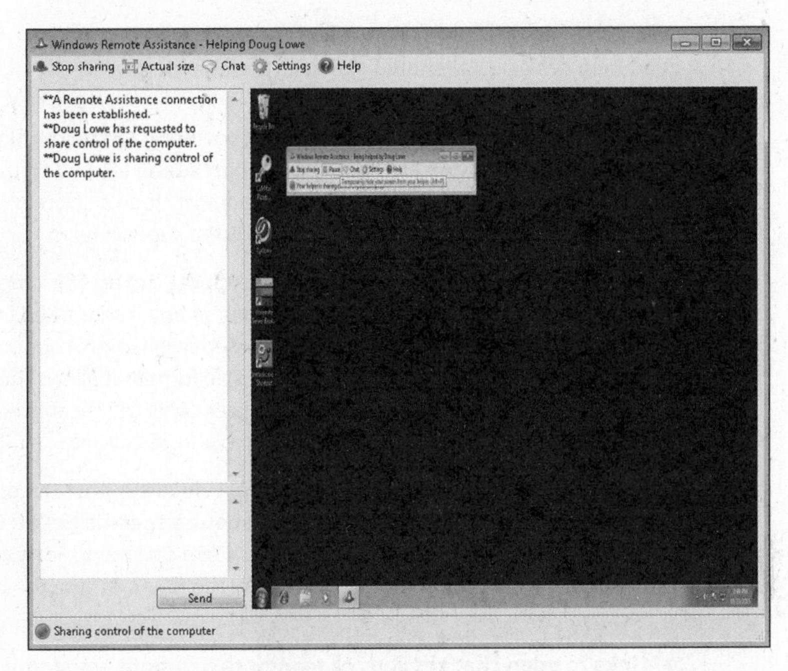

FIGURE 15-10: Using the chat window.

Creating a Knowledge Base

An important aspect of any help desk is creating and maintaining a comprehensive knowledge base of common issues along with their solutions. Whenever a recurring issues is discovered, a member of the help desk team who has good writing skills should be assigned to create a new article in the knowledge base. The article should clearly list the symptoms typically encountered along with a detailed, step-by-step solution that solves the problem.

The knowledge base should be shared throughout your company, perhaps on your company intranet. That way, users can search it on their own. Sharing a well-stocked knowledge base on your intranet can cut down on the number of support calls your help desk receives because users can search it to find solutions on their own.

Creating a Self-Service Help Portal

One of the best ways to improve the efficiency of your help desk is to create a self-service help desk portal where users can look for their solutions on their own, create and track support tickets, initiate chat sessions, and so on.

A good help desk portal should provide the following:

- **Contact information for the help desk outlining all the methods a user can use to communicate with the support team:** Instead of just providing an email address, the Contacts section should include a link that creates an email message automatically addressed to the support team.
- **A link that lets the user create a new support ticket:** The simplest way to do this would be to create a link that opens an email addressed to the support team, with the subject line prefilled to something like "New Support Request." But a better alternative is to provide a simple form that allows the user to fill in fields to describe the issue. When the user clicks OK, the form creates a new support ticket that can be assigned to a support technician.
- **A summary of the user's current and previous support tickets, with links to pull up detailed information about a specific ticket:** For a current ticket, the user should have the ability to add a comment. For a closed ticket, the user should have the ability to reopen the ticket, in case the problem recurs.
- **Links to open chat sessions or to initiate a remote assistance session.**
- **A search field that enables the user to search the knowledge base.**
- **Information about current or upcoming outages, recent upgrades, or other noteworthy stuff.**

If your company has a staff of web developers (or if you happen to be a web developer yourself), you can develop the help desk portal. Otherwise, you can find plenty of commercial options that will meet this need. (For more information, see the section "Using Help Desk Management Software" later in this chapter.)

Using Satisfaction Surveys

It's always a good idea to follow up on every support request with a brief survey asking the user to rate his or her satisfaction with the help desk's support. The easiest way to do this is to follow up with an email that contains a link to a survey page. Note that the survey should be short — just a few questions that can be shown on a single page.

Here are some suggested questions for the survey:

- How satisfied are you with the support you received from the help desk for this issue?
- Was your problem resolved in a timely manner?
- Was the issue resolved during the initial contact (phone call, chat, or email)?
- How understanding of your situation was the support technician?
- Did the technician ensure that your issue was resolved before closing the ticket?
- After the issue was resolved, did the technician ask if there were any other issues he or she could assist you with?
- Do you have any comments?

A simple Internet search for terms such as *help desk satisfaction survey* will turn up tons of software that can help with satisfaction surveys. One of the best known is Survey Monkey (`www.surveymonkey.com`). Survey Monkey even has a help desk satisfaction survey template to help you get started (see Figure 15-11).

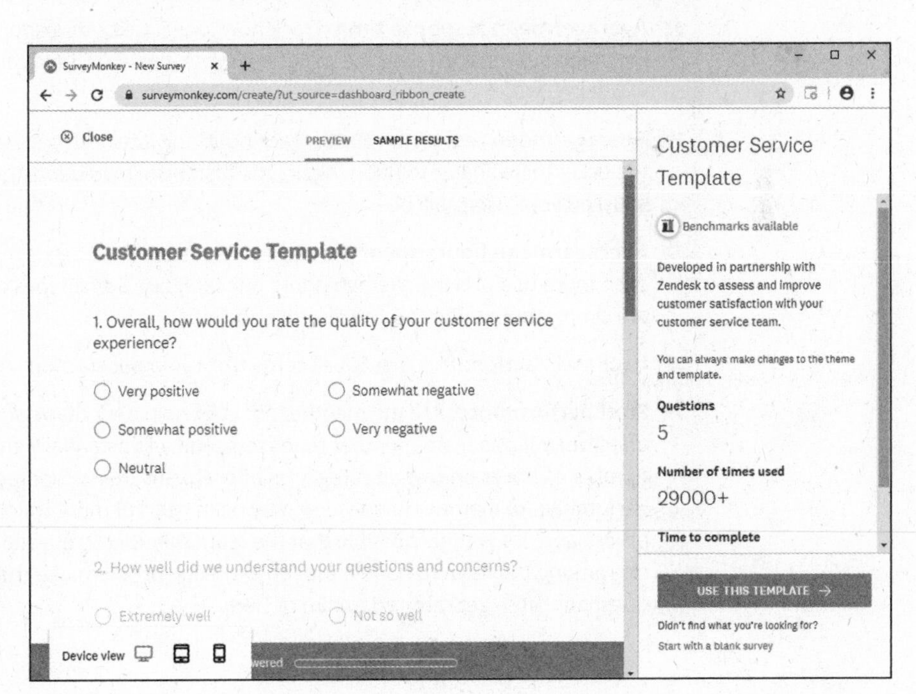

FIGURE 15-11: Survey Monkey's help desk satisfaction survey template.

Tracking Help Desk Performance

If you want to know how efficient and effective your help desk team is, you need to track a number of factors and periodically evaluate how you're doing. Here are some of the most important indicators to track on a monthly basis:

- **Total ticket count:** The most important baseline number you need to track is the total number of tickets your help desk receives. Unless you know how many tickets enter your help desk, you won't be able to make any sense of the other statistics. Keeping an eye on the growth of this number can help you justify the need for additional staff.
- **Tickets successfully resolved:** Hopefully, this number equals the number of tickets received. But it's important to note the differential, because sometimes you'll encounter issues that just can't be resolved. You want to pay close attention to these, to determine whether the problem is simply intractable, the user was unreasonable, or your staff just gave up.
- **Ticket categories:** Assuming you've created a good set of categories, tracking the overall percentage of tickets in each category will help you understand how to train and staff your help desk. (Can you say "pie chart"?)
- **Average/mean response time:** Track how long it takes to respond to incoming tickets. The faster the initial response, the less frustrated your users will be.
- **Average/mean resolution time:** Track how long it takes from the receipt of the ticket to its final resolution. Again, the faster the resolution, the more satisfied your users will be.
- **Average/mean hours spent per ticket:** The overall average is useful, but even more useful is the average hours per category. Some types of problems are simply more difficult to resolve than others.
- **User satisfaction:** This metric will come from your satisfaction surveys.
- **Staff performance:** Are the members of your help desk team working effectively? If one team member tends to spend a substantially greater number of hours on certain categories of problems than other team members, the team member in question may be in need of more training. (On the other hand, it's entirely possible that the team member in question is the best one among the team to handle the most difficult problems, so the tough questions often get referred to him or her!)

Using Help Desk Management Software

As you can tell from this chapter, running an efficient help desk is a big deal. And having an effective help desk is vital to your company's overall performance. The sooner the help desk can resolve your users' issues, the sooner your users can get back to productive work.

Instead of attempting to cobble together all the various pieces needed to run a help desk, consider acquiring a comprehensive package that includes all aspects of IT service management. Such software isn't cheap, but it can save your company money in the long run.

Good help desk management software should include the following features:

- **Ticket management:** Comprehensive management of trouble tickets from their inception to their resolution, with the ability to sort and filter by status, category, user, technician, and other factors. The ticketing system should allow for custom fields so that you can integrate your company's unique needs into the system.
- **Self-service portal:** The software should make it easy for you to stand up a self-service portal that enables your users to find solutions without engaging the help desk.
- **Knowledge base:** A good, customizable knowledge base is a must.
- **Reporting:** The software should keep track of important performance metrics and should have reporting features that let you track how the help desk is doing.
- **Asset management:** A definite plus. Some products offer this as an extra-charge feature.
- **Deployment flexibility:** Ideally, you should have the option to deploy the software on-site or use the software as a cloud-based service.

Search the Internet for *help-desk software* and you'll find many options to choose from. Here are just a few of the better known products — note that this list is by no means complete:

- **Zendesk** (`www.zendesk.com`)**:** An excellent cloud-based solution that charges a monthly subscription fee for each member of your help desk team. Most organizations opt for the Professional plan ($19.99 per month, as of this writing) or the Enterprise plan ($49 per month, as of this writing). ***Remember:*** The cost is per help-desk team member, not per end user. Zendesk is consistently one of the top-ranked help-desk solution providers, so it should be one that you carefully consider.

- **RescueAssist Service Desk (`http://get.gotoassist.com`):** A comprehensive help desk solution that includes powerful remote assistance, from the makers of the popular GoToMeeting conferencing software.
- **Salesforce (`www.salesforce.com`):** If your company is already a Salesforce customer, consider using its help desk features to automate your help desk.

» **Enabling Group Policy on a Windows Server**

» **Creating, filtering, and forcing Group Policy Objects**

Chapter 16 Using Group Policy

Group Policy is a feature of Windows operating systems that lets you control how certain aspects of Windows and other Microsoft software work throughout your network. Many features that you may expect to find in a management console, such as Active Directory Users and Computers, are controlled by Group Policy instead. You must use Group Policy to control how often users must change their passwords, for example, and how complicated their passwords must be. As a result, Group Policy is an important tool for any Windows network administrator.

Unfortunately, Group Policy can be a confusing beast. In fact, it's one of the most confusing aspects of Windows network administration. So, don't be put off if you find this chapter more confusing than other chapters in this part. Group Policy becomes clear after you spend some time actually working with it.

Understanding Group Policy

Here it is in a nutshell: A *policy* defines configuration options for specific Windows features. Each policy specifies how some aspect of Windows or some other Microsoft software should be configured. For example, a policy might specify the home page that's initially displayed when any user launches Internet Explorer. When a user logs on to the domain, that policy is retrieved and applied to the user's Internet Explorer configuration.

The way a specific policy is put into play is through a *Group Policy Object* (GPO). A GPO is a collection of one or more policies that are applied to either computers or users. For example, a single GPO might set the default Internet Explorer home page and also set Internet Explorer security options. When that GPO is applied to a computer or user, all the policies included in the GPO are applied.

A GPO that applies to a computer will be enforced for any user of the computer, and a GPO that applies to a user will be enforced for that user no matter what computer he or she logs on to. As a network administrator, you'll be concerned mostly with policies that apply to users. But computer policies are useful from time to time as well.

To use Group Policy, you have to know how to do two things: (1) create individual GPOs that specify which policies you want to apply, and then (2) apply — or *link* — those GPOs to user and computers. Both tasks can be a little tricky.

The trick to creating GPOs is finding the particular policy or policies you want to employ. Trying to find a specific policy among the thousands of available policies can be frustrating. Suppose that you want to force all network users to change their passwords every 30 days. Your users will hate you, but you decide to do it anyway. The good news is that you know there's a policy to do that. The bad news is that you have no idea where to find it. You'll find help with this aspect of working with Group Policy in the section titled "Creating Group Policy Objects," later in this chapter.

After you've created a GPO, you're faced with the task of linking it to the users or computers you want it to apply to. Creating a GPO that applies to all users or computers is simple enough. But things get more complicated if you want to be more selective — for example, if you want the GPO to apply only to users in a particular organizational unit (OU) or to users that belong to a particular group. You'll find help for this aspect of working with Group Policy in the section "Filtering Group Policy Objects," later in this chapter.

Enabling Group Policy Management on Windows Server 2019

Before you can work with Group Policy on Windows Server 2019, you must enable Group Policy on the server. The procedure is simple enough and needs to be done only once for each server. Here are the steps:

1. **In the Server Manager, click Add Roles and Features.**
2. **Follow the wizard until you get to the Select Features page, which is shown in Figure 16-1.**

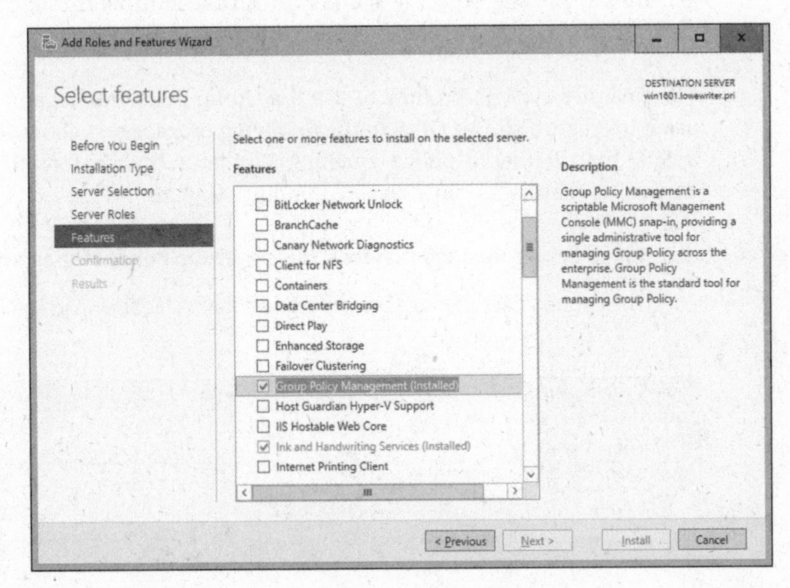

FIGURE 16-1: Enabling Group Policy management on Windows Server 2019.

3. **If the Group Policy Management check box is not already checked, select it.**
4. **Click Next.**
5. **When the confirmation page appears, click Install.**

   Be patient; installation may take a few minutes.
6. **Click Close.**

   You're done!

After you've completed this procedure, a new command titled Group Policy Management appears on the Tools menu in the Server Manager.

Creating Group Policy Objects

The easiest way to create GPOs is to use the Group Policy Management Console, which you can run from the Server Manager by choosing Tools⇨Group Policy Management.

A single GPO can consist of one setting or many individual Group Policy settings. The Group Policy Management Console presents the thousands of Group Policy settings that are available for your use in several categories. The more you work with Group Policy, the more these categories will begin to make sense. When you get started, you can expect to spend a lot of time hunting through the lists of policies to find the specific one you're looking for.

The easiest way to learn how to use the Group Policy Management Console is to use it to create a simple GPO. In the following procedure, I show you how to create a GPO that defines a policy enabling Windows Update for all computers in a domain so that users can't disable Windows Update.

1. **In the Server Manager, choose Tools ⇨ Group Policy Management.**

   The Group Policy Management console appears, as shown in Figure 16-2.

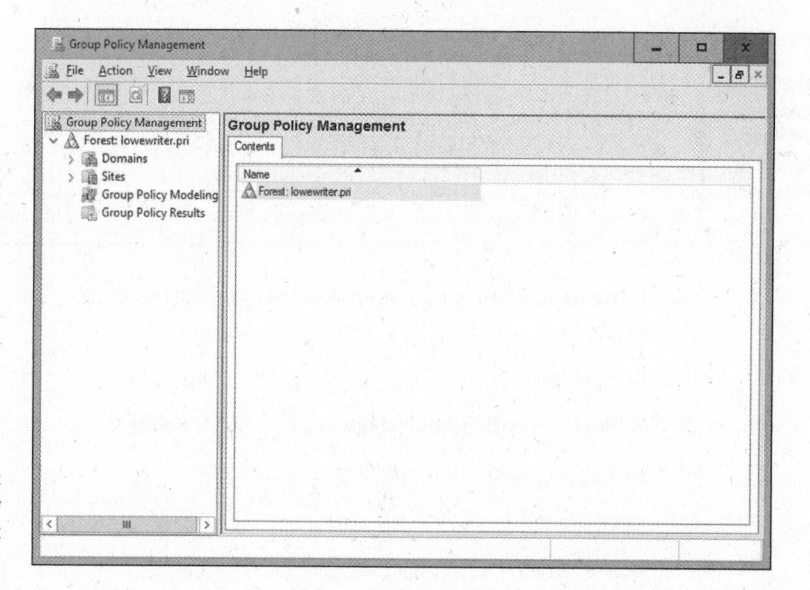

FIGURE 16-2: The Group Policy Management console.

2. **In the Navigation pane, drill down through the Domains node to the node for your domain, and then select the Group Policy Objects node for your domain.**
3. **Right-click the Group Policy Objects node and then choose New from the contextual menu that appears.**

   This command brings up the dialog box shown in Figure 16-3.

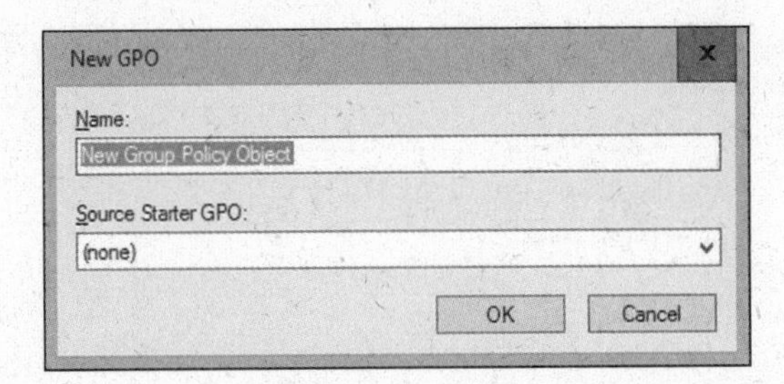

FIGURE 16-3: Creating a new GPO.

4. **Type a name for the GPO and then click OK.**

   For this example, type something like Windows Update for a GPO that will manage the Windows Update feature.

   When you click OK, the GPO is created and appears in the Group Policy Objects section of the Group Policy Management window.

5. **Double-click the new GPO.**

   The GPO opens, as shown in Figure 16-4. Note that at this stage, the Location section of the GPO doesn't list any Active Directory objects. As a result, this GPO is not yet linked to any Active Directory domains or groups. I get to that topic in a moment. First, create the policy settings for the new GPO.

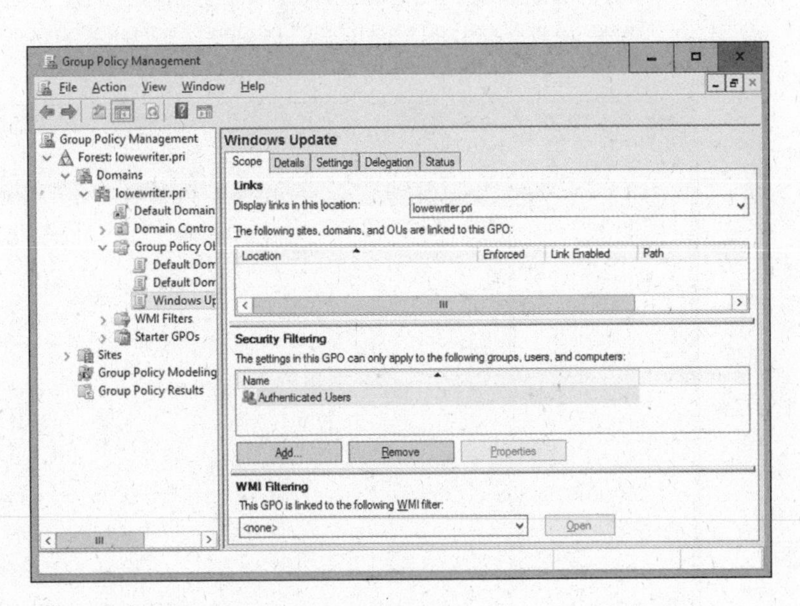

FIGURE 16-4: A new GPO.

6. **Click the Settings tab.**

   The message "Generating Report" appears for a moment, and then the policy settings are displayed, as shown in Figure 16-5.

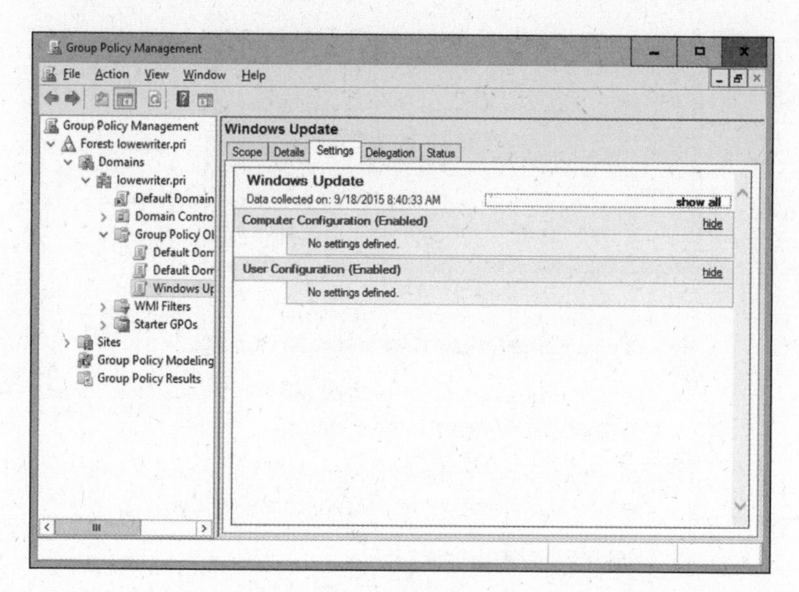

FIGURE 16-5: Policy settings.

7. **Right-click Computer Configuration and then choose Edit from the contextual menu.**

   This command opens the Group Policy Management Editor, as shown in Figure 16-6, where you can edit the Computer Configuration policies.

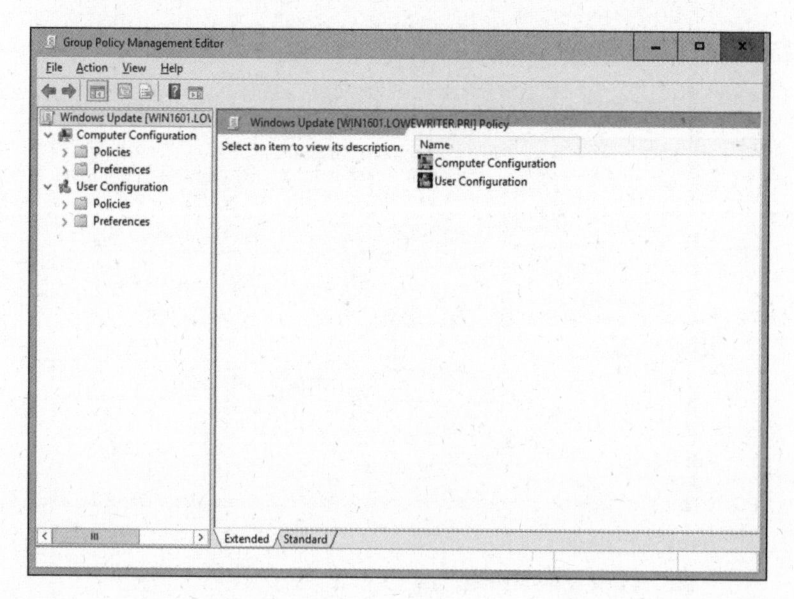

FIGURE 16-6: Editing a GPO.

8. **In the Navigation pane, navigate to Computer Configuration ⇨ Administrative Templates ⇨ Windows Components ⇨ Windows Update.**

   This step brings up the Windows Update policy, as shown in Figure 16-7.

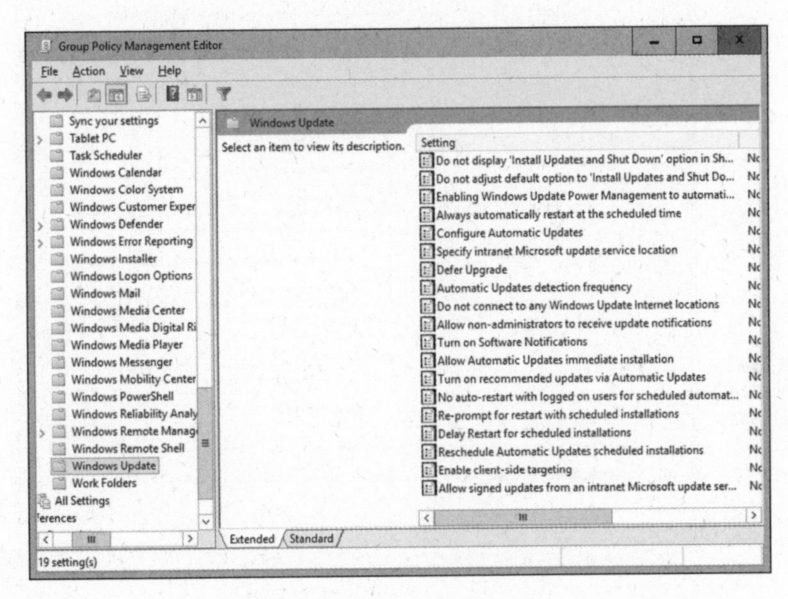

FIGURE 16-7: The Windows Update policy.

9. **Double-click Configure Automatic Updates.**

   This step brings up the Configure Automatic Updates dialog box, as shown in Figure 16-8.

10. **Select Enabled to enable the policy.**
11. **Configure the Windows Update settings however you want.**

    For this example, I configure Windows Update so that updates are automatically downloaded every day at 3 a.m.

12. **Click OK.**

    You return to the Group Policy Management Editor.

13. **Close the Group Policy Management Editor window.**

    This step returns you to the Group Policy Management settings window.

14. **Right-click Computer Configuration, and choose Refresh from the contextual menu.**

    The Windows Update policy is visible, as shown in Figure 16-9. (To show the full details of the policy, I expanded the Administrative Templates and Windows Components/Windows Update sections of the policy report.)

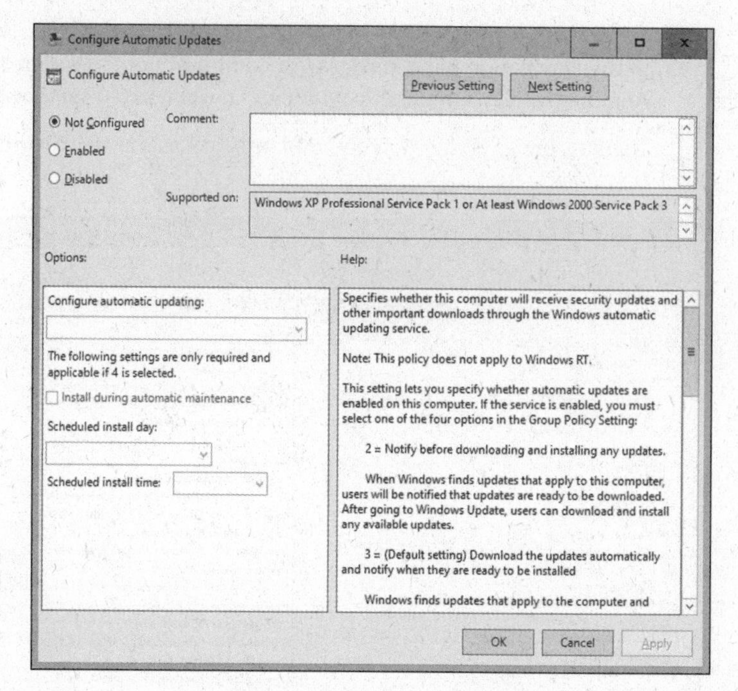

FIGURE 16-8: The Configure Automatic Updates dialog box.

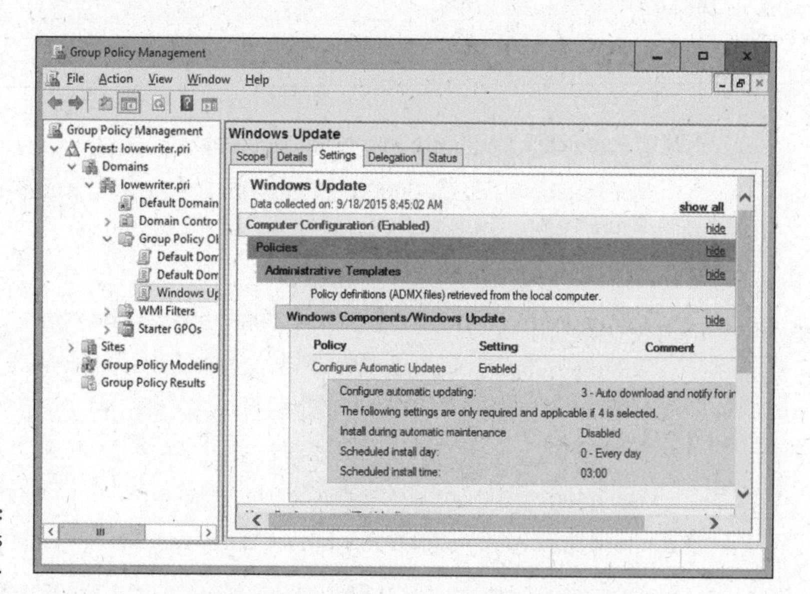

FIGURE 16-9: The Windows Update policy.

15. **In the Navigation pane of the Group Policy Management window, drag the new Windows Update GPO to the top-level domain (in this case, `lowewriter.pri`).**

When you release the mouse button, the dialog box shown in Figure 16-10 appears.

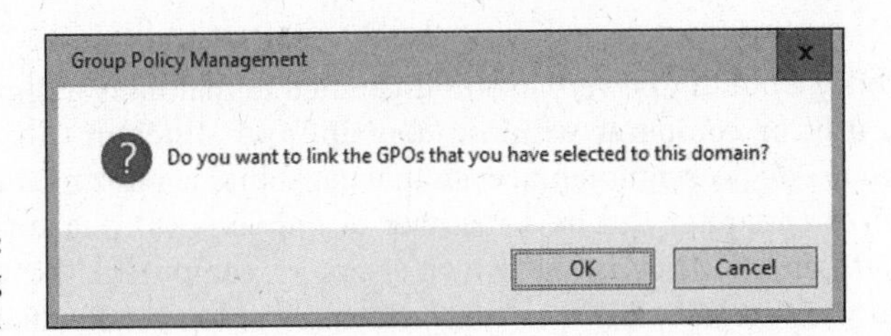

FIGURE 16-10: Confirming the scope.

16. **Click OK.**

    The domain is added to the scope, as shown in Figure 16-11.

17. **Close the Group Policy Management window.**

    The new GPO is now active.

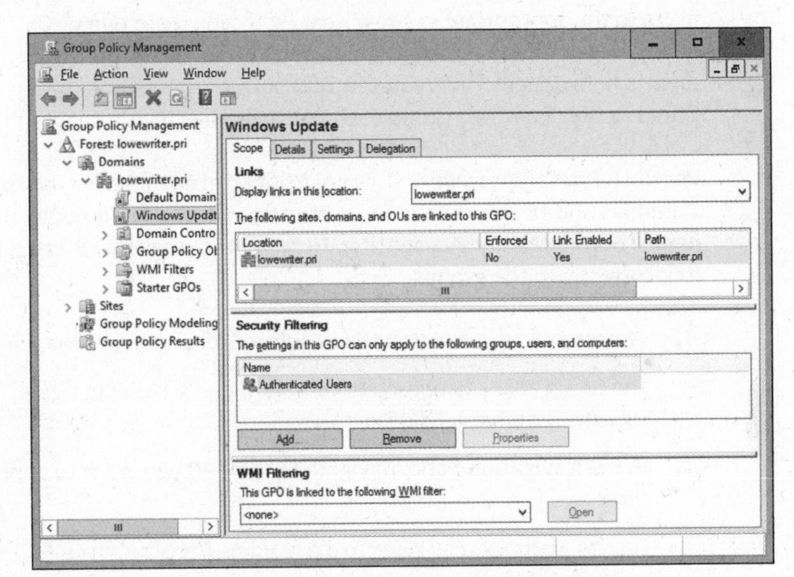

FIGURE 16-11: The GPO is finished.

Filtering Group Policy Objects

One of the most confusing aspects of Group Policy is that even though it applies to users and computers, you don't associate GPOs with users or computers. Instead, you link GPOs to sites, domains, or organizational units (OUs). At first glance, this aspect may seem to limit the usefulness of Group Policy. For most simple networks, you'll work with Group Policy mostly at the domain level and occasionally at the OU level. Site-level GPOs are used only for very large or complex networks.

Group Policy wouldn't be very useful if you had to manually assign the same GPO to every user or computer within a domain. And although OUs can help break down Group Policy assignments, even that capability is limiting, because a particular user or computer can be a member of only one OU. Fortunately, GPOs can have *filters* that further refine which users or computers the GPO applies to. Although you can filter GPOs so that they apply only to individual users or computers, you're more likely to use groups to apply your GPOs.

Suppose that you want to use Group Policy to assign two different default home pages for Internet Explorer. For the Marketing department, you want the default home page to be `www.dummies.com`, but for the Accounting department, you'd like the default home page to be `www.beancounters.com`. You can easily accomplish this task by creating two groups named Marketing and Accounting in Active Directory Users and Computers, and assigning the marketing and accounting users to the appropriate groups. Next, you can create two GPOs: one for the Marketing department's home page and the other to assign the Accounting department's home page. Then you can link both of these GPOs to the domain and use filters to specify which group each policy applies to.

For the following procedure, I've created two group policies, named IE Home Page Dummies and IE Home Page Beancounter, as well as two Active Directory groups, named Marketing and Accounting. Here are the steps for filtering these policies to link correctly to the groups:

1. **Choose Start ⇨ Administrative Tools ⇨ Group Policy Management.**

   The Group Policy Management console appears. (Refer to Figure 16-2 for a refresher on what it looks like.)

2. **In the Navigation pane, navigate to the GPO you want to apply the filter to.**

   For this example, I navigated to the IE Home Page Dummies GPO, as shown in Figure 16-12.

3. **In the Security Filtering section, click Authenticated Users and then click Remove.**

   This step removes Authenticated Users so that the GPO won't be applied to all users.

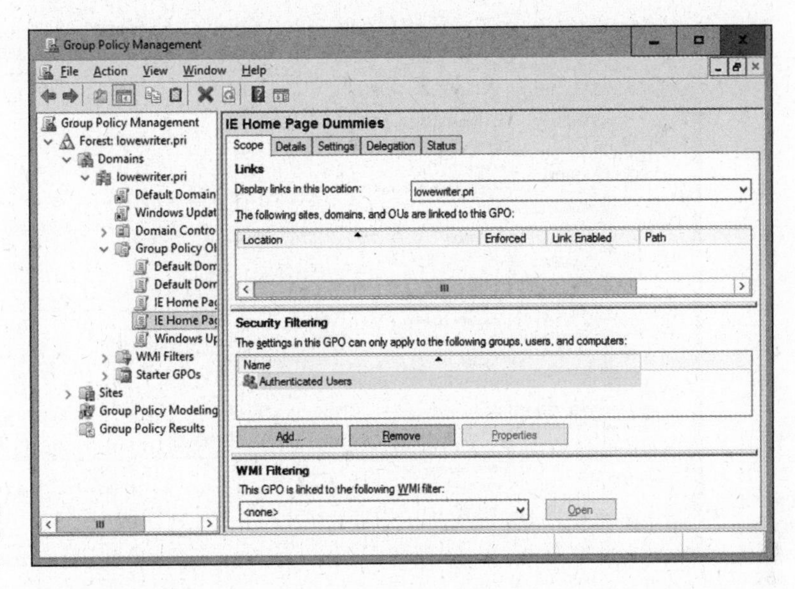

FIGURE 16-12: The IE Home Page Dummies GPO.

4. **Click Add.**

   This step brings up the Select User, Computer, or Group dialog box, as shown in Figure 16-13.

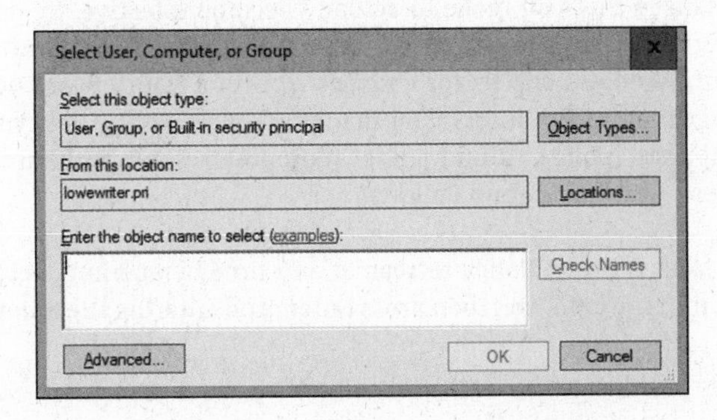

FIGURE 16-13: The Select User, Computer, or Group dialog box.

5. **Type** Marketing **in the text box and then click OK.**

   The GPO is updated to indicate that it applies to members of the Marketing group, as shown in Figure 16-14.

6. **Repeat Steps 2 through 5 for the IE Home Page Beancounter GPO, applying it to the Accounting group.**

   You're done!

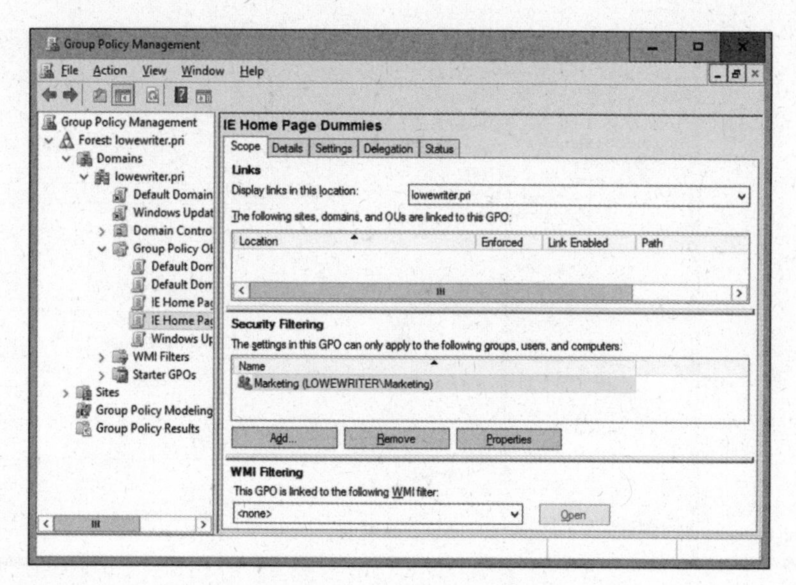

FIGURE 16-14: A GPO that uses a filter.

Forcing Group Policy Updates

One of the more frustrating aspects of working with Group Policy is waiting for policy changes that you make to actually become effective. By default, Windows applies Group Policy automatically whenever a user logs on. After the user has logged on, Windows checks for changes to Group Policy every 90 minutes. (You can change that default interval by changing the Set Group Policy Refresh Interval for Computers policy, found under Computer Configuration ⇨ Administrative Templates ⇨ System ⇨ Group Policy.)

You can force a Group Policy refresh at any time by opening a command prompt (press Start, type **cmd**, and then press Enter) and entering the following command:

```
GPUPDATE
```

This command forces an immediate refresh of all GPOs.

Note that this command works only on the computer on which you run it; it doesn't affect other computers on your network. You can force a Group Policy update on all computers in an Active Directory organizational unit by right-clicking the organizational unit in the Group Policy Management Console and choosing the Group Policy Update command.

WARNING

As with all Group Policy changes, make sure you've tested your change on a limited number of computers before forcing an update on all computers!

» Using license servers

» Exploring the deployment options

» Keeping up to date with patches and service packs

Chapter 17

Managing Software Deployment

An important task of any network administrator is managing the various bits and pieces of software that are used by your users throughout the network. Most, if not all, of your network users will have a version of Microsoft Office installed on their computers. Depending on the type of business, other software may be widely used. For example, accounting firms require accounting software, engineering firms require engineering software, and the list goes on. . . .

Long gone are the days when you could purchase one copy of a computer program and freely install it on every computer on your network. Most software has a built-in feature — commonly called *copy protection* — designed to prevent such abuse. But even in the absence of copy protection, nearly all software is sold with a license agreement that dictates how many computers you can install and use the software on. As a result, managing software licenses is an important part of network management.

Some software programs have a license feature that uses a server computer to regulate the number of users who can run the software at the same time. As the network administrator, your job is to set up the license server and keep it running.

Another important aspect of managing software on the network is figuring out the most expedient way to install the software on multiple computers. The last thing you want to do is manually run the software's Setup program individually on each computer in your network. Instead, you'll want to use the network itself to aid in the deployment of the software.

Finally, you'll want to ensure that all the software programs installed throughout your network are kept up to date with the latest patches and updates from the software vendors.

This chapter elaborates on these aspects of network software management.

Understanding Software Licenses

Contrary to popular belief, you don't really buy software. Instead, you buy the right to *use* the software. When you purchase a computer program at a store, all you really own after you complete the purchase is the box the software comes in, the discs the software is recorded on, and a license that grants you the right to use the software according to the terms offered by the software vendor. The software itself is still owned by the vendor.

That means that you're obligated to follow the terms of the license agreement that accompanies the software. Very few people actually read the complete text of a software agreement before they purchase and use software. If you do, you'll find that a typical agreement contains restrictions, such as the following:

- **You're allowed to install the software on one, and only one, computer.** Some license agreements have specific exceptions to this rule, allowing you to install the software on a single computer at work and a single computer at home, or on a single desktop computer and a single laptop computer, provided that both computers are used by the same person. However, most software licenses stick to the one-computer rule.
- **The license agreement probably allows you to make a backup copy of the discs.** The number of backup copies you can make, though, is probably limited to one or two.
- **You aren't allowed to reverse-engineer the software.** In other words, you can't use programming tools to dissect the software in an effort to learn the secrets of how it works.
- **Some software restricts the kinds of applications it can be used for.** For example, you might purchase a student or home version of a program

that prohibits commercial use. And some software — for example, Oracle's Java — prohibits its use by nuclear facilities.

- **Some software has export restrictions that prevent you from taking it out of the country.**
- **Nearly all software licenses limit the liability of the software vendor to replacing defective installation discs.** In other words, the software vendor isn't responsible for any damage that might be caused by bugs in the software. In a few cases, these license restrictions have been set aside in court, and companies have been held liable for damage caused by defective software. For the most part, though, you use software at your own risk.

In many cases, software vendors give you a choice of several different types of licenses to choose from. When you purchase software for use on a network, you need to be aware of the differences between these license types so you can decide which type of license to get. Here are the most common types:

- **Retail:** The software you buy directly from the software vendor, a local store, or an online store. A retail software license usually grants you the right for a single user to install and use the software. Depending on the agreement, the license may allow that user to install the software on two computers — one at work and one at home. The key point is that only one user may use the software. (However, it is usually acceptable to install the software on a computer that's shared by several users. In that case, more than one user can use the software, provided they use it one at a time.)

  The main benefit of a retail license is that it stays with the user when the user upgrades his or her computer. In other words, if you get a new computer, you can remove the software from your old computer and install it on your new computer.

- **Original equipment manufacturer (OEM):** The software that's installed by a computer manufacturer on a new computer. For example, if you purchase a computer from Dell and order Microsoft Office Professional along with the computer, you're getting an OEM license. The most important thing to know about an OEM license is that it applies only to the specific computer for which you purchased the software. You are never allowed to install the software on any computer other than the one for which you purchased the software.

  So, if one day, in a fit of rage, you throw your computer out the fifth-floor window of your office and the computer smashes into little pieces in the parking lot below, your OEM version of Office is essentially lost forever. When you buy a replacement computer, you'll have to buy a new OEM license of Office for the new computer. You can't install the old software on the new computer.

» **Volume:** Allows you to install and use the software on more than one computer. The simplest type of volume license simply specifies how many computers you can install the software on. For example, you may purchase a 20-user version of a program that allows you to install the software on 20 computers. Usually, you're on the honor system to make sure that you don't exceed the quantity.

TIP

Set up some type of system to keep track of this type of software license. For example, you could create a Microsoft Excel spreadsheet in which you record the name of each person for whom you install the software.

Volume licenses can become considerably more complicated. For example, Microsoft offers several different types of volume license programs, each with different pricing, features, and benefits. Table 17-1 summarizes the features of the more popular license programs. For more information, refer to `www.microsoft.com/licensing`.

» **Subscription:** A *subscription* isn't really a separate type of license but rather an optional add-on to a volume license. The added subscription fee entitles you to technical support and free product upgrades during the term of the subscription, which is usually annual. For some types of products, the subscription also includes periodic downloads of new data. For example, antivirus software usually includes a subscription that regularly updates your virus signature data. Without the subscription, the antivirus software would quickly become ineffective.

Note that many types of software that were traditionally offered under retail, OEM, or volume licensing plans are evolving toward subscription plans. Vendors such as Microsoft (for its Office 365), Adobe (for its Creative Design Suite), and Autodesk (for its AutoCAD family of products) are moving in that direction, as are many others.

TABLE 17-1 **Microsoft Volume License Plans**

Plan	Features
Open License	Purchase as few as five end-user licenses.
Open Value	Purchase as few as five end-user licenses and receive free upgrades during the subscription term (three years).
Select License	This is a licensing program designed for companies with 250 or more employees.
Enterprise	This is an alternative to the Select License program that's designed to cost-effectively provide Windows, Office, and certain other programs throughout an organization of at least 250 employees.

Using a License Server

Some programs let you purchase network licenses that enable you to install the software on as many computers as you want, but regulate the number of people who can use the software at any given time. To control how many people use the software, a special license server is set up. Whenever a user starts the program, the program checks with the license server to see whether a license is available. If so, the program is allowed to start, and the number of available licenses on the license server is reduced by one. Later, when the user quits the program, the license is returned to the server.

A commonly used license server software is FlexNet Publisher, by Flexera Software. (This program used to be named FlexLM, and many programs that depend on it still distribute it as FlexLM.) It's used by AutoCAD, as well as by many other network software applications. FlexNet Publisher uses special license files issued by a software vendor to indicate how many licenses of a given product you purchased. Although the license file is a simple text file, its contents are cryptic and generated by a program that only the software vendor has access to. Here's an example of a typical license file for AutoCAD:

```
SERVER server1 000ecd0fe359
USE_SERVER
VENDOR adskflex port=2080
INCREMENT 57000ARDES_2010_0F adskflex 1.000 permanent 6 \
VENDOR_STRING=commercial:permanent BORROW=4320
SUPERSEDE \
DUP_GROUP=UH ISSUED=07-May-2007 SN=339-71570316 SIGN="102D \
85EC 1DFE D083 B85A 46BB AFB1 33AE 00BD 975C 8F5C 5ABC 4C2F \
F88C 9120 0FB1 E122 BA97 BCAE CC90 899F 99BB 23C9 CAB5 613F \
E7BB CA28 7DBF 8F51 3B21" SIGN2="033A 6451 5EEB 3CA4 98B8 F92C \
184A D2BC BA97 BCAE CC90 899F 2EF6 0B45 A707 B897 11E3 096E 0288 \
787C 997B 0E2E F88C 9120 0FB1 782C 00BD 975C 8F5C 74B9 8BC1"
```

(Don't get any wild ideas here. I changed the numbers in this license file so that it won't actually work. I'm not crazy enough to publish an actual valid AutoCAD license file!)

One drawback to opting for software that uses a license server is that you have to take special steps to run the software when the server isn't available. For example, what if you have AutoCAD installed on a notebook computer, but you want to use it while you're away from the office? In that case, you have two options:

» **Use virtual private network (VPN) software to connect to the network.** After you're connected with the VPN, the license server will be available so you can use the software. (Read about VPNs in Chapter 24.)

» **Borrow a license.** When you borrow a license, you can use the software for a limited period of time while you're disconnected from the network. Of course, the borrowed license is subtracted from the number of available licenses on the server.

In most cases, the license server is a mission-critical application — as important as any other function on your network. If the license server goes down, all users who depend on it will be unable to work. Don't worry — they'll let you know. They'll be lining up outside your door demanding to know when you can get the license server up and running so they can get back to work.

Because the license server provides such an important function, treat it with special care. Make sure that the license server software runs on a stable, well-maintained server computer. Don't load up the license server computer with a bunch of other server functions.

And make sure that it's backed up. If possible, install the license server software on a second server computer as a backup. That way, if the main license server computer goes down and you can't get it back up and running, you can quickly switch over to the backup license server.

Deploying Network Software

After you acquire the correct license for your software, the next task of the network administrator is to deploy the software — that is, install the software on your users' computers and configure the software so that it runs efficiently on your network. The following sections describe several approaches to deploying software to your network.

Deploying software manually

Most software is shipped on CD or DVD media along with a Setup program that you run to install the software. The Setup program usually asks you a series of questions, such as where you want the program installed, whether you want to install all of the program's features or just the most commonly used features, and so on. You may also be required to enter a serial number, registration number, license key, or other code that proves you purchased the software. When all these questions are answered, the Setup program then installs the program.

If only a few of your network users will be using a particular program, the Setup program may be the most convenient way to deploy the program. Just take the installation media with you to the computer you want to install the program on, insert the disc into the CD/DVD drive, and run the Setup program.

WARNING

When you finish manually installing software from a CD or DVD, don't forget to remove the disc from the drive! It's easy to leave the disc in the drive, and if the user rarely or never uses the drive, it might be weeks or months before anyone discovers that the disc is missing. By that time, you'll be hard-pressed to remember where it is.

Running Setup from a network share

If you plan on installing a program on more than two or three computers on your network, you'll find it much easier to run the Setup program from a network share rather than from the original CDs or DVDs. To do so, follow these steps:

1. **Create a network share and a folder within the share where you can store the Setup program and other files required to install the program.**

   I usually set up a share named Software and then create a separate folder in this share for each program I want to make available from the network. You should enable Read access for all network users, but allow full access only for yourself and your fellow administrators.

   Read more about creating shares and setting permissions in Chapter 13.

2. **Copy the entire contents of the program's CD or DVD to the folder you create in Step 1.**

   To do so, insert the CD or DVD into your computer's CD/DVD drive. Then, use Windows Explorer to select the entire contents of the disc and drag it to the folder you created in Step 1.

   Alternatively, you can click Start, type **cmd**, and then press Enter to open a command prompt. Then, enter a command, such as the following:

   ```
   xcopy d:\*.* \\server1\software\someprogram\*.* /s
   ```

   In this example, `d:` is the drive letter of your CD/DVD drive, *`server1`* is the name of your file server, and *`software`* and *`someprogram`* are the names of the share and folder you created in Step 1.

3. **To install the program on a client computer, open a Windows Explorer window, navigate to the share and folder you create in Step 1, and double-click the `Setup.exe` file.**

   The Setup program launches.

4. **Follow the instructions displayed by the Setup program.**

   When the Setup program is finished, the software is ready to use.

Copying the Setup program to a network share spares you the annoyance of carrying the installation discs to each computer you want to install the software on. It doesn't spare you the annoyance of purchasing a valid license for each computer, though! It's illegal to install the software on more computers than the license you acquired from the vendor allows.

Installing silently

Copying the contents of a program's installation media to a network share does spare you the annoyance of carrying the installation discs from computer to computer, but you still have to run the Setup program and answer all its annoying questions on every computer. Wouldn't it be great if there were a way to automate the Setup program so that after you run it, it runs without any further interaction from you? With many programs, you can.

In some cases, the Setup program itself has a command-line switch that causes it to run silently. You can usually find out what command-line switches are available by entering the following at a command prompt:

```
setup /?
```

With luck, you'll find that the Setup program itself has a switch, such as `/quiet` or `/silent`, that installs the program with no interaction, using the program's default settings.

If the Setup program doesn't offer any command-line switches, don't despair. The following procedure describes a technique that often lets you silently install the software:

1. **Open an Explorer window and navigate to** `C:\Users\`*`name`*`\AppData\Local\Temp`

   Then, delete the entire contents of this folder.

   This is the `Temp` folder where various programs deposit temporary files. Windows may not allow you to delete every file in this folder, but it's a good idea to begin this procedure by emptying the `Temp` folder as much as possible.

2. **Run the Setup program and follow the installation steps right up to the final step.**

WARNING

When you get to the confirmation screen that says the program is about ready to install the software, stop! Do *not* click the OK or Finish button.

3. **Return to the** `Temp` **folder you opened in Step 1, and then poke around until you find the** `.msi` **file created by the Setup program you ran in Step 2.**

   The `:msi` file is the actual Windows Installer program that Setup runs to install the program. It may have a cryptic name, such as `84993882.msi`.

4. **Copy the** `.msi` **file to the network share from which you want to install the program on your client computers.**

   For example, `\\server1\software\someprogram`.

5. **(Optional) Rename the** `.msi` **file to** `setup.msi`.

   This step is optional, but I prefer to run `setup.msi` rather than `84993882.msi`.

6. **Use Notepad to create a batch file to run the** `.msi` **file with the** `/quiet` **switch.**

   To create the batch file, follow these steps:

   1. Right-click in the folder where the `.msi` file is stored.
   2. Choose New ➪ Text Document.
   3. Change the name of the text document to `setup.bat`.
   4. Right-click the `setup.bat` file and choose Edit.
   5. Add the following line to the file:

      setup.msi /quiet
   6. Save the file.

   You can now install the software by navigating to the folder you created the `setup.bat` file in and double-clicking the `setup.bat` file.

Creating an administrative installation image

Some software, such as Microsoft Office and AutoCAD, comes with tools that let you create a fully configured silent setup program that you can then use to silently install the software. For Microsoft software, this silent setup program is called an *administrative installation image.* (Note that the OEM versions of Office don't include this feature. You need to purchase a volume license to create an administrative installation.)

To create an administrative image, you simply run the configuration tool supplied by the vendor. The configuration tool lets you choose the installation options you want to have applied when the software is installed. Then it creates a network setup program on a network share that you specify. You can then install the software on a client computer by opening an Explorer window, navigating to the network share where you saved the network setup program, and running the network setup program.

Pushing out software with Group Policy

One final option you should consider for network software deployment is using Windows Group Policy to automatically install software to network users. Group Policy is a Windows Server feature that lets you create policies that are assigned to users. You use the Windows Group Policy feature to specify that certain users should have certain software programs available to them.

Note that group policies aren't actually assigned to individual users, but to organizational units (OUs), which are used to categorize users in Active Directory. Thus, you might create a policy to specify that everyone in the Accounting Department OU should have Excel.

Then, whenever anyone in the Accounting Department logs on to Windows, Windows checks to make sure that Excel is installed on the user's computer. If Excel is *not* installed, Windows advertises Excel on the computer. *Advertising* software on a computer means that a small portion of the software is downloaded to the computer — just enough to display an icon for the program on the Start menu and to associate Excel with the Excel file extensions (`.xls`, `.xlsx`).

If the user clicks the Start menu icon for the advertised application or attempts to open a document that's associated with the advertised application, the application is automatically installed on the user's computer. The user will have to wait a few minutes while the application is installed, but the installation is automatic.

For more information about setting up Group Policy software installation, search the web for *Group Policy software.*

Keeping Software Up to Date

One of the annoyances that every network manager faces is applying software patches to keep the operating system and other software up to date. A *software patch* is a minor update that fixes the small glitches that crop up from time to time, such as minor security or performance issues. These glitches aren't significant enough to merit a new version of the software, but they're important enough to require fixing. Most patches correct security flaws that computer hackers have uncovered in their relentless attempts to wreak havoc on the computer world.

Periodically, all the recently released patches are combined into a service pack. Although the most diligent network administrators apply all patches when they're released, many administrators just wait for the service packs.

Windows includes the Windows Update feature that automatically installs patches and service packs when they become available. These patches apply not just to Windows but to other Microsoft software as well. To use Windows Update, click the Start button, type **Windows Update**, and press Enter. The Windows Update settings appear, as shown in Figure 17-1.

From the Windows Update window, you can click the Check for Updates button to see if you have any updates that need to be installed. Any required updates will automatically be downloaded and installed.

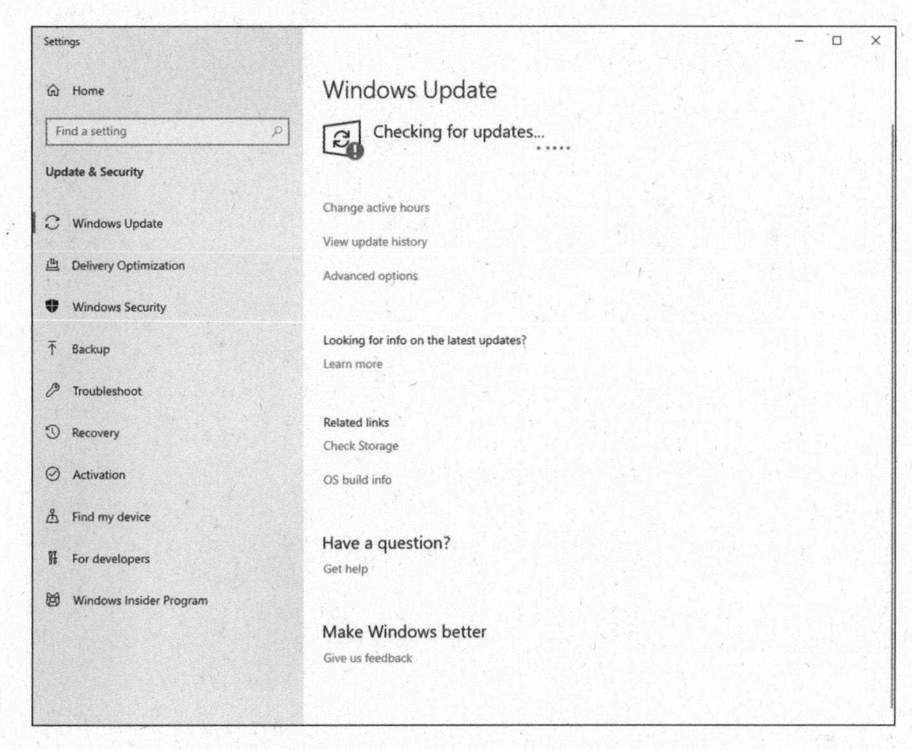

FIGURE 17-1: Windows Update.

» Configuring Windows Mobile devices for Exchange access

» Considering security implications

Chapter 18
Managing Mobile Devices

A computer consultant once purchased a used BlackBerry device on eBay for $15.50. When he put in a new battery and turned on the device, he discovered that it contained confidential emails and personal contact information for executives of a well-known financial institution.

Oops!

It turns out that a former executive with the company sold his old BlackBerry on eBay a few months after he left the firm. He'd assumed that because he'd removed the battery, everything on the BlackBerry had been erased.

The point of this true story is that mobile devices such as smartphones and tablet computers pose a special set of challenges for network administrators. These challenges are now being faced even by administrators of small networks. Just a few years ago, only large companies had BlackBerry or other mobile devices that integrated with Exchange email, for example. Now it isn't uncommon for companies with just a few employees to have mobile devices connected to the company network.

This chapter is a brief introduction to mobile devices and the operating systems they run, with an emphasis on iPhone and Android devices. You find out more about how these devices can interact with Exchange email and the steps you can take to ensure their security.

The Many Types of Mobile Devices

Once upon a time, there were mobile phones and PDAs. A mobile phone was just that: a handheld telephone you could take with you. The good ones had nice features such as a call log, an address book, and perhaps a crude game, but not much else. PDAs — *Personal Digital Assistants* — were little handheld computers designed to replace the old-fashioned Day-Timer books people used to carry around with them to keep track of their appointment calendars and address books.

All that changed when cellular providers began adding data capabilities to their networks. Now cellphones can have complete mobile Internet access. This fact has resulted in the addition of sophisticated PDA features to mobile phones and phone features to PDAs so that the distinctions are blurred.

A *mobile device* can be any one of a wide assortment of devices that you can hold in one hand and that are connected through a wireless network. The term *handheld* is a similar generic name for such devices. The following list describes some of the most common specifics of mobile devices:

- **Mobile phone:** Primary purpose is to enable phone service. Most mobile phones also include text messaging, address books, appointment calendars, and games; they may also provide Internet access.
- **Smartphone:** A smartphone is a cellphone that also functions as a handheld computer. Smartphones feature touchscreens instead of physical buttons or keys to press. Besides the features ordinarily found on a mobile phone, smartphones also offer email, calendar, contacts, task lists, and web access, as well as apps that can be purchased and installed on the phone.
- **Android:** Android is an open-source operating system (OS) for smartphones, developed by Google. Android is far and away the most popular platform for smartphones, being used on more than 80 percent of the smartphones sold since 2015.
- **iOS:** iOS is the OS used on Apple's popular iPhone and iPad mobile devices. Although outnumbered by Android devices, many people consider iOS devices to be more innovative than Android devices. The main thing that holds iOS back in market share is cost: Apple devices are considerably more expensive than their Android equivalents.
- **BlackBerry:** BlackBerry was once the king of the smartphone game. For many years, BlackBerry had a virtual monopoly on the mobile devices market because it was the first mobile device that could synchronize well with Microsoft Exchange. Now that Android and Apple devices do that just

as well as (actually, much better than) BlackBerry, BlackBerry devices have fallen out of vogue. However, BlackBerry is still around and there are still plenty of BlackBerry users out there. (Note that newer BlackBerry phones run Android rather than the old proprietary BlackBerry OS.)

Considering Security for Mobile Devices

As a network administrator, one of your main responsibilities regarding mobile devices is to keep them secure. Unfortunately, that's a significant challenge. Here are some reasons why:

- **Mobile devices connect to your network via other networks that are out of your control.** You can go to great lengths to set up firewalls, encryption, and a host of other security features, but mobile devices connect via public networks whose administrators may not be as conscientious as you.
- **Mobile devices are easy to lose.** A user might leave her smartphone at a restaurant or hotel, or it might fall out of her pocket on the subway.
- **Mobile devices run operating systems that aren't as security conscious as Windows.**
- **Users who wouldn't dare install renegade software on their desktop computers think nothing of downloading free games or other applications to their handheld devices.** Who knows what kinds of viruses or Trojans these downloads carry?
- **Inevitably, someone will buy his own handheld device and connect it to your network without your knowledge or permission.**

Here are some recommendations for beefing up security for your mobile devices:

- Establish clear, consistent policies for mobile devices, and enforce them.
- Make sure employees understand that they aren't allowed to bring their own devices into your network. Allow only company-owned devices to connect.
- Train your users in the security risks associated with using mobile devices.
- Implement antivirus protection for your mobile devices.

Managing iOS Devices

In 2007, the Apple iPhone, one of the most innovative little gadgets in many, many years, hit the technology market. In just a few short years, the iPhone captured a huge slice of a market previously dominated almost exclusively by RIM and its BlackBerry devices. Since then, the iPhone's share of the mobile-phone market has grown beyond that of the former king, BlackBerry.

The success of the iPhone was due in large part to the genius of its operating system, called iOS. In 2010, Apple released the iPad, a tablet computer that runs the same iOS as the iPhone. And in 2012, Apple introduced a smaller version of the iPad: the iPad mini. Together, these devices are commonly known as *iOS devices.*

Understanding the iPhone

The iPhone is essentially a combination of four devices:

- A cellphone
- An iPod with a memory capacity of 16GB to 512GB
- A digital camera
- An Internet device with its own web browser (Safari) and applications, such as email, calendar, and contact management

The most immediately noticeable feature of the iPhone is its lack of a keyboard. Instead, nearly the entire front surface of the iPhone is a high-resolution, touch-sensitive LCD display. The display is not only the main output device of the iPhone, but also its main input device. The display can become a keypad input for dialing a telephone number or a keyboard for entering text. You can also use various finger gestures, such as tapping icons to start programs or pinching to zoom in the display.

The iPhone has several other innovative features:

- An *accelerometer* tracks the motion of the iPhone in three directions. The main use of the accelerometer is to adjust the orientation of the display from landscape to portrait based on how the user is holding the phone. Some other applications — mostly games — use the accelerometer as well.
- A Wi-Fi interface lets the iPhone connect to local Wi-Fi networks for faster Internet access.

- GPS capability provides location awareness for many applications, including Google Maps.
- The virtual private network (VPN) client lets you connect to your internal network.

Of all the unique features of the iPhone, probably the most important is its huge collection of third-party applications that can be downloaded from a special web portal, the App Store. Many of these applications are free or cost just a few dollars. (Many are just 99 cents or $1.99.) As of this writing, more than 1.8 million applications — everything from business productivity to games — were available from the App Store.

Understanding the iPad

The iPad is essentially an iPhone without the phone but with a larger screen. The iPhone comes with a 3.5-inch screen; the iPad has a 9.7-inch screen; and its smaller cousin, the iPad mini, has a 7.9-inch screen and the iPad Pro has a whopping 12.9-inch screen.

Apart from these basic differences, an iPad is nearly identical to an iPhone. Any application that can run on an iPhone can also run on an iPad, and many applications are designed to take special advantage of the iPad's larger screen.

All the information that follows in this chapter applies equally to iPhones and iPads.

Integrating iOS devices with Exchange

An iOS device can integrate with Microsoft Exchange email via the Exchange ActiveSync feature, which is enabled by default on all versions of Exchange since 2010.

To verify the Exchange ActiveSync feature for an individual mailbox, follow these steps:

1. **Choose Start ⇨ Administrative Tools ⇨ Active Directory Users and Computers.**

   The Active Directory Users and Computers console opens.

2. **Expand the domain and then locate the user you want to enable mobile access for.**

3. **Right-click the user and then choose Properties from the contextual menu.**

4. **Click the Exchange Features tab.**

   The Exchange Features options are displayed, as shown in Figure 18-1.

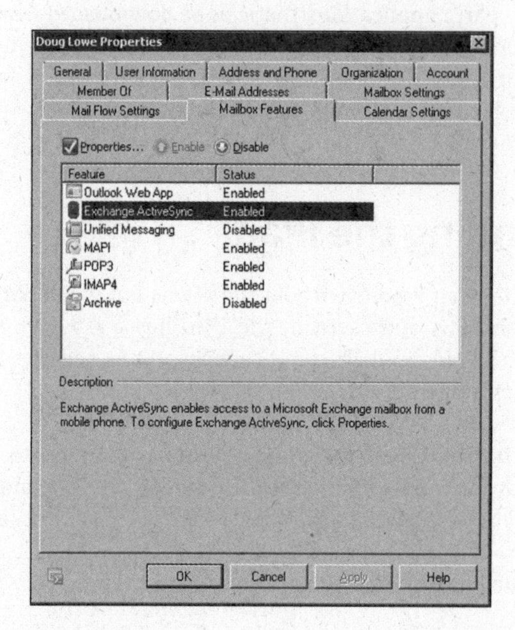

FIGURE 18-1: Enabling Exchange ActiveSync for a user.

5. **Ensure that the Exchange ActiveSync option is enabled.**

   If the options aren't already enabled, right-click each option and choose Enable from the contextual menu.

6. **Click OK.**

7. **Repeat Steps 5 and 6 for any other users you want to enable mobile access for.**

8. **Close Active Directory Users and Computers.**

That's all there is to it. After you enable these features, any users running Windows Mobile can synchronize their handheld devices with their Exchange mailboxes.

Configuring an iOS device for Exchange email

After ActiveSync is enabled for the mailbox, you can configure an iPhone or iPad to tap into the Exchange account by following these steps:

1. **On the iPhone or iPad, tap Settings and then tap Mail, Contacts, Calendars.**

   The screen shown in Figure 18-2 appears. This screen lists any existing email accounts that may already be configured on the phone and also lets you add a new account.

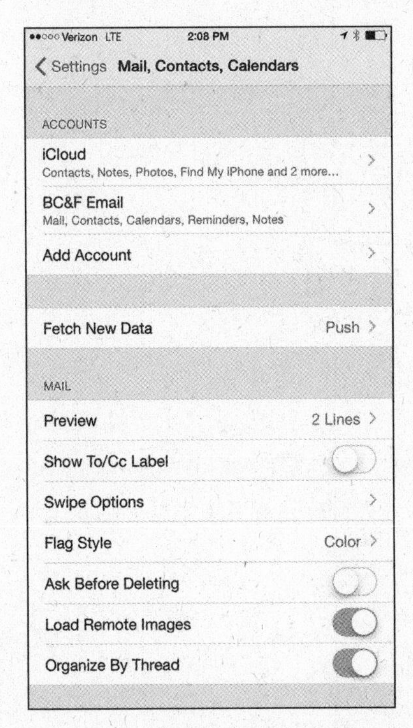

FIGURE 18-2: Adding an email account.

2. **Tap Add Account.**

   The screen shown in Figure 18-3 appears, allowing you to choose the type of email account you want to add.

3. **Tap Exchange.**

   The screen shown in Figure 18-4 appears, where you can enter basic information for your Exchange account.

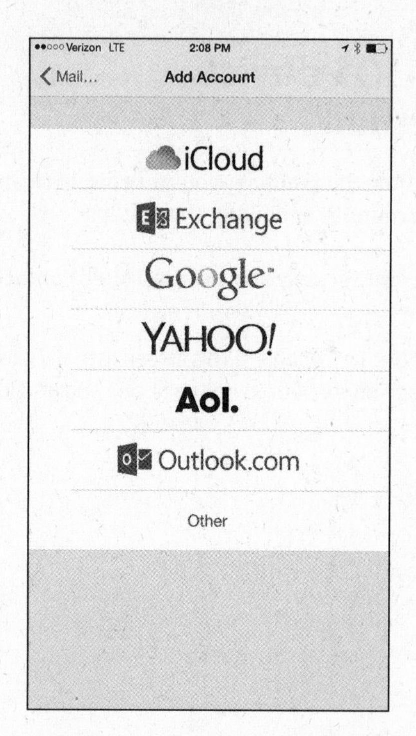

FIGURE 18-3: The iPhone can support many types of email accounts.

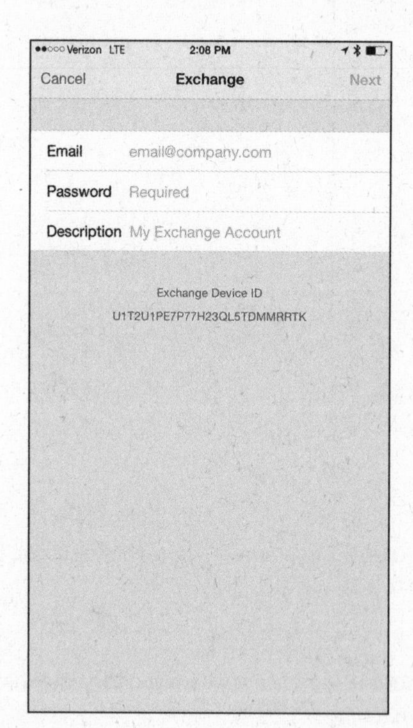

FIGURE 18-4: Enter your email address and password.

4. **Enter your email address, password, and a description of the account.**
5. **Tap Next.**

   The screen shown in Figure 18-5 appears.

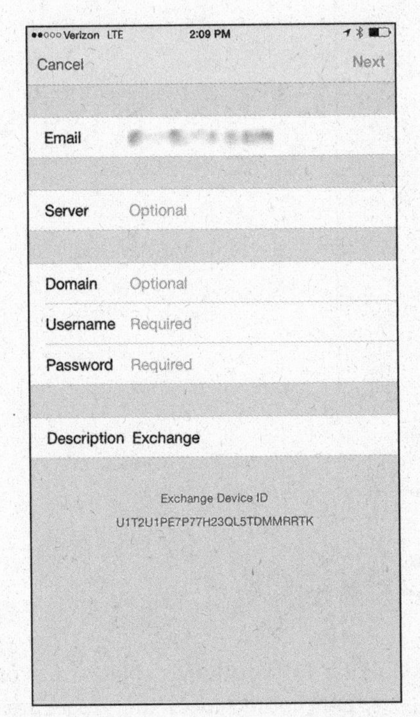

FIGURE 18-5: Enter your Exchange server information.

6. **Enter either the DNS name or the IP address of your Exchange server in the Server field.**

   I entered **smtp.lowewriter.com** for my Exchange server, for example.
7. **Enter your domain name, your Windows username, and your Windows password in the appropriate fields.**
8. **Tap Next.**

   The screen shown in Figure 18-6 appears. Here, you select which mailbox features you want to synchronize: Mail, Contacts, Calendars, Reminders, and/or Notes.
9. **Select the features you want to synchronize and then tap Done.**

   The email account is created.

After the email account has been configured, the user can access it via the Mail icon on the iPhone's home screen.

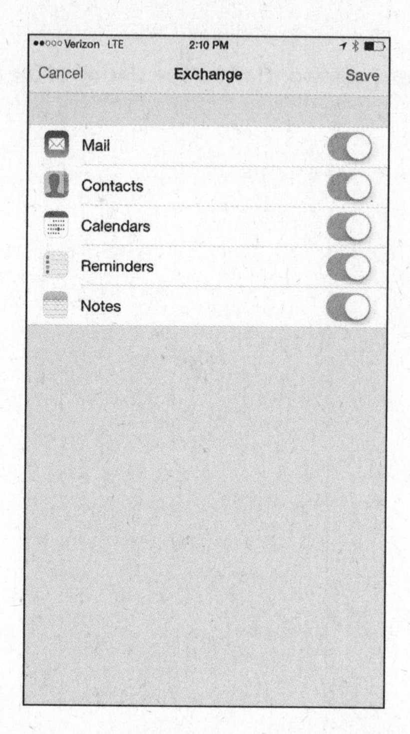

FIGURE 18-6: Select features to synchronize.

Managing Android Devices

This section is a brief introduction to the Android platform. You find out a bit about what Android actually is, and you discover the procedures for setting up Exchange email access on an Android phone.

Crucial differences exist between Android phones and iPhones. The most important difference — in many ways, the *only* important difference — is that Android phones are based on an open source OS derived from Linux, which can be extended and adapted to work on a wide variety of hardware devices from different vendors. With the iPhone, you're locked into Apple hardware. With an Android phone, though, you can buy hardware from a variety of manufacturers.

Looking at the Android OS

Most people associate the Android OS with Google, and it's true that Google is the driving force behind Android. The Android OS is an open source OS managed by the Open Handset Alliance (OHA). Google still plays a major role in the development of Android, but more than 50 companies are involved in the OHA, including hardware manufacturers (such as HTC, Intel, and Motorola), software companies (such as Google and eBay), and mobile-phone operators (such as T-Mobile and Sprint-Nextel).

Technically speaking, Android is more than just an OS. It's also a complete *software stack*, which comprises several key components that work together to create the complete Android platform:

- **The OS core,** which is based on the popular Linux OS
- **A middleware layer,** which provides drivers and other support code to enable the OS core to work with the hardware devices that make up a complete phone, such as a touch-sensitive display, the cellphone radio, the speaker and microphone, Bluetooth or Wi-Fi networking components, and so on
- **A set of core applications** that the user interacts with to make phone calls, read email, send text messages, take pictures, and so on
- **A Software Developers Kit (SDK)** that lets third-party software developers create their own applications to run on an Android phone, as well as a marketplace where the applications can be marketed and sold, much as the App Store lets iPhone developers market and sell applications for the iPhone

Besides the basic features provided by all operating systems, here are a few bonus features of the Android software stack:

- An optimized graphical display engine that can produce sophisticated 2-D and 3-D graphics
- GPS capabilities that provide location awareness that can be integrated with applications such as Google Maps
- Compass and accelerometer capabilities that can determine whether the phone is in motion and in which direction it's pointed
- A built-in SQL database server for data storage
- Support for several network technologies, including 3G, 4G, Bluetooth, and Wi-Fi
- Built-in media support, including common formats for still images, audio, and video files

Perusing Android's core applications

The Android OS comes preconfigured with several standard applications, which provide the functionality that most people demand from a modern smartphone. These applications include

- **Dialer:** Provides the basic cellphone function that lets users make calls.
- **Browser:** A built-in web browser that's similar to Google's Chrome browser.

- **Messaging:** Provides text (SMS) and multimedia (MMS) messaging.
- **Email:** A basic email client that works best with Google's Gmail but that can be configured to work with other email servers, including Exchange.
- **Contacts:** Provides a contacts list that integrates with the Dialer and Email applications.
- **Camera:** Lets you use the phone's camera hardware (if any) to take pictures.
- **Calculator:** A simple calculator application.
- **Alarm Clock:** A basic alarm clock. You can set up to three different alarms.
- **Maps:** An integrated version of Google Maps.
- **YouTube:** An integrated version of YouTube.
- **Music:** An MP3 player similar to the iPod. You can purchase and download music files from Amazon.
- **Google Play:** Lets you purchase and download third-party applications for the Android phone.
- **Settings:** Lets you control various settings for the phone.

Integrating Android with Exchange

Android's core Email application can integrate with Microsoft Exchange email. To do that, you must enable Exchange Mobile Services and then enable ActiveSync for the user's mailbox.

After you enable Exchange Mobile Services and ActiveSync on your Exchange server, you can easily configure the Android phone for email access. Just run the Email application on the Android phone, and follow the configuration steps, which ask you for basic information such as your email address, username, password, and Exchange mail server.

5 Securing Your Network

IN THIS PART . . .

Learn what cyber security is all about.

Deal with firewalls and antivirus software.

Protect your users' email accounts.

Back up your data.

Plan for disaster.

IN THIS CHAPTER

» Assessing the risk for security

» Determining your basic security philosophy

» Physically securing your network equipment

» Considering user account security

» Looking at other network security techniques

» Making sure your users are secure

Chapter 19 Welcome to Cybersecurity Network

As an IT professional, cybersecurity is the thing most likely to keep you awake at night. Consider the following scenarios:

- Your phone starts ringing like crazy at 3 o'clock one afternoon because no one anywhere on the network can access any of their files. You soon discover that your network has been infiltrated by *ransomware,* nefarious software that has encrypted every byte of data on your network, rendering it useless to your users until you pay a ransom to recover the data.
- Your company becomes a headline on CNN because a security breach has resulted in the theft of your customers' credit card information.
- On his last day of work, a disgruntled employee copies your company contact list and other vital intellectual property to a flash drive and walks away with it along with his Swingline stapler. A few months later, your company loses its biggest contract to the company where this jerk now works.

There is no way you can absolutely prevent such scenarios from ever happening, but with proper security, you can greatly reduce their likelihood. Fortunately, all modern operating systems have built-in security provisions for network security, deterring someone from stealing your files even if he does break down the door.

This chapter provides a brief overview of what's involved when it comes to securing your network. You'll find additional details in the remaining chapters of this part. Chapter 20 dives into the details of using firewalls and antivirus software to protect your users from the dangers of the Internet. Chapter 21 explores methods for preventing the bad guys from penetrating your network via email. And Chapters 22 and 23 look at the tools you'll need to recover in the event of a successful cyberattack or other disaster.

Do You Need Security?

Most small networks are in small businesses or departments where everyone knows and trusts everyone else. Folks don't lock up their desks when they take a coffee break, and although everyone knows where the petty cash box is, money never disappears.

Cybersecurity isn't necessary in an idyllic setting like this one, is it? You bet it is. Here's why any network should be set up with at least some concern for security:

- Even in the friendliest office environment, some information is and should be confidential. If this information is stored on the network, you want to store it in a directory that's available only to authorized users.
- Not all security breaches are malicious. A network user may be routinely scanning through files and come across a filename that isn't familiar. The user may then call up the file, only to discover that it contains confidential personnel information, juicy office gossip, or your résumé. Curiosity, rather than malice, is often the source of security breaches.
- Sure, everyone at the office is trustworthy now. However, what if someone becomes disgruntled, a screw pops loose, and he decides to trash the network files before jumping out the window? What if someone decides to print a few $1,000 checks before packing off to Tahiti?
- Sometimes the mere opportunity for fraud or theft can be too much for some people to resist. Give people free access to the payroll files, and they may decide to vote themselves a raise when no one is looking.

If you think that your network doesn't contain any data worth stealing, think again. For example, your personnel records probably contain more than enough information for an identity thief: names, addresses, phone numbers, Social Security numbers, and so on. Also, your customer files may contain your customers' credit card numbers.

- Hackers who break into your network may be looking to plant a Trojan horse program on your server, enabling them to use your server for their own purposes. For example, someone may use your server to send thousands of unsolicited spam email messages. The spam won't be traced back to the hackers; it'll be traced back to *you.*
- Not everyone on the network knows enough about how Windows and the network work to be trusted with full access to your network's data and systems. A careless mouse click can wipe out a directory of network files. One of the best reasons for activating your network's security features is to protect the network from mistakes made by users who don't know what they're doing.

The Three Pillars of Cybersecurity

There are three pillars that you must consider as part of your cybersecurity plan. They're like the legs of a three-legged stool — if one fails, the whole thing comes crashing down. The three pillars of cybersecurity are

- **Prevention technology:** The first pillar of cybersecurity is technology that you can deploy to prevent bad actors from penetrating your network and stealing or damaging your data. This technology includes firewalls that block unwelcome access, antivirus programs that detect malicious software, patch management tools that keep your software up to date, and antispam programs that keep suspicious email from reaching your users' inboxes.

  Chapter 20 addresses firewalls and antivirus software, while Chapter 21 shows you how to employ antispam software.
- **Recovery technology:** The second pillar of cybersecurity is necessary because the first pillar isn't always successful. Successful cyberattacks are inevitable, so you need to have technology and plans in place to quickly recover from them when you do. You can learn more about recovery technology in Chapters 22 and 23.

» **The human firewall:** The most important element of cybersecurity is what I call the *human firewall.* Technology can only go so far in preventing successful cyberattacks. Most successful attacks are the result of human error, when users open email attachments or click web links that they should have known were dangerous. So, in addition to providing technology to prevent attacks, you also need to make sure your users know how to spot and avoid suspicious email attachments and web links. (For more information, see the section "Securing the Human Firewall" later in this chapter.)

Two Approaches to Security

When you're planning how to implement security on your network, first consider which of two basic approaches to security you'll take:

» **Open door:** You grant everyone access to everything by default and then place restrictions just on those resources to which you want to limit access.

» **Closed door:** You begin by denying access to everything and then grant specific users access to the specific resources that they need.

In most cases, an open door policy is easier to implement. Typically, only a small portion of the data on a network really needs security, such as confidential employee records, or secrets, such as the Coke recipe. The rest of the information on a network can be safely made available to everyone who can access the network.

If you choose a closed door approach, you set up each user so that he has access to nothing. Then, you grant each user access only to those specific files or folders that he needs.

A closed door approach results in tighter security but can lead to the Cone of Silence Syndrome: Like how Max and the Chief can't hear each other but still talk while they're under the Cone of Silence, your network users will constantly complain that they can't access the information that they need. As a result, you'll find yourself often adjusting users' access rights. Choose a closed door approach only if your network contains a lot of sensitive information, and only if you're willing to invest time administrating your network's security policy.

You can think of an open door approach as an *entitlement model,* in which the basic assumption is that users are entitled to network access. In contrast, the closed-door policy is a *permissions model,* in which the basic assumption is that users aren't entitled to anything but must get permissions for every network resource that they access.

TECHNICAL STUFF

If you've never heard of the Cone of Silence, go to YouTube (`www.youtube.com`) and search for "Cone of Silence." You'll find several clips from the original *Get Smart* series.

Physical Security: Locking Your Doors

The first level of security in any computer network is *physical security.* I'm amazed when I walk into the reception area of an accounting firm and see an unattended computer sitting on the receptionist's desk. Often, the receptionist has logged on to the system and then walked away from the desk, leaving the computer unattended.

Physical security is important for workstations but vital for servers. Any good hacker can quickly defeat all but the most paranoid security measures if they can gain physical access to a server. To protect the server, follow these guidelines:

- Lock the computer room.
- Give the key only to people you trust.
- Keep track of who has the keys.
- Mount the servers on cases or racks that have locks.
- Keep a trained guard dog in the computer room and feed it only enough to keep it hungry and mad. (Just kidding.)

REMEMBER

There's a big difference between a door with a lock and a locked door. And locks are quite worthless if you don't use them.

Client computers should be physically secure:

- Instruct users to not leave their computers unattended while they're logged on and unlocked. Users should always lock their computers whenever they leave their desks, even if only for a moment. (To lock a computer, simultaneously press the Windows key and L.)
- In high-traffic areas (such as the receptionist's desk), users should secure their computers with the keylock, if the computer has one.
- Users should lock their office doors when they leave.

WARNING

Here are some other threats to physical security that you may not have considered:

- The nightly cleaning crew probably has complete access to your facility. How do you know that the person who vacuums your office every night doesn't really work for your chief competitor or doesn't consider computer hacking to be a sideline hobby? You don't, so consider the cleaning crew to be a threat.
- What about your trash? Paper shredders aren't just for Enron accountants. Your trash can contain all sorts of useful information: sales reports, security logs, printed copies of the company's security policy, even hand-written passwords. For the best security, every piece of paper that leaves your building via the trash bin should first go through a shredder.
- Where do you store your backup tapes? Don't just stack them up next to the server. Not only does that make them easy to steal, it also defeats one of the main purposes of backing up your data in the first place: securing your server from physical threats, such as fires. If a fire burns down your computer room and the backup tapes are sitting unprotected next to the server, your company may go out of business and you'll certainly be out of a job. Store the backup tapes securely in a fireproof safe and keep a copy off-site, too.
- I've seen some networks in which the servers are in a locked computer room, but the hubs or switches are in an unsecured closet. Remember that every unused port on a hub or a switch represents an open door to your network. The hubs and switches should be secured just like the servers.

Securing User Accounts

Next to physical security, the careful use of user accounts is the most important type of security for your network. Properly configured user accounts can prevent unauthorized users from accessing the network, even if they gain physical access to the network. The following sections describe some of the steps that you can take to strengthen your network's use of user accounts.

Obfuscating your usernames

Huh? When it comes to security, *obfuscation* simply means picking obscure usernames. For example, most network administrators assign usernames based on some combination of the user's first and last name, such as `BarnyM` or `baMiller`. However, a hacker can easily guess such a user ID if he or she knows the name of at least one employee. After the hacker knows a username, he or she can focus on breaking the password.

You can slow down a hacker by using names that are more obscure. Here are some suggestions on how to do that:

- Add a random three-digit number to the end of the name. For example: BarnyM320 or baMiller977.
- Throw a number or two into the middle of the name. For example: Bar6nyM or ba9Miller2.
- Make sure that usernames are different from email addresses. For example, if a user's email address is `baMiller@Mydomain.com`, do *not* use baMiller as the user's account name. Use a more obscure name.

WARNING

Do *not* rely on obfuscation to keep people out of your network! Security by obfuscation doesn't work. A resourceful hacker can discover the most obscure names. Obfuscation can *slow* intruders, not stop them. If you slow intruders down, you're more likely to discover them before they crack your network.

Using passwords wisely

One of the most important aspects of network security is the use of passwords.

REMEMBER

Usernames aren't usually considered *secret.* Even if you use obscure names, even casual hackers will eventually figure them out.

Passwords, on the other hand, are top secret. Your network password is the one thing that keeps an impostor from logging on to the network by using your username and therefore receiving the same access rights that you ordinarily have. *Guard your password with your life.*

Here are some tips for creating good passwords:

- Don't use obvious passwords, such as your last name, your kid's name, or your dog's name.
- Don't pick passwords based on your hobbies. A friend of mine is a boater, and his password is the name of his boat. Anyone who knows him can quickly guess his password. Five lashes for naming your password after your boat.
- Store your password in your head — not on paper.

  WARNING

  Especially bad: Writing your password down on a sticky note and sticking it on your computer's monitor.

» Most network operating systems enable you to set an expiration time for passwords. For example, you can specify that passwords expire after 30 days. When a user's password expires, the user must change it. Your users may consider this process a hassle, but it helps to limit the risk of someone swiping a password and then trying to break into your computer system later.

» You can configure user accounts so that when they change passwords, they can't reuse a *recent* password. For example, you can specify that the new password can't be identical to any of the user's past three passwords.

» You can also configure security policies so that passwords must include a mixture of uppercase letters, lowercase letters, numerals, and special symbols. Thus, passwords like DIMWIT or DUFUS are out. Passwords like 87dIM@wit or duF39&US are in.

WARNING

» Some administrators of small networks opt against passwords altogether because they feel that security isn't an issue on their network. Or short of that, they choose obvious passwords, assign every user the same password, or print the passwords on giant posters and hang them throughout the building. Ignoring basic password security is rarely a good idea, even in small networks. You should consider not using passwords only if your network is very small (say, two or three computers), if you don't keep sensitive data on a file server, or if the main reason for the network is to share access to a printer rather than sharing files. (Even if you don't use passwords, imposing basic security precautions, like limiting access that certain users have to certain network directories, is still possible. Just remember that if passwords aren't used, nothing prevents a user from signing on by using someone else's username.)

Generating passwords For Dummies

How do you come up with passwords that no one can guess but that you can remember? Most security experts say that the best passwords don't correspond to any words in the English language but consist of a random sequence of letters, numbers, and special characters. Yet, how in the heck are you supposed to memorize a password like Dks4%DJ2? Especially when you have to change it three weeks later to something like 3pQ&X(d8.

TIP

Here's a compromise solution that enables you to create passwords that consist of two four-letter words back to back. Take your favorite book (if it's this one, you need to get a life) and turn to any page at random. Find the first four- or five-letter word on the page. Suppose that word is *When*. Then repeat the process to find another four- or five-letter word; say you pick the word *Most* the second time. Now combine the words to make your password: WhenMost. I think you'll agree that WhenMost is easier to remember than 3PQ&X(D8 and is probably just about as hard to guess. I probably wouldn't want the folks at the Los Alamos Nuclear Laboratory using this scheme, but it's good enough for most of us.

Here are additional thoughts on concocting passwords from your favorite book:

- If the words end up being the same, pick another word. And pick different words if the combination seems too commonplace, such as WestWind or FootBall.
- For an interesting variation, insert a couple of numerals or special characters between the words. You end up with passwords like into#cat, ball3%and, or tree47wing. If you want, use the page number of the second word as a separator. For example, if the words are *know* and *click* and the second word comes from page 435, use know435click.
- To further confuse your friends and enemies, use medieval passwords by picking words from Chaucer's *Canterbury Tales.* Chaucer is a great source for passwords because he lived before the days of word processors with spell-checkers. He wrote *seyd* instead of *said, gret* instead of *great, welk* instead of *walked, litel* instead of *little.* And he used lots of seven-letter and eight-letter words suitable for passwords, such as *glotenye* (gluttony), *benygne* (benign), and *opynyoun* (opinion). And he got A's in English.

TIP

- If you use any of these password schemes and someone breaks into your network, don't blame me. You're the one who's too lazy to memorize D#Sc$h4@bb3xaz5.
- If you do decide to go with passwords, such as KdI22UR3xdkL, you can find random password generators on the Internet. Just go to a search engine, such as Google, and search for Password Generator. You'll find web pages that generate random passwords based on criteria that you specify, such as how long the password should be, whether it should include letters, numbers, punctuation, uppercase and lowercase letters, and so on.

TIP

Recent research is suggesting that much of what we've believed about password security for the last 30 or so years may actually be counterproductive. Why? Two reasons:

- The requirement to change passwords frequently and making them too complicated to memorize simply encourages users to write their passwords down, which makes them easy to steal.
- A common way that passwords are compromised is by theft of the encrypted form of the password database, which can then be attacked using simple dictionary methods. Even the most complex passwords can be cracked using a dictionary attack if the password is relatively short; the most important factor in making passwords difficult to crack is not complexity but length.

As a result, the National Institute for Standards and Technology (NIST) recommends new guidelines for creating secure passwords:

- Encourage longer passwords.
- Drop the complexity requirement. Instead, encourage users to create passwords that they can easily remember. A simple sentence or phrase consisting of ordinary words will suffice, as long as the sentence or phrase is long. For example, "My password is a simple sentence" would make a good password.
- Drop the requirement to change passwords periodically; it only encourages users to write down their passwords.

Old ways are difficult to change, and it will take a while for these new guidelines to catch on. Personally, I wouldn't drop the requirement to change passwords periodically without also increasing the minimum length to at least 12 characters.

Secure the Administrator account

It stands to reason that at least one network user must have the authority to use the network without any of the restrictions imposed on other users. This user is the *administrator*. The administrator is responsible for setting up the network's security system. To do that, the administrator must be exempt from all security restrictions.

WARNING

Many networks automatically create an administrator user account when you install the network software. The username and password for this initial administrator are published in the network's documentation and are the same for all networks that use the same network operating system. One of the first things that you must do after getting your network up and running is to change the password for this standard administrator account. Otherwise, your elaborate security precautions are a complete waste of time. Anyone who knows the default administrator username and password can access your system with full administrator rights and privileges, thus bypassing the security restrictions that you so carefully set up.

WARNING

Don't forget the password for the administrator account! If a network user forgets his or her password, you can log on as the supervisor and change that user's password. If you forget the administrator's password, though, you're stuck.

Managing User Security

User accounts are the backbone of network security administration. Through the use of user accounts, you can determine who can access your network as well as what network resources each user can and can't access. You can restrict access to the network to just specific computers or to certain hours of the day. In addition, you can lock out users who no longer need to access your network. The following sections describe the basics of setting up user security for your network.

User accounts

Every user who accesses a network must have a *user account.* User accounts allow the network administrator to determine who can access the network and what network resources each user can access. In addition, the user account can be customized to provide many convenient features for users, such as a personalized Start menu or a display of recently used documents.

Every user account is associated with a *username* (sometimes called a *user ID*), which the user must enter when logging on to the network. Each account also has other information associated with it. In particular:

- **The user's password:** This also includes the password policy, such as how often the user has to change his or her password, how complicated the password must be, and so on.
- **The user's contact information:** This includes full name, phone number, email address, mailing address, and other related information.
- **Account restrictions:** This includes restrictions that allow the user to log on only during certain times of the day. This feature can restrict your users to normal working hours so that they can't sneak in at 2 a.m. to do unauthorized work. This feature also discourages your users from working overtime because they can't access the network after hours, so use it judiciously. You can also specify that the user can log on only at certain computers.
- **Account status:** You can temporarily disable a user account so the user can't log on.
- **Home directory:** This specifies a shared network folder where the user can store documents.
- **Dial-in permissions:** These authorize the user to access the network remotely via a dialup connection.
- **Group memberships:** These grant the user certain rights based on groups to which she belongs.

TIP

For more information, see the section, "Group therapy," later in this chapter.

Built-in accounts

Most network operating systems come preconfigured with two built-in accounts, Administrator and Guest. In addition, some server services, such as web or database servers, create their own user accounts under which to run. The following sections describe the characteristics of these accounts.

- **The Administrator account:** The Administrator account is the King of the Network. This user account isn't subject to any of the account restrictions to which mere mortal accounts must succumb. If you log on as the administrator, you can do anything. For this reason, avoid using the Administrator account for routine tasks. Log in as the Administrator only when you really need to.

  TIP

  Because the Administrator account has unlimited access to your network, it's imperative that you secure it immediately after you install the server. When the operating system Setup program asks for a password for the Administrator account, start with a good random mix of uppercase and lowercase letters, numbers, and symbols. Don't pick some easy-to-remember password to get started, thinking you'll change it to something more cryptic later. You'll forget, and in the meantime, someone will break in and reformat the server's C: drive or steal your customer's credit card numbers.

- **The Guest account:** Another commonly created default account is the *Guest account.* This account is set up with a blank password and — if any — access rights. The Guest account is designed to allow anyone to step up to a computer and log on, but after they do, it then prevents them from doing anything. Sounds like a waste of time to me. I suggest you disable the Guest account.

- **Service accounts:** Some network users aren't actual people. I don't mean that some of your users are subhuman. Rather, some users are actually software processes that require access to secure resources, and therefore, require user accounts. These user accounts are usually created automatically for you when you install or configure server software.

  For example, when you install Microsoft's web server (IIS), an Internet user account called IUSR is created. The complete name for this account is `IUSR_<servername>`. So if the server is named WEB1, the account is named `IUSR_WEB1`. IIS uses this account to allow anonymous Internet users to access the files of your website.

  TIP

  Don't mess with these accounts unless you know what you're doing. For example, if you delete or rename the IUSR account, you must reconfigure IIS to use the changed account. If you don't, IIS will deny access to anyone trying to reach your site. (Assuming that you *do* know what you're doing, renaming these accounts can increase your network's security. However, don't start playing with these accounts until you've researched the ramifications.)

User rights

User accounts and passwords are the front line of defense in the game of network security. After a user accesses the network by typing a valid user ID and password, the second line of security defense — *rights* — comes into play.

In the harsh realities of network life, all users are created equal, but some users are more equal than others. The Preamble to the Declaration of Network Independence contains the statement "We hold these truths to be self-evident, that *some* users are endowed by the network administrator with certain inalienable rights. . . ."

The rights that you can assign to network users depend on which network operating system you use. These are some of the possible user rights for Windows servers:

- **Log on locally:** The user can log on to the server computer directly from the server's keyboard.
- **Change system time:** The user can change the time and date registered by the server.
- **Shut down the system:** The user can perform an orderly shutdown of the server.
- **Back up files and directories:** The user can perform a backup of files and directories on the server.
- **Restore files and directories:** The user can restore backed-up files.
- **Take ownership of files and other objects:** The user can take over files and other network resources that belong to other users.

NetWare has a similar set of user rights.

Permissions (who gets what)

User rights control what a user can do on a network-wide basis. *Permissions* enable you to fine-tune your network security by controlling access to specific network resources, such as files or printers, for individual users or groups. For example, you can set up permissions to allow users into the accounting department to access files in the server's `\ACCTG` directory. Permissions can also enable some users to read certain files but not modify or delete them.

NETWORK RIGHTS WE WANT TO SEE

The network rights allowed by most operating systems are pretty boring. Here are a few rights I wish would be allowed:

- **Cheat:** Provides a special option that enables you to see what cards the other players are holding when you're playing Hearts.
- **Spy:** Eavesdrops on other users' Internet sessions so you can find out what websites they're viewing.
- **Grumble:** Automatically sends email messages to other users that explain how busy, tired, or upset you are.
- **Set pay:** Grants you special access to the payroll system so that you can give yourself a pay raise.
- **Sue:** In America, everyone has the right to sue. So this right should be granted automatically to all users.
- **Fire:** Wouldn't it be great if the network could grant you the right to play Donald Trump and fire your annoying co-workers?

Each network operating system manages permissions in a different way. Whatever the details, the effect is that you can give permission to each user to access certain files, folders, or drives in certain ways. For example, you might grant a user full access to some files but grant read-only access to other files.

TIP

Any permissions you specify for a folder apply automatically to any of that folder's subfolders, unless you explicitly specify different permissions for the subfolder.

TECHNICAL STUFF

You can use Windows permissions only for files or folders that are created on drives formatted as NTFS or ReFS volumes. If you insist on using FAT or FAT32 for your Windows shared drives, you can't protect individual files or folders on the drives. This is one of the main reasons for using NTFS for your Windows servers.

Group therapy

A *group account* is an account that doesn't represent an individual user. Instead, it represents a group of users who use the network in a similar way. Instead of granting access rights to each of these users individually, you can grant the rights to the group and then assign individual users to the group. When you assign a user to a group, that user inherits the rights specified for the group.

For example, suppose that you create a group named Accounting for the accounting staff and then allow members of the Accounting group access to the network's accounting files and applications. Then, instead of granting each accounting user access to those files and applications, you simply make each accounting user a member of the Accounting group.

Here are a few additional details about groups:

- Groups are one of the keys to network management nirvana. As much as possible, avoid managing network users individually. Instead, clump them into groups and manage the groups. When all 50 users in the accounting department need access to a new file share, would you rather update 50 user accounts or just 1 group account?
- A user can belong to more than one group. Then, the user inherits the rights of each group. For example, you can have groups set up for Accounting, Sales, Marketing, and Finance. A user who needs to access both Accounting and Finance information can be made a member of both groups. Likewise, a user who needs access to both Sales and Marketing information can be made a member of both the Sales and Marketing groups.
- You can grant or revoke specific rights to individual users to override the group settings. For example, you may grant a few extra permissions for the manager of the accounting department. You may also impose a few extra restrictions on certain users.

User profiles

User profiles are a Windows feature that keeps track of an individual user's preferences for his or her Windows configuration. For a non-networked computer, profiles enable two or more users to use the same computer, each with his or her own desktop settings, such as wallpaper, colors, Start menu options, and so on.

The real benefit of user profiles becomes apparent when profiles are used on a network. A user's profile can be stored on a server computer and accessed whenever that user logs on to the network from any Windows computer on the network.

The following are some of the elements of Windows that are governed by settings in the user profile:

- Desktop settings from the Display Properties dialog box, including wallpaper, screen savers, and color schemes
- Start menu programs and Windows toolbar options

- Favorites, which provide easy access to the files and folders that the user accesses often
- Network settings, including drive mappings, network printers, and recently visited network locations
- Application settings, such as option settings for Microsoft Word
- The Documents folder

Logon scripts

A *logon script* is a batch file that runs automatically whenever a user logs on. Logon scripts can perform several important logon tasks for you, such as mapping network drives, starting applications, synchronizing the client computer's time-of-day clock, and so on. Logon scripts reside on the server. Each user account can specify whether to use a logon script and which script to use.

This sample logon script maps a few network drives and synchronizes the time:

```
net use m: \\MYSERVER\Acct
net use n: \\MYSERVER\Admin
net use o: \\MYSERVER\Dev
net time \\MYSERVER /set /yes
```

Logon scripts are a little out of vogue because most of what a logon script does can be done via user profiles. Still, many administrators prefer the simplicity of logon scripts, so they're still used even on Windows 2019 Server systems.

Securing the Human Firewall

Security techniques, such as physical security, user account security, server security, and locking down your servers are child's play compared with the most difficult job of network security: securing your network's users. All the best-laid security plans will go for naught if your users write their passwords on sticky notes and post them on their computers.

The key to securing your network users is to create a written network security policy and to stick to it. Have a meeting with everyone to go over the security policy to make sure that everyone understands the rules. Also, make sure to have consequences when violations occur.

Here are some suggestions for some basic security rules that can be incorporated into your security policy:

- Never write down your password or give it to someone else.
- Accounts shouldn't be shared. Never use someone else's account to access a resource that you can't access under your own account. If you need access to some network resource that isn't available to you, formally request access under your own account.
- Likewise, never give your account information to a co-worker so that he or she can access a needed resource. Your co-worker should instead formally request access under his or her own account.
- Don't install any software or hardware on your computer without first obtaining permission. This especially includes wireless access devices or modems.
- Don't enable file and printer sharing on workstations without first getting permission.
- Never attempt to disable or bypass the network's security features.

- » Examining the different types of firewalls
- » Using the built-in Windows firewall
- » Looking at virus protection
- » Patching your computers

Chapter 20
Hardening Your Network

If your network is connected to the Internet, a whole host of security issues bubble to the surface. You probably connected your network to the Internet so that your network's users could get out to the Internet. Unfortunately, however, your Internet connection is a two-way street. Not only does it enable your network's users to step outside the bounds of your network to access the Internet, but it also enables others to step in and access your network.

And step in they will. The world is filled with hackers who are looking for networks like yours to break into. They may do it just for the fun of it, or they may do it to steal your customer's credit card numbers or to coerce your mail server into sending thousands of spam messages on their behalf. Whatever their motive, rest assured that your network will be broken into if you leave it unprotected.

This chapter presents an overview of three basic techniques for securing your network's Internet connection: controlling access via a firewall, detecting viruses with antivirus software, and fixing security flaws with software patches.

Firewalls

A *firewall* is a security-conscious router that sits between the Internet and your network with a single-minded task: preventing *them* from getting to *you*. The firewall acts as a security guard between the Internet and your local area network (LAN).

All network traffic into and out of the LAN must pass through the firewall, which prevents unauthorized access to the network.

WARNING

Some type of firewall is a must-have if your network has a connection to the Internet, whether that connection is residential or business-class broadband (cable modem or DSL) or enterprise-grade fiber. Without a firewall, hackers will quickly discover your unprotected network. Within a few hours your network will be toast.

The most common way to set up a firewall is to purchase a *firewall appliance,* which is basically a self-contained router with built-in firewall features. Most firewall appliances include a web-based interface that enables you to connect to the firewall from any computer on your network using a browser. You can then customize the firewall settings to suit your needs.

Alternatively, you can set up a server computer to function as a firewall computer. The server can run just about any network operating system, but most dedicated firewall systems run Linux. However, this alternative is less commonly used.

Whether you use a firewall appliance or a firewall computer, the firewall must be located between your network and the Internet, as shown in Figure 20-1. Here, one end of the firewall is connected to a network hub, which is, in turn, connected to the other computers on the network. The other end of the firewall is connected to the Internet. As a result, all traffic from the LAN to the Internet and vice versa must travel through the firewall.

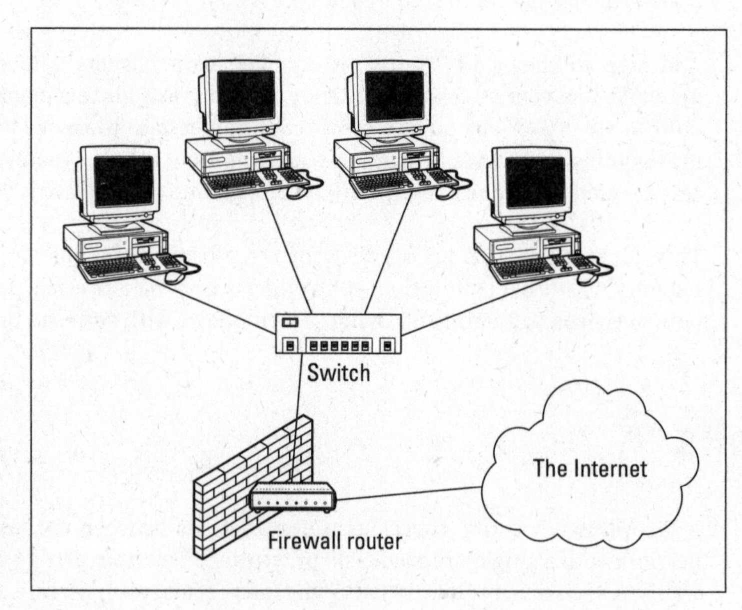

FIGURE 20-1: A firewall router creates a secure link between a network and the Internet.

THE BUILT-IN WINDOWS FIREWALL

All versions of Windows since Windows XP come with a built-in packet-filtering firewall. If you don't have a separate firewall router, you can use this built-in firewall to provide a basic level of protection. See Chapter 9 for the steps to follow to configure the Windows Firewall.

Do *not* rely on the built-in Windows firewall as your sole source of firewall protection. Although it is a good second line of defense, the built-in Windows firewall is not nearly as capable as a dedicated firewall appliance. Your computers will be much safer behind a bona-fide firewall.

The term *perimeter* is sometimes used to describe the location of a firewall on your network. In short, a firewall is like a perimeter fence that completely surrounds your property and forces all visitors to enter through the front gate.

WARNING

In large networks — especially campus-wide or even metropolitan networks — it's sometimes hard to figure out exactly where the perimeter is located. If your network has two or more wide area network (WAN) connections, make sure that every one of those connections connects to a firewall and not directly to the network. You can do this by providing a separate firewall for each WAN connection or by using a firewall with more than one WAN port.

The Many Types of Firewalls

Firewalls employ four basic techniques to keep unwelcome visitors out of your network. The following sections describe these basic firewall techniques.

Packet filtering

A *packet-filtering* firewall examines each packet that crosses the firewall and tests the packet according to a set of rules that you set up. If the packet passes the test, it's allowed to pass. If the packet doesn't pass, it's rejected.

Packet filters are the least expensive type of firewall. As a result, packet-filtering firewalls are very common. However, packet filtering has a number of flaws that knowledgeable hackers can exploit. As a result, packet filtering by itself doesn't make for a fully effective firewall.

Packet filters work by inspecting the source and destination IP and port addresses contained in each TCP/IP packet. *TCP/IP ports* are numbers that are assigned to specific services that help to identify for which service each packet is intended. For example, the port number for the HTTP protocol is 80. As a result, any incoming packets headed for an HTTP server will specify port 80 as the destination port.

Port numbers are often specified with a colon following an IP address. For example, the HTTP service on a server whose IP address is 192.168.10.133 would be 192.168.10.133:80.

Literally thousands of established ports are in use. Table 20-1 lists a few of the most popular ports.

TABLE 20-1 Some Well-Known TCP/IP Ports

Port	Description
20	File Transfer Protocol (FTP)
21	File Transfer Protocol (FTP)
22	Secure Shell Protocol (SSH)
23	Telnet
25	Simple Mail Transfer Protocol (SMTP)
53	Domain Name Server (DNS)
80	World Wide Web (HTTP)
110	Post Office Protocol (POP3)
119	Network News Transfer Protocol (NNTP)
137	NetBIOS Name Service
138	NetBIOS Datagram Service
139	NetBIOS Session Service
143	Internet Message Access Protocol (IMAP)
161	Simple Network Management Protocol (SNMP)
194	Internet Relay Chat (IRC)
389	Lightweight Directory Access Protocol (LDAP)
396	NetWare over IP
443	HTTP over TLS/SSL (HTTPS)

The rules that you set up for the packet filter either permit or deny packets that specify certain IP addresses or ports. For example, you may permit packets that are intended for your mail server or your web server and deny all other packets. Or, you may set up a rule that specifically denies packets that are heading for the ports used by NetBIOS. This rule keeps Internet hackers from trying to access NetBIOS server resources, such as files or printers.

One of the biggest weaknesses of packet filtering is that it pretty much trusts that the packets themselves are telling the truth when they say who they're from and who they're going to. Hackers exploit this weakness by using a hacking technique called *IP spoofing*, in which they insert fake IP addresses in packets that they send to your network.

Another weakness of packet filtering is that it examines each packet in isolation, without considering what packets have gone through the firewall before and what packets may follow. In other words, packet filtering is *stateless*. Rest assured that hackers have figured out how to exploit the stateless nature of packet filtering to get through firewalls.

In spite of these weaknesses, packet filter firewalls have several advantages that explain why they're commonly used:

- **Packet filters are very efficient.** They hold up each inbound and outbound packet for only a few milliseconds while they look inside the packet to determine the destination and source ports and addresses. After these addresses and ports have been determined, the packet filter quickly applies its rules and either sends the packet along or rejects it. In contrast, other firewall techniques have a more noticeable performance overhead.
- **Packet filters are almost completely transparent to users.** The only time a user will be aware that a packet filter firewall is being used is when the firewall rejects packets. Other firewall techniques require that clients and/or servers be specially configured to work with the firewall.
- **Packet filters are inexpensive.** Most routers include built-in packet filtering.

Stateful packet inspection (SPI)

Stateful packet inspection (SPI) is a step up in intelligence from simple packet filtering. A firewall with SPI looks at packets in groups rather than individually. It keeps track of which packets have passed through the firewall and can detect patterns that indicate unauthorized access. In some cases, the firewall may hold on to packets as they arrive until the firewall has gathered enough information to make a decision about whether the packets should be authorized or rejected.

Stateful packet inspection was once found only on expensive, enterprise-level routers. Now, however, SPI firewalls are affordable enough for small- or medium-sized networks to use.

Circuit-level gateway

A *circuit-level gateway* manages connections between clients and servers based on TCP/IP addresses and port numbers. After the connection is established, the gateway doesn't interfere with packets flowing between the systems.

For example, you could use a Telnet circuit-level gateway to allow Telnet connections (port 23) to a particular server and prohibit other types of connections to that server. After the connection is established, the circuit-level gateway allows packets to flow freely over the connection. As a result, the circuit-level gateway can't prevent a Telnet user from running specific programs or using specific commands.

Application gateway

An *application gateway* is a firewall system that's more intelligent than a packet-filtering, stateful packet inspection, or circuit-level gateway firewall. Packet filters treat all TCP/IP packets the same. In contrast, application gateways know the details about the applications that generate the packets that pass through the firewall. For example, a web application gateway is aware of the details of HTTP packets. As a result, it can examine more than just the source and destination addresses and ports to determine whether the packets should be allowed to pass through the firewall.

In addition, application gateways work as proxy servers. Simply put, a *proxy server* is a server that sits between a client computer and a real server. The proxy server intercepts packets that are intended for the real server and processes them. The proxy server can examine the packet and decide to pass it on to the real server, or it can reject the packet. Or the proxy server may be able to respond to the packet itself, without involving the real server at all.

For example, web proxies often store copies of commonly used web pages in a local cache. When a user requests a web page from a remote web server, the proxy server intercepts the request and checks to see whether it already has a copy of the page in its cache. If so, the web proxy returns the page directly to the user. If not, the proxy passes the request on to the real server.

Application gateways are aware of the details of how various types of TCP/IP servers handle sequences of TCP/IP packets, so they can make more intelligent decisions about whether an incoming packet is legitimate or is part of an attack. As a result, application gateways are more secure than simple packet-filtering firewalls, which can deal with only one packet at a time.

The improved security of application gateways, however, comes at a price. Application gateways are more expensive than packet filters, both in terms of their purchase price and in the cost of configuring and maintaining them. In addition, application gateways slow down the network performance because they do more detailed checking of packets before allowing them to pass.

Next-generation firewall

Many modern firewalls use the term *next generation* to describe new types of advanced threat-protection intelligence that are designed to watch for types of packet behavior that indicates the likelihood of malicious attack. A firewall that includes these new protections is called a *next-generation firewall,* usually abbreviated *NGFW.*

A next generation firewall performs all the functions of a standard firewall and more. Using a technique called *deep packet inspection,* next-generation firewalls look beyond the surface of data packets as they enter your network to find threats that simpler types of firewalls would overlook. Next generation firewalls can often stop malware before it ever gets into your network.

Virus Protection

Viruses are one of the most misunderstood computer phenomena around these days. What is a virus? How does it work? How does it spread from computer to computer? I'm glad you asked.

What is a virus?

Make no mistake — viruses are real. Now that most people are connected to the Internet, viruses have really taken off. Every computer user is susceptible to attacks by computer viruses, and using a network increases your vulnerability because it exposes all network users to the risk of being infected by a virus that lands on any one network user's computer.

Viruses don't just spontaneously appear out of nowhere. *Viruses* are computer programs that are created by malicious programmers who've lost a few screws and should be locked up.

What makes a virus a virus is its capability to make copies of itself that can be spread to other computers. These copies, in turn, make still more copies that spread to still more computers, and so on, ad nauseam.

Then, the virus waits patiently until something triggers it — perhaps when you type a particular command or press a certain key, when a certain date arrives, or when the virus creator sends the virus a message. What the virus does when it strikes also depends on what the virus creator wants the virus to do. Some viruses harmlessly display a "gotcha" message. Some send email to everyone it finds in your address book. Some wipe out all the data on your hard drive. Ouch.

A few years back, viruses moved from one computer to another by latching themselves onto floppy disks. Whenever you borrowed a floppy disk from a buddy, you ran the risk of infecting your own computer with a virus that may have stowed away on the disk.

Nowadays, virus programmers have discovered that email is a much more efficient method to spread their viruses. Typically, a virus masquerades as a useful or interesting email attachment, such as instructions on how to make $1,000,000 in your spare time, pictures of naked celebrities, or a Valentine's Day greeting from your long-lost sweetheart. When a curious but unsuspecting user double-clicks the attachment, the virus springs to life, copying itself onto the user's computer and, in some cases, sending copies of itself to all the names in the user's address book.

After the virus has worked its way onto a networked computer, the virus can then figure out how to spread itself to other computers on the network.

Here are some more tidbits about protecting your network from virus attacks:

» The term *virus* is often used to refer not only to true virus programs (which can replicate themselves) but also to any other type of program that's designed to harm your computer. These programs include so-called *Trojan horse* programs that usually look like games but are, in reality, hard drive formatters.

» A *worm* is similar to a virus, but it doesn't actually infect other files. Instead, it just copies itself onto other computers on a network. After a worm has copied itself onto your computer, there's no telling what it may do there. For example, a worm may scan your hard drive for interesting information, such as passwords or credit card numbers, and then email them to the worm's author.

- Computer virus experts have identified several thousand "strains" of viruses. Many of them have colorful names, such as the I Love You virus, the Stoned virus, and the Michelangelo virus.
- Antivirus programs can recognize known viruses and remove them from your system, and they can spot the telltale signs of unknown viruses. Unfortunately, the idiots who write viruses aren't idiots (in the intellectual sense), so they're constantly developing new techniques to evade detection by antivirus programs. New viruses are frequently discovered, and antivirus programs are periodically updated to detect and remove them.

Antivirus programs

The best way to protect your network from virus infection is to use an antivirus program. These programs have a catalog of several thousand known viruses that they can detect and remove. In addition, they can spot the types of changes that viruses typically make to your computer's files, thus decreasing the likelihood that some previously unknown virus will go undetected.

Newer versions of Windows (8 and later) include built-in antivirus protection. For older versions, you can download an excellent free antivirus solution from Microsoft called Microsoft Security Essentials. Popular alternatives to Microsoft's built-in or free antivirus protection include Norton AntiVirus, Webroot SecureAnywhere Antivirus, and Kaspersky Antivirus.

The people who make antivirus programs have their fingers on the pulse of the virus world and often release updates to their software to combat the latest viruses. Because virus writers are constantly developing new viruses, your antivirus software is next to worthless unless you keep it up-to-date by downloading the latest updates.

The following are several approaches to deploying antivirus protection on your network:

- You can install antivirus software on each network user's computer. This technique would be the most effective if you could count on all your users to keep their antivirus software up-to-date. Because that's an unlikely proposition, you may want to adopt a more reliable approach to virus protection.
- Managed antivirus services place antivirus client software on each client computer in your network. Then, an antivirus server automatically updates the clients on a regular basis to make sure that they're kept up to date.

- Server-based antivirus software protects your network servers from viruses. For example, you can install antivirus software on your mail server to scan all incoming mail for viruses and remove them before your network users ever see them.
- Some firewall appliances include antivirus enforcement checks that don't allow your users to access the Internet unless their antivirus software is up to date. This type of firewall provides the best antivirus protection available.

Safe computing

Besides using an antivirus program, you can take a few additional precautions to ensure virus-free computing. If you haven't talked to your kids about these safe-computing practices, you had better do so soon.

- Regularly back up your data. If a virus hits you and your antivirus software can't repair the damage, you may need the backup to recover your data. Make sure that you restore from a backup that was created before you were infected by the virus!
- If you buy software from a store and discover that the seal has been broken on the disk package, take the software back. Don't try to install it on your computer. You don't hear about tainted software as often as you hear about tainted beef, but if you buy software that's been opened, it may well be laced with a virus infection.
- Use your antivirus software to scan your disk for virus infection after your computer has been to a repair shop or worked on by a consultant. These guys don't intend harm, but they occasionally spread viruses accidentally, simply because they work on so many strange computers.
- Don't open email attachments from people you don't know or attachments you weren't expecting.
- Use your antivirus software to scan any CD-ROM or flash drive that doesn't belong to you before you access any of its files.

Patching Things Up

One of the annoyances that every network manager faces is applying software patches to keep the operating system and other software up to date. A software *patch* is a minor update that fixes the small glitches that crop up from time to time, such as minor security or performance issues. These glitches aren't significant enough to merit a new version of the software, but they're important enough to require fixing. Most of the patches correct security flaws that computer hackers have uncovered in their relentless attempts to prove that they are smarter than the security programmers at Microsoft.

Periodically, all the recently released patches are combined into a *service pack*. Although the most diligent network administrators apply all patches when they're released, many administrators just wait for the service packs:

TIP

- » For all versions of Windows, you can use the built-in Windows Update feature to apply patches to keep your operating system and other Microsoft software up to date. Windows Update scans your computer's software and creates a list of software patches and other components that you can download and install. To use Windows Update, open the Control Panel, click System and Security, and then click Windows Update.
- » You can configure Windows Update to automatically notify you of updates so you don't have to remember to check for new patches.
- » You can subscribe to a service that automatically sends you email to let you know of new patches and updates.

TIP

Keeping a large network patched can be one of the major challenges of network administration. If you have more than a few dozen computers on your network, consider investing in server-based software that's designed to simplify the process. For example, Ivanti Patch (`www.ivanti.com`) is a server-based program that collects software patches from a variety of manufacturers and lets you create distributions that are automatically pushed out to client computers. With software like Ivanti Patch, you don't have to rely on end users to download and install patches, and you don't have to visit each computer individually to install patches.

» Setting up software to catch spam

» Tweaking your spam filters

» Reducing the amount of spam you get

Chapter 21
Securing Your Email

Spam, spam, spam, spam, spam, spam, and spam.

So goes the famous *Monty Python* sketch, in which a woman at a restaurant just wants to order something that doesn't have spam in it.

That pretty much sums up the situations with most people's inboxes these days. The legitimate emails get lost among the spam emails. Wouldn't you like to look at an inbox that wasn't filled with spam?

Nobody likes spam. You don't like it, and your users don't like it either. And believe me, they'll let you know if they're getting too much spam in their inboxes. They'll hold you personally responsible for every email with an offensive subject line, every email that tries to sell them stuff they aren't interested in, and every email that attempts to get them to provide their bank account password or credit card number.

As a network administrator, part of your job is protecting your users from spam. The holy grail of antispam is a solution that never allows a single piece of spam into anyone's inbox, but at the same time never mistakenly identifies a single legitimate piece of email as spam.

Good luck. This level of perfection doesn't exist. The best thing you can hope for is to find the right balance: a happy medium that lets only a small amount of actual spam through to users' inboxes and only occasionally misidentifies legitimate email as spam.

In this chapter, I explain what you need to know to find and deploy such a solution. I fill you in on the various kinds of spam, where spam comes from, how spammers get people's email addresses, and — most important — the many effective techniques you can employ to keep spam out of your users' inboxes.

Defining Spam

The most basic definition of *spam* is "any email that arrives in your inbox that you didn't ask for." Spam is unsolicited email. It's email that isn't welcome, email that you aren't expecting. It's email from people you don't know or haven't heard of, usually trying to sell you something you aren't interested in or couldn't possibly need, and often trying to trick you into parting with your money, your valuable personal information, or both.

One of the defining characteristics of spam is that it's sent out in bulk, often to thousands or even millions of recipients all at once. Most spam is not particularly well targeted. Instead of taking the time to figure out who might be interested in a particular product, spammers find it easier and cheaper to pitch their products to every email address they can get their hands on.

Spam is often compared to junk mail of the physical kind — the brochures, catalogs, and other solicitations that show up in your mailbox every day. In fact, spam is often called "junk email."

However, there is a crucial difference between physical junk mail and junk email. With physical junk mail, the sender must pay the cost of postage. As a result, even though junk mail can be annoying, most junk mail is carefully targeted. Junk mailers don't want to waste their money on postage to send mail to people who aren't interested in what they have to sell. They carefully measure response rates to ensure that their mailings are profitable.

In contrast, it costs very little money to send huge numbers of emails. To be sure, spam is expensive. But the bulk of the cost of spam is borne by the recipients, who must spend time and money to receive, store, and manage the unwelcome email, and by the network providers, who must build out their networks with ever greater capacity and speed to accommodate the huge volumes of spam emails that their networks must carry.

Estimates vary, but most studies indicate that as much as three-quarters of all the email sent via the Internet is spam. At the time that I wrote this, there were indications that spam was actually becoming less common, accounting for closer to half of all the emails sent. But some organizations report that 80 percent or 90 percent of the email that they receive is actually spam.

WARNING

One thing is sure: Spam is not just annoying; it's dangerous. Besides filling up your users' inboxes with unwanted email, spam emails often carry attachments that harbor viruses or other malware, or entice your users to click links that take them to websites that can infect your network. If your network is ever taken down by a virus, there's a very good chance that the virus entered your network by way of spam.

So, understanding spam and taking precautions to block it are important parts of any network administrator's job.

Sampling the Many Flavors of Spam

Spam is unsolicited and/or unwanted email. That's a pretty broad definition, but there are several distinct categories of spam:

- **Advertisements:** Most spam is advertising from companies you've never heard of, trying to sell you products you aren't interested in. The most common type of product pitched by spam emails are pharmaceuticals, but spam also commonly promotes food supplements, knockoffs of expensive products such as watches or purses, weight-loss products, and so on.
- **Phishing emails:** Among the most annoying and dangerous types of spam are phishing emails, which try to get the recipient to divulge private information such as credit card account numbers or passwords. Phishing email masquerades as legitimate email from a bank or other well-known institution and often includes a link to a phony website that resembles the institution's actual website. For example, you might get an email informing you that there was a suspicious charge on your credit card, with a link you can click to log in to verify that the charge is legitimate. When you click the link, you're taken to a page that looks exactly like your credit card company's actual website. However, the phony page exists solely to harvest your username and password.

  Another type of phishing email includes an attachment that claims to be an unpaid invoice or a failed parcel delivery notice. The attachment contains a Trojan that attempts to infect your computer with malware.
- **Scams:** The most common type of email scam is called an *advance-fee scam,* in which you're promised a large reward or prize in the future for advancing a relatively small amount of money now, in the form of a wire transfer or money order. You may have heard of or actually received the classic scam known as the Nigerian prince scam, in which a person claiming to be a Nigerian prince needs your help to transfer a huge amount of money (for example, $40 million) but can't use an African bank account. The prince needs to use your personal bank account, and will pay you a percentage — perhaps

$1 million — for your help. But you must first open a Nigerian account with a minimum balance — of perhaps $1,000 or $10,000 — to facilitate the transfer. All you have to do is wire the money, and they'll take care of the rest.

There are many variations of this story, but they all have one thing in common: They're too good to be true. They offer you a huge amount of money later, in exchange for a relatively small amount of money now.

- **Ads for pornographic websites:** Such websites are notorious for being top sources of viruses and other malware.
- **Get-rich-quick schemes:** Pyramid schemes, multilevel marketing schemes, phony real estate schemes, you name it — they're all in a category of spam that promises to make you rich.
- **Backscatter:** Backscatter is a particularly annoying phenomenon in which your inbox becomes flooded with dozens or perhaps hundreds of nondelivery reports (NDRs), indicating that an email that you allegedly sent didn't arrive. When you examine the NDRs, you can easily determine that you never sent an email to the intended recipient. What's actually going on here is that your email address has been used as the From address in a spam campaign, and you're receiving the NDRs from the mail servers of those spam emails that were not deliverable.

TIP

Though technically not spam, many uses consider advertisements and newsletters from companies they *have* dealt with in the past to be a form of spam. An important element of the definition of *spam* is the word *unsolicited.* When you register at a company's website, you're effectively inviting that company to send you email.

Using Antispam Software

The most effective way to eliminate spam from your users' inboxes is to use antispam software. *Antispam software* examines all incoming email with the intent of distinguishing between spam and legitimate email. Depending on how the software is configured, email identified as spam is deleted, moved to a separate location, or simply marked as possible spam by adding a tag to the email's subject line.

Antispam software works by analyzing every piece of incoming email using sophisticated techniques that determine the likelihood that the email is, indeed, spam. When a certain threshold of probability is reached, the email is deemed to be spam and deleted, moved, or tagged. If the threshold is not reached, the email is passed on to the user as usual.

TIP

Microsoft Exchange mailboxes include a Junk folder that is often the ultimate destination of email identified as spam. You should always check your Junk folder whenever you can't find an email you're expecting.

Not all antispam programs use the Junk folder. Some programs store spam email outside of the user's mailbox, in a separate location on the network or perhaps on the cloud. These programs usually deliver a daily email (often called a *digest*) that lists the emails that were identified as spam. You should review this email whenever you can't find an email you're expecting.

Determining whether an email is spam is not an exact science. As a result, *false positives* (in which a legitimate piece of email is mistakenly identified as spam) and *false negatives* (in which a spam email is not detected as spam and makes it into the user's inbox) are not uncommon. False positives can result in your users not receiving emails they're expecting. False negatives can leave users scratching their heads wondering how in the world the spam filter didn't catch the spam. Sometimes email that to a human is obviously spam slips right by the antispam software.

The challenge of any antispam tool is finding the right balance of not too many false positives and not too many false negatives. Most antispam tools let you tune the filters to some degree, setting them to be more or less permissive — that is, erring on the side of more false negatives or more false positives. The stricter the filters are set, the more false positives you'll have. Loosening the filters will result in more false negatives.

TIP

The possibility of false negatives is one of the main reasons that it's rarely a good idea to configure an antispam program to simply delete spam. Most programs can be configured to delete only the most obvious spam emails — the ones that can be identified as spam with 100 percent certainty. Email that is probably spam but with less than 100 percent certainty should be marked as spam but not deleted.

Understanding Spam Filters

Antispam programs use a variety of different techniques to determine the probability of a given piece of email being spam. These techniques are employed by *filters*, which examine each piece of email; each filter uses a specific technique.

Here are some of the most commonly used filter types:

- **Keyword checking:** The most obvious way to identify spam is to look for certain words that appear either in the email's subject line or in the email body. For example, a keyword checking filter might look for profanity, sexual terms, and other words or phrases such as "Get rich quick!"

Although this is the most obvious way to identify spam, it's also the least reliable. Spammers learned long ago to leave common words out of their spams to avoid these types of filters. Often they intentionally misspell words or substitute numbers or symbols for letters, such as the numeral 0 for the letter *o*, or the symbol ! for the letter *l*.

The biggest problem with keyword checking is that it often leads to false positives. Friends and relatives might intentionally or inadvertently use any of the banned words in their emails. Sometimes, the banned words appear in the middle of otherwise completely innocent words. For example, if you list *Cialis* as a keyword that you want blocked, you'll also block the words *specialist* or *socialist*.

For these reasons, keyword filters are typically used only for the most obvious and offensive words and phrases, if they're used at all.

- **Bayesian analysis:** One of the most trusted forms of spam filtering is *Bayesian analysis,* which works by assuming that certain words occur more often in spam email than in other email. This sounds a lot like keyword checking, but Bayesian analysis is much more sophisticated than simple keyword checking. The Bayesian filter maintains an index of words that are likely to be encountered in spam emails. Each word in this index has a probability associated with it, and each word in the email being analyzed is looked up in this index to determine the overall probability of the email being spam. If the probability calculated from this index exceeds a certain threshold, the email is marked as spam.

  Here's where the magic of Bayesian analysis comes in: The index is self-learning, based on the user's actual email. Whenever the filter misidentifies an email, the user trains the filter by telling the filter that it was incorrect. The user typically does this by clicking a button labeled "This is spam" or "This is not spam." When the user clicks either of these buttons, the filter adjusts the probability associated with the words that led it to make the wrong conclusion. So, when the filter encounters a similar email in the future, it's more likely to make the correct determination.

- **Sender Policy Framework (SPF):** Surprisingly, SMTP (the Internet email protocol) has very poor built-in security. In particular, any email server can easily send email that claims to be from any domain. This makes it easy to forge the From address in an email. SPF lets you designate via DNS which specific email servers are allowed to send email from your domain. An antispam SPF filter works by looking up the sending email server against the SPF records in the DNS of the domain specified by the email's From address.

- **Blacklisting:** Another trusted form of spam filtering is a *blacklist* (also known as *blocklist*), which uses a list of known spammers to block email from sources that aren't trustworthy. There are two types of blacklists: private and public. A private blacklist is a list that you set up yourself to designate sources you don't want to accept email from. A public blacklist is a list that is maintained by a company or organization and is available for others to use.

Note that simply blacklisting a sender email address isn't much help. That's because the sender email address is easy to forge. Instead, blacklists track individual email servers that are known to be sources of spam.

Unfortunately, spammers don't usually set up their own servers to send out their spam. Instead, they hijack other servers to do their dirty work. Legitimate email servers can be hijacked by spammers and, thus, become spam sources, often without the knowledge of their owners. This raises the unfortunate possibility that your own email server might be taken over by a spammer, and you might find your email server listed on a public blacklist. If that happens, you won't be able to send email to anyone who uses that blacklist until you've corrected the problem that allowed your server to be hijacked and petitioned the blacklist owners to have your server removed.

- **Whitelisting:** One of the most important elements of any antispam solution is a *whitelist,* which ensures that email from known senders will never be blocked. Typically, the whitelist consists of a list of email addresses that you trust. When the antispam tool has confirmed that the From address in the email has not been forged (perhaps by use of an SPF filter), the whitelist filters looks up the address in the whitelist database. If the address is found, the email is immediately marked as legitimate email, and no other filters are applied. So, if the email is marked as legitimate by the whitelist filter, the other filters are not used.

Most whitelist filters will let you whitelist entire domains, as well as individual email addresses. You most certainly do *not* want to whitelist domains of large email providers such as `gmail.com` or `comcast.net`. But you should whitelist the domains of all your business partners and clients to ensure that emails from new employees at these key companies are never marked as spam.

Some antispam programs automatically add the recipient addresses of all outgoing emails to the whitelist. In other words, anyone that you send an email to is automatically added to the whitelist. Over time, this feature can drastically reduce the occurrence of false positives.

Use the whitelist to preemptively allow important email that you're expecting from new customers, vendors, or service providers. For example, if you switch payroll providers, find out in advance what email addresses the new provider will be using so that your payroll staff doesn't miss important emails.

- **Graylisting:** Graylisting is an effective antispam technique that exploits the fact that if a legitimate email server can't successfully deliver an email on its first attempt, the server will try again later, typically in 30 minutes. A graylist filter automatically rejects the first attempt to deliver a message but keeps track of the details of the message it rejected. Then, when the same message is received a second time, the graylist filter accepts the message and makes note of the sender so that future messages from the sender are accepted on the first attempt.

Graylisting works because spammers usually configure their servers to not bother with the second attempt. Thus, the graylist filter knows that if a second copy of the email arrives after the initial rejection, the mail is probably legitimate.

The drawback of graylisting is that the first time you receive an email from a new sender, the email will be delayed. Many users find that the benefit of graylisting is not worth the cost of the delayed emails, so they simply disable the graylist filter.

Looking at Three Types of Antispam Software

The many different antispam programs that are available fall into three broad categories: on-premises, appliance, and cloud based (hosted). The following sections describe the relative merits of each of these approaches to providing antispam for your organization.

On-premises antispam

An on-premises antispam program runs on a server on your network and interacts directly with your email server. Email that arrives at your server is passed over to the antispam program, which evaluates the email to determine whether it's spam or legitimate mail. The antispam software uses a variety of techniques to identify spam and can usually be configured for optimal performance. Email that is identified as legitimate is handed back to the email server for normal processing. Depending on how you configure the software, email that is identified as spam may be sent to your users' Junk folders or stored in some other location.

In smaller organizations, the antispam software can run on the same server as the email server (for example, Microsoft Exchange). In larger organizations, the antispam software can be configured to run on its own dedicated server, separate from the mail server(s).

Here are some of the advantages of using an on-premises antispam product:

- **You have complete control over the configuration and operation of the software.** Most on-premises antispam software is highly configurable, often providing a dozen or more distinct filtering methods, which you can customize in many different ways. (For more information, see the section "Understanding Spam Filters," earlier in this chapter.)

- **On-premises antispam software is usually tightly integrated not only with Microsoft Exchange but also with Microsoft Outlook.** Spam email typically appears in the users' Junk folders, and the software often provides an Outlook add-in that makes it easy for users to mark incorrectly identified email.
- **On-premises software is relatively inexpensive.** Typically, you pay an upfront fee to purchase the license, as well as an annual maintenance fee to receive regular updates not only to the software but also to the spam filters.

Here are the main disadvantages of on-premises antispam software:

- **You're responsible for installing, patching, configuring, updating, and otherwise maintaining the software.**
- **Because the relationship between the email server and the antispam software is complicated, on-premises antispam software periodically malfunctions.** Such a malfunction usually halts mail flow throughout your organization. It then becomes your responsibility to correct the problem so that mail begins flowing again. (This usually happens just at the moment when your boss is expecting an important email, and you'll find yourself diagnosing and fixing the problem while your boss watches over your shoulder.)
- **On-premises antispam software increases the workload on your servers, requiring additional resources in the form of processor time, RAM, disk storage, and network bandwidth.**

Antispam appliances

An *antispam appliance* is essentially an on-premises server in a dedicated box that you install at your location. The appliance is usually a self-contained Linux-based computer running antispam software that is pre-installed on the appliance. This makes the appliance essentially plug-and-play — you just set it up, connect it to your network, turn it on, and configure it using a simple web-based interface. When the appliance is up and running, it can provide many, if not all, of the features of on-premises antispam software.

Here are some of the main advantages of using an antispam appliance:

- Because the appliance includes its own hardware and pre-installed operating system, you don't have to worry about purchasing hardware separately, installing an operating system, installing software, or any of the other tasks associated with setting up a server.
- After it's set up, an appliance will pretty much take care of itself. You'll need to check on it once in a while, but appliances are designed to be self-sufficient.

- The appliance may provide other security features, such as antivirus and firewall protection. Thus, a single appliance can handle many of your network's security and protection needs.

Using an antispam appliance is not without its disadvantages:

- Eventually, you'll outgrow the appliance. For example, if the number of users on your network doubles, you may run out of disk space.
- If the appliance fails, you may have trouble getting it back up and running. When a normal Windows server fails, you can usually troubleshoot the problem. But because of the self-contained nature of an appliance, troubleshooting it can be difficult when it's nonresponsive.

Cloud-based antispam services

A cloud-based antispam service (also called *hosted antispam*) is an Internet-based service that filters your email before it ever arrives at your mail server. When you use hosted antispam, you reconfigure your public DNS so that your mail server (the MX record) points to the cloud-based antispam server rather that to your mail server. That way, all email sent to your organization is first processed by the servers at the antispam service before it ever arrives at your mail server. Only those emails that are deemed to be legitimate are forwarded to your mail server; spam emails are stored in the cloud, where they can be reviewed and retrieved by your users if necessary.

Typically, you pay for hosted antispam based on how many users you have. For example, you might pay a monthly fee of $2 per user. As your organization grows, you simply purchase additional subscriptions.

Here are some of the main advantages of using cloud-based antispam:

- You get to skip the hassle of installing and configuring software, integrating the software with Exchange, maintaining and patching the software, and all the other chores associated with hosting your own server on your own premises. Your monthly subscription charges cover the cost of someone else doing all that work.
- Because you don't have to buy software or hardware, there is no initial investment. You simply subscribe to the service and pay the monthly service charges. (As an added bonus, if you're dissatisfied with the service, you can easily move to a different one. Switching to a different antispam appliance or on-premises solution is a much more complicated and expensive affair.)

- A cloud-based antispam solution scales easily with your organization. If you double the number of users, you simply pay twice as much per month. You don't have to worry about running out of disk space, RAM, clock cycles, or network bandwidth.
- Cloud-based antispam takes a huge load off your network and your mail server. Because someone else filters your spam for you, spam never enters your network. In most organizations, email is one of the most taxing applications running on the network. Using cloud-based antispam can easily cut incoming network traffic in half; in some cases, it might cut traffic by as much as 90 percent.

As you would expect, there are drawbacks to using cloud-based antispam:

- You give up some control. Cloud-based services usually have fewer configuration options than on-premises software. For example, you'll probably have fewer options for customizing the spam filters.
- If the service goes down, so does your incoming email, and you won't be able to do anything about it except call technical support. Oh, and you can count on getting a busy signal, because when the service goes down, you aren't the only one affected — it's you and all the other customers. (Of course, this gives such services plenty of motivation to ensure that they fix the problem right away.)

Minimizing Spam

TIP

No antispam program is perfect, so you need to understand and expect that a certain amount of spam will get through to your inbox. Here are some tips that you (and your users) should keep in mind to minimize the amount of spam that gets through undetected:

- **Never trust email that requests your password or credit card.** A bank will *never* send you an email notifying you of a potential problem and containing a link to its online portal's login page. Nor will a credit card company ever send you an email alerting you to potential fraud and containing a link to a page that requests your credit card number to verify the transaction. Such emails may look very convincing, but you can rest assured they're fraudulent.

  If you're in doubt, do *not* click the link. Instead, open a browser window and navigate to the address you know for a fact to be the legitimate login page for your bank or credit card company's web portal.

- **Never open attachments in spam.** Attachments in a spam email almost certainly contain malware. Often, the malware in a spam email harvests all the contacts from your computer and sends them to the spammer, or hijacks your computer so the spammer can use it to send spam email.
- **Do not reply to spam.** If you reply to spam email, you merely confirm to the spammers that they've found a legitimate email address. You'll get even more spam.
- **Use your antispam program's "This is spam" feature.** If your antispam program has a "This is spam" or similar button, be sure to use it. Doing so alerts the antispam program that it has missed a spam message, which helps improve the filters the antispam program uses to detect spam.
- **Unsubscribe from legitimate but unwanted emails.** Much of what many users consider to be spam is actually mail from legitimate organizations. If the spam is from a reputable organization, it probably isn't really spam; you probably at one time signed up to receive emails from the organization. Click the unsubscribe link on these types of emails to remove yourself from the mailing list.

WARNING

Spammers often include an unsubscribe link on their spam emails. If the email is actually spam, clicking the unsubscribe link is akin to replying to the spam — it simply confirms to the spammers that they've found a legitimate email address, and you'll just get more spam. Worse yet, the link may take you to a malicious website that will attempt to install malware on your computer. So, before you click the unsubscribe link, make sure that the email is indeed from a legitimate sender.

- **Protect your email address.** Be careful who you give your email address to, especially when you fill out forms online. Make sure you give your email address only to trusted websites. And read the fine print when you sign up for an account — you'll often find check boxes that allow you to opt out of mailings such as newsletters or announcements about product updates and so on.
- **Use an alternate email address.** One useful technique to manage the amount of spam you get is to set up a free email account with a provider such as Gmail. Then use this email account for websites that require an email address for registration when you don't want to use your real email address. You can delete or change the alternate email address if it becomes the target of spam.
- **Don't publish your email address.** If you have a personal website or are on social media, don't publish your email address there. Spammers love to harvest email addresses from public websites and social media.

» Learning the Golden Rule of backups

» Understanding the different types of backups

» Considering the difference between file-level backups and backing up virtual machines

» Dealing with tape

Chapter 22 Backing Up Your Data

If you're the hapless network manager, the safety of the data on your network is your responsibility. In fact, it's your primary responsibility. You get paid to lie awake at night worrying about your data. Will it be there tomorrow? If it's not, can you get it back? And — most important — if you can't get it back, will you have a job tomorrow?

This chapter covers the ins and outs of creating a solid backup plan. No one gives out merit badges for this stuff, but someone should.

3-2-1: The Golden Rule of Backups

Having data backed up is the cornerstone of any disaster recovery plan. Without backups, a simple hard drive failure can set your company back days or even weeks while it tries to reconstruct lost data. In fact, without backups, your company's very existence is in jeopardy.

But you can't get by with haphazard, once-in-a-while backups. Instead, you need to create a comprehensive plan for backing up your data. Then, you need to follow through, making sure that your backups run according to the plan.

To get your plan started, I suggest you begin with the most widely accepted rule for creating a good backup plan, called the *3-2-1* rule:

- **Keep three copies of your data.** One copy is your primary copy, the one that is accessed every day. The other two copies are backup copies.
- **Use two different types of media.** If you keep all three copies on the same type of media, all three copies will be subject to the same risks. Therefore, you should use at least two types of media. Local disk storage is used for the primary copy, but at least one of the other two copies should be something other than disk, such as cloud-based storage or tape.
- **Keep one copy off-site.** If all three copies are stored together, a physical calamity such as a fire can destroy all three copies. Therefore, at least one of the backup copies should be stored somewhere else.

The 3-2-1 rule has been around for decades and has served us well, but I think it needs an update in light of today's environment. Here's my new version of the 3-2-1 rule:

- **Keep three backup copies of your data.** One primary and two backup copies are no longer sufficient in today's world, where threats to our data are constant and sophisticated. So I recommend you plan for three backup copies in addition to your primary data.

  TIP

  A bit later in this chapter, I discuss the concept of backup *restore points,* which refer to how many date-specific copies of your data are contained in a backup. When I say you should keep three backups, I don't mean to keep three restore points. Instead, you should keep three distinct sets of backups, each stored on different media and in different locations. Each of those three backups can and should contain more than a single restore point.
- **Keep two copies off-site.** Your main backup copy should be kept close to your primary data for fast recovery. It isn't so important that you have two distinct types of media, but it is imperative that you keep backups off-site to protect your data in the event of a physical disaster. One option for off-site backups is the cloud. The other is a backup target that is located at a different site, such as at a branch office.
- **Keep one copy offline.** Off-site is not the same as offline. These days, your off-site backup copy is likely to be either an off-site backup appliance or the cloud. As safe as these backup locations may seem, they are not completely safe from the efforts of a highly motivated and high skilled cyberattack. If all your backups are online, a hacker can break into them and delete your backups. And that would be the ultimate nightmare scenario: That you come in to work some day and find that your primary data has been erased, your

on-site backup has been erased, and the off-site backup, which you thought was secure, has also been erased.

For this reason, your third and final line of defense should be a form of backup that can be removed entirely from all networks, making it completely immune to cyberattack. In most cases, the best form of storage for this is tape.

WARNING

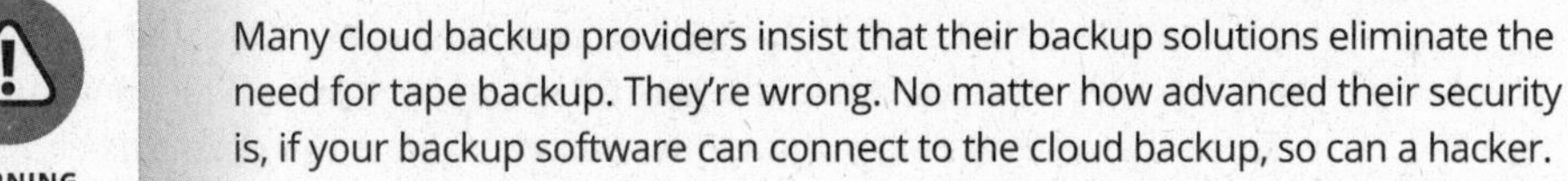

Many cloud backup providers insist that their backup solutions eliminate the need for tape backup. They're wrong. No matter how advanced their security is, if your backup software can connect to the cloud backup, so can a hacker.

Here's an example of a backup plan that implements the 3-2-1 rule:

- The primary backup is written to a network attached storage (NAS) device with at least twice the capacity of the actual amount of data being backed up. (For more information about NAS, please refer to Chapter 13.)
- The off-site backup is written to a second, identical NAS device that is located in a separate building.
- The offline backup is written to a tape device, and the tapes are removed when the backup is complete.

How Often Should You Back Up Your Data?

Another crucial factor in a backup plan is how often you should back up your data. The best way to answer *that* question is to ask *this* question: How much work can your company afford to lose?

If your company can afford to lose an entire week's worth of work, you can get by with a simple backup plan that backs up your data just once a week. If the threshold is just one day, you'll need a more robust system that can back up your data every day.

If the loss of even one day's work is intolerable, you can create a backup plan that will copy your data every hour, or even more often if necessary.

TIP

As a simple way to calculate the cost of loss, ask yourself how much it costs for your company to not be able to work for an hour, a day, and a week. For example, suppose your company is a consulting firm with 20 employees, with an average billable rate of $100 per employee. The math is simple: If no one can work for one hour, the lost income is $2,000. For an eight-hour day, the lost income is $16,000. For an entire week, the loss is $80,000. This calculation will help guide you in determining how much money you should invest in a backup solution.

An important factor to consider when determining how often you should back up your data is determining how long each backup takes to complete. This period of time is called the *backup window.* For example, if it takes two hours to back up your data, you obviously can't back up every hour.

Fortunately, you can often work around a short backup window by choosing a backup method that doesn't back up all your data every time, but just backs up the data that has changed since the last time you backed up. For more information, refer to the section "Examining File-Based Backups" later in this chapter.)

TIP

You don't have to use the same backup interval for all three layers of your backup plan. For example, your primary backup might run every hour, your off-site just once a day (typically overnight), and your off-site backup once a week (typically over the weekend).

Choosing Where to Back Up Your Data

If you plan on backing up the data on your network server's hard drives, you obviously need some type of media on which to back up the data. This media is called the *backup target* because it's the location to which your backup data is sent. (The data being backed up is called the *backup source.*)

The most common backup target options are

- **Storage Area Network (SAN):** Disk storage that is directly attached to your servers. Your network's primary data most likely exists on a SAN. (For more information about SANs, refer to Chapter 13.)

  You should *not,* under any circumstances, back up your data to the same SAN that the live data is on. The reason is simple: If the SAN fails, you'll lose not only your live data, but also your backup.

  However, you can set up a second SAN to receive your backups. Then your backup data will be on a separate device from your live data.

- **NAS:** A device that connects directly to your network. A NAS device is often used as a backup device because it's less expensive than a SAN. Depending on your needs, you can acquire an enterprise-grade NAS device that is rack-mounted, or you can purchase an inexpensive consumer-grade NAS device that is portable. (For more information about NAS, refer to Chapter 13.)
- **Backup appliance:** A disk device that is specifically designed to serve as a backup target. Backup appliances are popular because they require very little engineering work to get them set up: You just plug it in, turn it on, and then configure your backup software to send backup data to it.

Backup appliances usually include advanced features to help them pack as much backup data as possible into their disk storage. For example, most backup appliances use *de-duplication* to avoid storing identical data blocks twice. With de-duplication technology, a backup appliance can almost always shrink your data by a factor of at least 2:1, and often more like 3:1 or even 4:1 or more.

- **Cloud backup:** An increasingly popular option is to use a third-party service to back up your data to a remote location via the Internet. Cloud backup has the advantage of already being offsite.
- **Tape:** Magnetic tape, the oldest storage medium for backups, is still one of the most widely used types. One of the biggest advantages of tape backups is that tape cartridges are small and can thus be easily transported to an offsite location. (For more information about tape backup, see the section "Backing Up to Tape," coming up in just a bit.)

Establishing Two Key Backup Objectives

When determining how often to back up your data and what type of media to store your backups on, you should consider establishing two backup objectives:

- **Recovery point objective (RPO):** The point in time you want to be able to recover your data to. The RPO reflects the frequency of your backups: If you back up once per day, your RPO is one day.
- **Recovery time objective (RTO):** The amount of time it will take to recover your data in the event of a loss.

  In reality, it isn't possible to set a single RTO that applies to all types of losses, because the time required for recovery depends on the nature and extent of the data loss incident and which of your backup sets you'll be recovering from. If you're recovering from your primary backups, which reside on high-speed disk and are close to your primary storage, recovery will be relatively fast — measured in hours. On the other hand, if you're recovering from a cyberattack that corrupted your live data, your primary backup and your off-site backup, you'll have to recover your data from your offline backup, which is likely stored on much slower tape — possibly measured in days. And if you're recovering from a fire that destroyed your server room, you'll need to replace all your server equipment before you can start the recovery.

  For more information about planning for recovery from the wide variety of risks your data faces, refer to Chapter 23.

Backing Up to Tape

Another benefit of using a tape backup is that you can run it unattended. In fact, you can schedule a tape backup to run automatically during off hours when no one is using the network. For unattended backups to work, though, you must ensure that you have enough tape capacity to back up your entire network server's hard drive without having to manually switch tapes. If your network server has only 100GB of data, you can easily back it up onto a single tape. If you have 1,000GB of data, however, invest in a tape drive that features a magazine changer that can hold several tapes and automatically cycle them in and out of the drive. That way, you can run your backups unattended.

The most popular type of tape drive in use today is linear tape open (LTO). LTO tape technology is sometimes called Ultrium, though the word *Ultrium* refers to the size and shape of the cartridges that house the LTO tape rather than the tape itself.

LTO tape has gone through eight generations since its introduction in the year 2000, with each generation storing more data on a single tape. LTO tape uses built-in data compression, so the capacity of a single tape is listed both as a raw (uncompressed) capacity and an estimated compressed capacity. (The compressed capacity is an estimate because the degree of compression achieved depends on the nature of your data.)

The generations, along with their capacities and write times, are as follows:

Generation	Raw Capacity	Compressed Capacity	Time to Write Full Tape (h:mm)
LTO-1	100GB	200GB	1:25
LTO-2	200GB	400GB	1:25
LTO-3	400GB	800GB	1:25
LTO-4	800GB	1.6TB	1:50
LTO-5	1.5TB	3.0TB	3:10
LTO-6	2.5TB	6.25TB	4:35
LTO-7	6TB	15TB	5:55
LTO-8	12TB	30TB	9:15

As you can see, each generation roughly doubles the capacity of the each tape.

The total amount of time required to fill a single tape provides a useful guide for how long a tape backup job will require to complete. It may seem as if the newer tapes are slower because they take so much longer to fill the tape, but when you consider the much larger amount of data stored on the tape, they're actually much faster. For example, an LTO-5 tape takes just over 3 hours to copy 3TB of data, chugging along at about 1TB per hour. But the newest version, LTO-8, can copy 30TB of data to tape in 9 hours and 15 minutes; that's a whopping 3.25TB per hour, three times as fast as the LTO-5.

Keep in mind that many other factors can affect the speed at which data is written to the tape. Like any other network performance issue, the speed of an overall system is limited by the speed of its slowest link. No matter how fast your tape drive can theoretically write data to the tape, it won't achieve anywhere near this speed if it's connected to your servers over a slow network connection. So you should keep the tape drive as close as possible to the data it's backing up, and use the fastest possible connection between the tape drive and the data.

TIP

If the total amount of data being backed up exceeds the capacity of a single tape, you should consider a robotic tape library that can automatically change tapes when one tape becomes full. That way, you can load as many tapes as your backup will require so that the entire tape backup job can complete without the need to manually swap tapes.

Understanding Backup Software

Windows Server comes with a rudimentary backup program called Windows Server Backup (WSB). However, this backup program is not sophisticated enough for any real-world use. Here are just a few of the limitations of WSB:

- WSB can back up only to a local disk drive or to a network share. It can't back up to removable media, including tape.
- If you back up to a network share, WSB automatically overwrites the previous backup every time it runs. That means that while a WSB backup is running with a network share as its target, you have no backup at all until the backup completes.
- WSB can back up only the single server it's installed on. You can't use a single WSB user interface to back up all the servers on your network. As a result, you'll have to configure WSB separately on every server to ensure that all servers are backed up, and you'll have to log in to every server and fire up WSB to ensure that the backups are running successfully.

Because WSB has so many limitations, most organizations should use a more sophisticated backup program that is designed specifically to back up server computers.

Backup programs do more than just copy data from your hard drive to tape. Backup programs use special compression techniques to squeeze your data so that you can cram more data onto fewer tapes. Compression factors of 2:1 are common, so you can usually squeeze 100GB of data onto a tape that would hold only 50GB of data without compression. (Tape drive manufacturers tend to state the capacity of their drives by using compressed data, assuming a 2:1 compression ratio. Thus, a 200GB tape has an uncompressed capacity of 100GB.)

WARNING

Whether you achieve a compression factor of 2:1 depends on the nature of the data you're backing up:

- **Documents:** If your network is used primarily for Microsoft Office applications and is filled with Word and Excel documents, you'll probably get better than 2:1 compression.
- **Graphics:** If your network data consists primarily of graphic image files, you probably won't get much compression. Most graphic image file formats are already compressed, so they can't be compressed much more by the backup software's compression methods.

Backup programs also help you keep track of which data has been backed up and which hasn't. They also offer options, such as incremental or differential backups that can streamline the backup process, as I describe in the next section.

REMEMBER

If your network has more than one server, invest in good backup software. One popular choice is Barracuda Backup, made by BarracudaWare (`www.barracuda.com`). Besides being able to handle multiple servers, one of the main advantages of backup software is that it can properly back up Microsoft Exchange server data.

Examining File-Based Backups

One popular option for backing up your data is to use a backup program that copies individual files from your servers to the backup media of your choice. This type of software often works by installing a program called an *agent* on each of your server computers. Then, you use a single centralized program to schedule and manage the individual server backups, which are performed by the agents installed on each server.

TIP

Backup programs allow you to select any combination of drives and folders to back up. As a result, you can customize the file selection for a backup operation to suit your needs. For example, you can set up one backup schedule that backs up all of a server's shared folders and drives, but then leaves out folders that rarely change, such as the operating system folders or installed program folders. You can then back up those folders on a less-regular basis. The drives and folders that you select for a backup operation are collectively called the *backup selection.*

When using file-based backups, you can perform four distinct types of backups: Full, Copy, Incremental, and Differential. The differences among these four types of backups involve a little technical detail known as the *archive bit,* which indicates whether a file has been modified since it was backed up. The archive bit is a little flag stored along with the filename, creation date, and other directory information. Any time a program modifies a file, the archive bit is set to the On position. That way, backup programs know that the file has been modified and needs to be backed up.

Each of the four types of backups uses the archive bit in a different way:

Backup Type	Selects Files Based on Archive Bit?	Resets Archive Bits after Backing Up?
Normal	No	Yes
Copy	No	No
Incremental	Yes	Yes
Differential	Yes	No

The archive bit would have made a good Abbott and Costello routine. ("All right, I wanna know who modified the archive bit." "What." "Who?" "No, What." "Wait a minute . . . just tell me what's the name of the guy who modified the archive bit!" "Right.")

Full backups

A *full backup* is the basic type of backup. In a full backup, all files in the backup selection are backed up regardless of whether the archive bit has been set. In other words, the files are backed up even if they haven't been modified since the last time they were backed up. When each file is backed up, its archive bit is reset, so subsequent backups that select files based on the archive bit setting won't back up the files.

When a full backup finishes, none of the files in the backup selection has its archive bit set. As a result, if you immediately follow a full backup with an incremental backup or a differential backup, files won't be selected for backup by the incremental or differential backup because no file will have its archive bit set.

One simple and common backup scheme is to schedule a full backup every night. That way, all your data is backed up on a daily basis.

Copy backups

A *copy backup* is similar to a full backup except that the archive bit isn't reset when each file is copied. As a result, copy backups don't disrupt the cycle of normal and incremental or differential backups.

Copy backups are usually used for occasional one-shot backups that fall between your normal backup cycle. If you're about to perform an operating system upgrade, for example, you should back up the server before proceeding. If you do a full backup, the archive bits are reset, and your regular backups are disrupted. If you do a copy backup, however, the archive bits of any modified files remain unchanged. As a result, your regular normal and incremental or differential backups are unaffected.

If you don't incorporate incremental or differential backups into your backup routine, the difference between a copy backup and a full backup is moot.

Incremental backups

An *incremental backup* backs up only those files that were modified since the last time you did a backup. Incremental backups are a lot faster than full backups because your users probably modify only a small portion of the files on the server on any given day. As a result, fewer files are backed up in an incremental backup than in a full backup.

When an incremental backup copies each file, it resets the file's archive bit. That way, the file won't be backed up again in a subsequent incremental backup unless a user modifies the file.

Here are some thoughts about using incremental backups:

- **One common scheme for using incremental backups is the following:**
  - A *full backup* once a week, typically Friday so the full backup can complete over the weekend.
  - An *incremental* backup on each remaining normal business day (typically Monday, Tuesday, Wednesday, and Thursday).
- **When you use incremental backups, the complete backup consists of the full backup tapes and all the incremental backup tapes that you've made since you did the full backup.**

  If the hard drive crashes, and you have to restore the data onto a new drive, you first restore the full backup, and then restore all the subsequent incremental backups.

» **Incremental backups complicate restoring individual files because the most recent copy of the file may be in the full backup or any of the incremental backups.**

TECHNICAL STUFF

Backup programs keep track of the location of the most recent version of each file to simplify the process.

Differential backups

A *differential backup* is similar to an incremental backup except that it doesn't reset the archive bit when files are backed up. As a result, each differential backup represents the difference between the last full backup and the current state of the hard drive.

To do a full restore from a differential backup, you first restore the last full backup and then restore the most recent differential backup.

Suppose that you do a full backup on Friday and differential backups on Monday, Tuesday, and Wednesday, and your server crashes Friday morning. On Friday afternoon, you install a new hard drive. To restore the data, you first restore the full backup from the previous weekend. Then you restore the differential backup from Thursday. The Tuesday and Wednesday differential backups aren't needed.

The main differences between incremental and differential backups are

» *Incremental* backups result in smaller and faster backups.

» *Differential* backups are easier to restore.

Backup and Virtualization

If your servers are virtualized using either VMware or Hyper-V, you should adopt an altogether different approach to backups. Instead of creating complicated schemes of weekly full backups and daily incremental backups that are based on backing up the hundreds of thousands (or even millions) of individual files on all your servers, a virtual backup solution can focus on backing up the files that represent entire virtual machines (VMs). These files are very large, but software exists that allows you to easily and quickly replicate these files onto other media.

Virtualization platforms such as VMware and Hyper-V have built-in capabilities to manage this replication, but you can also purchase third-party solutions that can turn this replication capability into a full-fledged backup solution. For example, the Swiss-based company Veeam (`www.veeam.com`) has a powerful backup solution that is specifically designed for virtual environments. With Veeam, you can do full and incremental backups of VMs in a way that lets you recover either individual files or entire machines. One of the best features of Veeam is that you can run a virtual server directly from a backup image, without the need to first do a time-consuming restore. This can cut your recovery time from hours to minutes. And, while continuing to run the machine from the backup image, you can simultaneously restore the machine to its primary media. After the restore is completed, Veeam will automatically switch over to the restored copy of the machine.

Virtual-aware backup programs have a different set of backup types than file-based backups. For example, Veeam has five basic types of backup jobs:

- **Full backup:** A *full backup* is simply a copy of an entire VM. The VM along with all of its data are written to a single file. Naturally, a full backup can take a long time if the VM is large. Note that the first time you back up a VM using any of the Veeam backup methods, a full backup is produced to provide a starting point for the other types of backups.
- **Forward incremental backup:** Often called just an *incremental backup,* a *forward incremental backup* copies just those disk blocks in a VM that have changed since the last time the VM was backed up. Like a file-based incremental backup, a forward incremental backup creates a set of increment files that must be combined with the original full backup to restore a VM.
- **Reverse incremental backup:** A *reverse incremental backup* is similar to a forward incremental backup, but with a twist: Instead of creating a separate backup file that contains the changes since the last backup, the reverse incremental backup incorporates all changed data into the most recent full backup, and at the same time creates a separate file that holds the previous versions of all the data blocks that have changed. When reverse increments are used, the current version of the VM is always contained in the full backup file. To restore a previous version of a VM, the full backup is first restored. Then one or more of the reverse increment files are applied to the full backup to "walk back" the data until it reaches the desired restore point.
- **Synthesized full backup:** When forward incremental backups are used, the Veeam software can create a new full backup without actually transferring all the data from the source VM. Instead, it makes a copy of the most recent full backup; then it applies all the incremental backups that have been made since the most recent full backup. The result is called a *synthesized full backup* because it's identical to a regular full backup but it was made without actually copying any data from the source VM.

Most Veeam backup plans use a combination of occasional full backups followed by a sequence of full increments and periodic synthetic fulls. For example, you might employ a four-week cycle that looks something like this:

- **Week 1:** A full backup on Friday, followed by daily forward increments Monday through Thursday
- **Week 2:** A synthetic full backup on Friday, followed by daily forward increments Monday through Thursday
- **Week 3:** A synthetic full backup on Friday, followed by daily forward increments Monday through Thursday
- **Week 4:** A synthetic full backup on Friday, followed by daily forward increments Monday through Thursday

At the end of this cycle, you'll have one full backup, three synthesized full backups, and 20 forward increments. This cycle results in 24 distinct recovery points from which you can restore data.

Veeam also offers a fifth type of backup type, called a *backup copy*. When you run a backup copy, an existing Veeam backup is copied to another location, creating a backup of your backup. Backup copy jobs are used to create off-site backups at a remote location or to cloud storage, and can also be used to create a tape backup.

Verifying Tape Reliability

Speaking of tape, from painful experience I've found that although tape drives are very reliable, they do run amok once in a while. The problem is that they don't always tell you when they're not working. A tape drive can spin along for hours, pretending to back up your data — but in reality, your data isn't being written reliably to the tape. In other words, a tape drive can trick you into thinking that your backups are working just fine. Then, when disaster strikes and you need your backup tapes, you may just discover that the tapes are worthless.

TIP

Don't panic! Here's a simple way to assure yourself that your tape drive is working. Just activate the "compare after backup" feature of your backup software. As soon as your backup program finishes backing up your data, it rewinds the tape, reads each backed-up file, and compares it with the original version on the hard drive. If all files compare, you know that your backups are trustworthy.

Here are some additional thoughts about the reliability of tapes:

- » The compare-after-backup feature doubles the time required to do a backup, but that doesn't matter if your entire backup fits on one tape. You can just run the backup after hours. Whether the backup and repair operation takes one hour or ten doesn't matter, as long as it's finished by the time the network users arrive at work the next morning.
- » If your backups require more than one tape, you may not want to run the compare-after-backup feature every day. Be sure to run it periodically, however, to check that your tape drive is working.
- » If your backup program reports errors, throw away the tape, and use a new tape.
- » Actually, you should ignore that last comment about waiting for your backup program to report errors. You should discard tapes *before* your backup program reports errors. Most experts recommend that you should use a tape only about 20 times before discarding it. If you use the same tape every day, replace it monthly. If you have tapes for each day of the week, replace them twice yearly. If you have more tapes than that, figure out a cycle that replaces tapes after about 20 uses.

Keeping Backup Equipment Clean and Reliable

An important aspect of backup reliability is proper maintenance of your tape drives. Every time you back up to tape, little bits and specks of the tape rub off onto the read and write heads inside the tape drive. Eventually, the heads become too dirty to read or write data reliably.

To counteract this problem, clean the tape heads regularly. The easiest way to clean them is to use a cleaning cartridge for the tape drive. The drive automatically recognizes when you insert a cleaning cartridge and then performs a routine that wipes the cleaning tape back and forth over the heads to clean them. When the cleaning routine is done, the tape is ejected. The whole process takes only about 30 seconds.

Because the maintenance requirements of drives differ, check each drive's user's manual to find out how and how often to clean the drive. As a general rule, clean drives once weekly.

The most annoying aspect of tape drive cleaning is that the cleaning cartridges have a limited life span, and unfortunately, if you insert a used-up cleaning cartridge, the drive accepts it and pretends to clean the drive. For this reason, keep track of how many times you use a cleaning cartridge and replace it as recommended by the manufacturer.

Setting Backup Security

Backups create an often-overlooked security exposure for your network: No matter how carefully you set up user accounts and enforce password policies, if any user (including a guest) can perform a backup of the system, that user may make an unauthorized backup. In addition, your backup tapes themselves are vulnerable to theft. As a result, make sure that your backup policies and procedures are secure by taking the following measures:

- **Set up a user account for the user who does backups.** Because this user account has backup permission for the entire server, guard its password carefully. Anyone who knows the username and password of the backup account can log on and bypass any security restrictions that you place on that user's normal user ID.
- **Counter potential security problems by restricting the backup user ID to a certain client and a certain time of the day.** If you're really clever (and paranoid), you can probably set up the backup user's account so that the only program it can run is the backup program.
- **Use encryption to protect the contents of your backup tapes.**
- **Secure the backup tapes in a safe location, such as . . . um, a safe.**

IN THIS CHAPTER

» Looking at the types of disasters that can impact your organization

» Considering the impact of a disaster

» Developing a plan to survive a disaster

Chapter 23
Planning for Disaster

On April Fools' Day about 30 years ago, my colleagues and I discovered that some loser had broken into the office the night before and pounded our computer equipment to death with a crowbar. (I'm not making this up.)

Sitting on a shelf right next to the mangled piles of what used to be a Wang minicomputer system was an undisturbed disk pack that contained the only complete backup of all the information that was on the destroyed computer. The vandal didn't realize that one more swing of the crowbar would have escalated this major inconvenience into a complete catastrophe. Sure, we were up a creek until we could get the computer replaced. And in those days, you couldn't just walk into your local Computers R Us and buy a new computer off the shelf — this was a Wang minicomputer system that had to be specially ordered. After we had the new computer, though, a simple restore from the backup disk brought us right back to where we were on March 31. Without that backup, getting back on track would have taken months.

I've been paranoid about disaster planning ever since. Before then, I thought that disaster planning meant doing good backups. That's a part of it, but I can never forget the day we came within one swing of the crowbar of losing everything. Vandals are probably much smarter now: They know to smash the backup tapes, as well as the computers themselves. Being prepared for disasters entails much more than just doing regular backups.

Nowadays, the trendy term for disaster planning is a *business continuity plan* (BCP). I suppose the term *disaster planning* sounded too negative, like we were planning for disasters to happen. The new term refocuses attention on the more positive aspect of preparing a plan that will enable a business to carry on with as little interruption as possible in the event of a disaster.

For more in-depth information about this topic, please refer to *IT Disaster Recovery Planning For Dummies,* by Peter Gregory (Wiley).

Assessing Different Types of Disasters

Disasters come in many shapes and sizes. Some types of disasters are more likely than others. For example, your building is more likely to be struck by lightning than to be hit by a comet. In some cases, the likelihood of a particular type of disaster depends on where you're located. For example, crippling snowstorms are more likely in New York than in Florida.

In addition, the impact of each type of disaster varies from company to company. What may be a disaster for one company may only be a mere inconvenience for another. For example, a law firm may tolerate a disruption in telephone service for a day or two — loss of communication via phone would be a major inconvenience but not a disaster. To a telemarketing firm, however, a day or two with the phones down is a more severe problem because the company's revenue depends on the phones.

One of the first steps in developing a business continuity plan is to assess the risk of the various types of disasters that may affect your organization. Weigh the likelihood of a disaster happening with the severity of the impact that the disaster would have. For example, a meteor crashing into your building would probably be pretty severe, but the odds of that happening are miniscule. On the other hand, the odds of your building being destroyed by fire are much higher, and the consequences of a devastating fire would be about the same as those from a meteor impact.

The following sections describe the most common types of risks that most companies face. Notice throughout this discussion that although many of these risks are related to computers and network technology, some are not. The scope of business continuity planning is much larger than just computer technology.

Environmental disasters

Environmental disasters are what most people think of first when they think of disaster recovery. Some types of environmental disasters are regional. Others can happen pretty much anywhere. Here are some examples:

- **Fire:** Fire is probably the first disaster that most people think of when they consider disaster planning. Fires can be caused by unsafe conditions; carelessness, such as electrical wiring that isn't up to code; natural causes, such as lightning strikes; or arson.
- **Earthquakes:** Not only can earthquakes cause structural damage to your building, but they can also disrupt the delivery of key services and utilities, such as water and power. Serious earthquakes are rare and unpredictable, but some areas experience them with more regularity than others. If your business is located in an area known for earthquakes, your BCP should consider how your company would deal with a devastating earthquake.
- **Weather:** Weather disasters can cause major disruption to your business. Moderate weather may close transportation systems so that your employees can't get to work. Severe weather may damage your building or interrupt delivery of services, such as electricity and water.
- **Water:** Flooding can wreak havoc with electrical equipment, such as computers. If floodwaters get into your computer room, chances are good that the computer equipment will be totally destroyed. Flooding can be caused not only by bad weather but also by burst pipes or malfunctioning sprinklers.
- **Lightning:** Lightning storms can cause electrical damage to your computers and other electronic equipment from lightning strikes, as well as surges in the local power supply.

Deliberate disasters

Some disasters are the result of deliberate actions by others. For example:

- **Intentional damage:** Vandalism or arson may damage or destroy your facilities or your computer systems. The vandalism or arson may be targeted at you specifically, by a disgruntled employee or customer, or it may be random. Either way, the effect is the same.

REMEMBER

Don't neglect the possibility of sabotage. A disgruntled employee who gets hold of an administrator's account and password can do all sorts of nasty things to your network.

» **Theft:** Theft is always a possibility. You may come to work someday to find that your servers or other computer equipment have been stolen.

» **Terrorism:** Terrorism used to be something that most Americans weren't concerned about, but September 11, 2001, changed all that. No matter where you live in the world, the possibility of a terrorist attack is real.

Disruption of services

You may not realize just how much your business depends on the delivery of services and utilities. A BCP should take into consideration how you'll deal with the loss of certain services:

» **Electricity:** Electrical power is crucial for computers and other types of equipment. During a power failure once (I live in California, so I'm used to it), I discovered that I can't even work with pencil and paper because all my pencil sharpeners are electric. Electrical outages are not uncommon, but the technology to deal with them is readily available. Uninterruptible power supply (UPS) equipment is reliable and inexpensive.

» **Communications:** Communication connections can be disrupted by many causes. A few years ago, a railroad overpass was constructed across the street from my office. One day, a backhoe cut through the phone lines, completely cutting off our phone service — including our Internet connection — for a day and a half.

» **Water:** An interruption in the water supply may not shut down your computers, but it can disrupt your business by forcing you to close your facility until the water supply is reestablished.

Equipment failure

Modern companies depend on many different types of equipment for their daily operations. The failure of any of these key systems can disrupt business until the systems are repaired:

» **Computer equipment failure can obviously affect business operations.**

» **Air-conditioning systems are crucial to regulate temperatures, especially in computer rooms.** Computer equipment can be damaged if the temperature climbs too high.

» **Elevators, automatic doors, and other equipment may also be necessary for your business.**

Other disasters

You should assess many other potential disasters. Here are just a few:

- Labor disputes
- Loss of key staff because of resignation, injury, sickness, or death
- Workplace violence
- Public health issues, such as epidemics, mold infestations, and so on
- Loss of a key supplier
- Nearby disaster, such as a fire or police action across the street that results in your business being temporarily blocked off

Analyzing the Impact of a Disaster

With a good understanding of the types of disasters that can affect your business, you can turn your attention to the impact that these disasters can have on your business. The first step is to identify the key business processes that can be impacted by different types of disasters. These business processes are different for each company. For example, here are a few of the key business processes for a publishing company:

- **Editorial,** such as managing projects through the process of technical editing, copyediting, and production
- **Acquisition,** such as determining product development strategies, recruiting authors, and signing projects
- **Human resources,** such as payroll, hiring, employee review, and recruiting
- **Marketing,** including sales tracking, developing marketing materials, sponsoring sales conferences, and exhibiting at trade events
- **Sales and billing,** such as filling customer orders, maintaining the company website, managing inventory, and handling payments
- **Executive and financial,** such as managing cash flow, securing credit, raising capital, deciding when to go public, and deciding when to buy a smaller publisher or sell out to a bigger publisher

The impact of a disruption to each of these processes will vary. One common way to assess the impact of business process loss is to rate the impact of various degrees of loss for each process. For example, you may rate the loss of each process for the following time frames:

- Zero to two hours
- Two to 24 hours
- One to two days
- Two days to one week
- More than one week

For some business processes, an interruption of two hours or even one day may be minor. For other processes, even the loss of a few hours may be very costly.

Developing a Business Continuity Plan

A BCP is simply a plan for how you'll continue operation of your key business processes should the normal operation of the process fail. For example, if your primary office location is shut down for a week because of a major fire across the street, you won't have to suspend operations if you have a business continuity plan in place.

The key to a BCP is redundancy of each component that is essential to your business processes. These components include

- **Facilities:** If your company already has multiple office locations, you may be able to temporarily squeeze into one of the other locations for the duration of the disaster. If not, you should secure arrangements in advance with a real estate broker so that you can quickly arrange an alternate location. By having an arrangement made in advance, you can move into an emergency location on a moment's notice.
- **Computer equipment:** It doesn't hurt to have a set of spare computers in storage somewhere so that you can dig them out to use in an emergency. Preferably, these computers would already have your critical software installed. The next best thing would be to have detailed plans available so that your IT staff can quickly install key software on new equipment to get your business up and running.

WARNING

Always keep a current set of backup media at an alternate location.

- **Phones:** Discuss emergency phone services in advance with your phone company. If you're forced to move to another location on 24-hour notice, how quickly can you get your phones up and running? And can you arrange to have your incoming toll-free calls forwarded to the new location?
- **Staff:** Unless you work for a government agency, you probably don't have redundant employees. However, you can make arrangements in advance with a temp agency to provide clerical and administrative help on short notice.
- **Stationery:** This sounds like a small detail, but you should store a supply of all your key stationery products (letterhead, envelopes, invoices, statements, and so on) in a safe location. That way, if your main location is suddenly unavailable, you don't have to wait a week to get new letterhead or invoices printed.
- **Hard copies:** Keep a backup copy of important printed material (customer billing files, sales records, and so on) at an alternate location.

Holding a Fire Drill

Remember in grade school when the fire alarm would go off and your teacher would tell you and the other kids to calmly put down your work and walk out to the designated safe zone in an orderly fashion? Drills are important so that if a real fire occurs, you don't run and scream and climb all over each other in order to be the first one to get out.

Any disaster recovery plan is incomplete unless you test it to see whether it works. Testing doesn't mean that you should burn your building down one day to see how long it takes you to get back up and running. You should, though, periodically simulate a disaster in order to prove to yourself and your staff that you can recover.

The most basic type of disaster recovery drill is a simple test of your network backup procedures. You should periodically attempt to restore key files from your backup tapes just to make sure that you can. You achieve several benefits by restoring files on a regular basis:

- Tapes are unreliable. The only way to be sure that your tapes are working is to periodically restore files from them.
- Backup programs are confusing to configure. I've seen people run backup jobs for years that don't include all the data they think they're backing up. Only when disaster strikes and they need to recover a key file do they discover that the file isn't included in the backup.

» Restoring files can be a little confusing, especially when you use a combination of normal and incremental or differential backups. Add to that the pressure of having the head of the company watching over your shoulder while you try to recover a lost file. If you regularly conduct file restore drills, you'll familiarize yourself with the restore features of your backup software in a low-pressure situation. Then you can easily restore files for real when the pressure's on.

You can also conduct walk-throughs of more serious disaster scenarios. For example, you can set aside a day to walk through moving your entire staff to an alternate location. You can double-check that all the backup equipment, documents, and data are available as planned. If something is missing, it's better to find out now rather than while the fire department is still putting water on the last remaining hot spots in what used to be your office.

6 More Ways to Network

IN THIS PART . . .

Connect to your network remotely.

Extend your network to the cloud.

Understand the hype about hybrid cloud.

IN THIS CHAPTER

» Accessing your email with Outlook Web Access

» Using a virtual private network

» Connecting with Remote Desktop Connection

Chapter 24
Accommodating Remote Users

A typical computer user takes work home to work on in the evening or over the weekend and bring back to the office the following weekday. This arrangement can work okay, except that exchanging information between your home computer and your office computer isn't easy.

One way to exchange files is to mark them for offline access, as I describe in Chapter 4. However, this approach has its drawbacks. What if someone goes to the office on Saturday and modifies the same file you're working on at home? What if you get home and discover that the file you need is on a folder you didn't mark for offline access?

This chapter introduces two features that can alleviate these problems. The first is Internet-based access to your email via Outlook Web App (OWA) in Microsoft Exchange. The second is a *virtual private network* (VPN), which lets you connect to your network from home as though you were at work so that you can safely access all your network resources as though you were locally connected to the network.

This chapter also introduces you to a third important feature for working with computers remotely: the built-in *Remote Desktop Connection* feature, which lets you connect to a remote computer and operate the computer as if you were sitting at it. This feature opens up all kinds of opportunities for working remotely. For example, suppose you have a computer at the office that has all your important work-related software and files on it. You can access that computer from your home computer by establishing a VPN connection to your work network and then using Remote Desktop Connection to connect to your work computer. Presto! You're working from home!

Using Outlook Web App

Most people who connect to their office networks from home really just need their email. If the only reason for accessing the office network is to get email, try this simple, easy tool: Outlook Web App, also known as *OWA*. This Microsoft Exchange Server feature can access your company email from any computer that has an Internet connection. The remote computer just needs a web browser and an Internet connection; no VPN or other special configuration is required.

The best part is that you don't have to do anything special to enable OWA; it's enabled by default when you install Microsoft Exchange. Although you can configure plenty of options to improve its use, OWA is functional right out of the box.

To access OWA from any web browser, just browse to the address designated for your organization's OWA. The default address is the DNS name of your mail server, followed by `/exchange`. For example, for the mail server `smtp.lowewriter.com`, the OWA address is `smtp.lowewriter.com/exchange`.

TECHNICAL STUFF

The connection must use the secure version of the normal HTTP web protocol. You must type **https://** before the OWA address. The complete address will be something like `https://smtp.lowewriter.com/exchange`.

When you browse to your OWA address, you're prompted to enter a name and password. Use your regular network logon name and password. OWA appears in the browser window, as shown in Figure 24-1.

If you're familiar with Outlook, you'll have no trouble using OWA. Almost all Outlook features are available, including your inbox, calendar, contacts, tasks, reminders, and even public folders. You can even set up an Out of Office reply.

One difference between OWA and Outlook is that there's no menu bar across the top. However, most of the functions that are available from the menu bar are available elsewhere in OWA. If you can't find a feature, look in the Options page, which you can reach by choosing Options ⇨ Show All Options near the top right of the page. Figure 24-2 shows the Options page. From here, you can create an Out of Office reply, set your signature, and change a variety of other options.

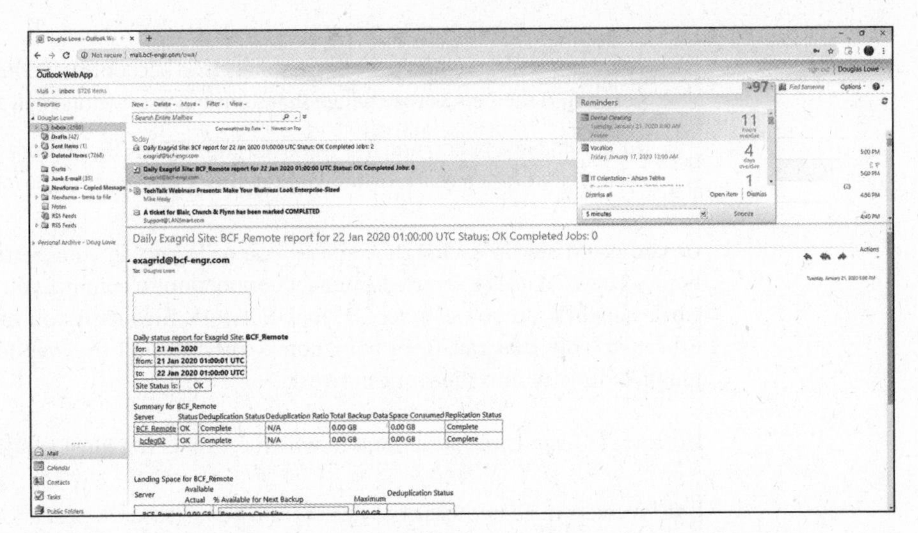

FIGURE 24-1: OWA looks a lot like Outlook.

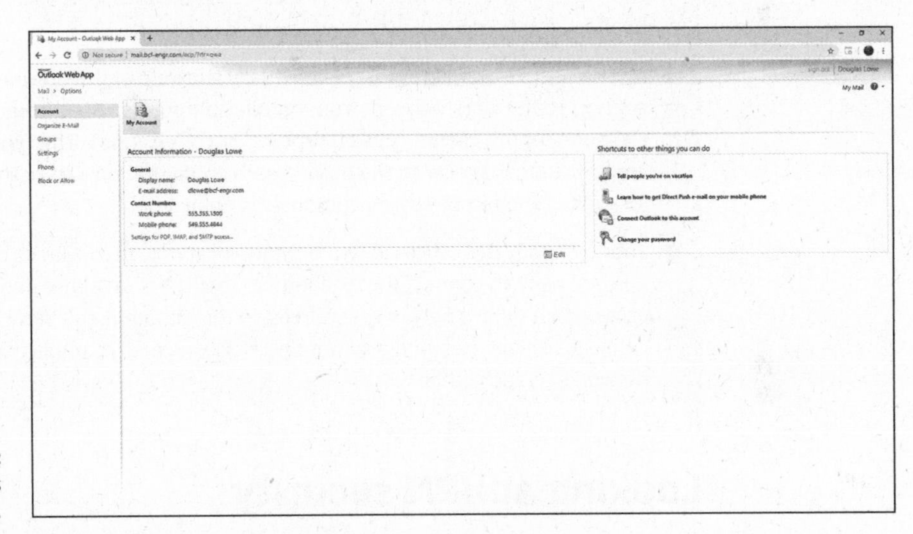

FIGURE 24-2: Set OWA options here.

Using a Virtual Private Network

A *virtual private network* (VPN) is a type of network connection that creates the illusion that you're directly connected to a network when in fact, you're not. For example, suppose you set up a LAN at your office, but you also occasionally work from home. But how will you access the files on your work computer from home?

- You could simply copy whatever files you need from your work computer onto a flash drive and take them home with you, work on the files, copy the updated files back to the flash drive, and take them back to work with you the next day.

» You could email the files to your personal email account, work on them at home, and then email the changed files back to your work email account.

» You could get a laptop and use the Windows Offline Files feature to automatically synchronize files from your work network with files on the laptop.

Or you could set up a VPN that allows you to log on to your work network from home. The VPN uses a secured Internet connection to connect you directly to your work network, so you can access your network files as if you had a really long Ethernet cable that ran from your home computer all the way to the office and plugged directly into the work network.

Here are at least three situations in which a VPN is the ideal solution:

» Workers need to occasionally work from home (as in the scenario just described). In this situation, a VPN connection establishes a secure connection between the home computer and the office network.

» Mobile users — who may not ever actually show up at the office — need to connect to the work network from mobile computers, often from locations like hotel rooms, clients' offices, airports, or coffee shops. This type of VPN configuration is similar to the home user's configuration except that the exact location of the remote user's computer is not fixed.

» Your company has offices in two or more locations, each with its own LAN, and you want to connect the locations so that users on either network can access each other's network resources. In this situation, the VPN doesn't connect a single user with a remote network; instead, it connects two remote networks to each other.

Looking at VPN security

The *V* in VPN stands for *virtual*, which means that a VPN creates the appearance of a local network connection when in fact the connection is made over a public network — the Internet. The term *tunnel* is sometimes used to describe a VPN because the VPN creates a tunnel between two locations, which can be entered only from either end. The data that travels through the tunnel from one end to the other is secure as long as it's within the tunnel — that is, within the protection provided by the VPN.

The *P* in VPN stands for *private*, which is the purpose of creating the tunnel. If the VPN didn't create effective security so that data can enter the tunnel only at one of the two ends, the VPN would be worthless; you may as well just open your network and your remote computer up to the Internet and let the hackers have their way.

Prior to VPN technology, the only way to provide private remote network connections was through direct-dial lines or dedicated private lines, which were (and still are) very expensive. For example, to set up a remote office, you could lease a private T1 line from the phone company to connect the two offices. This private T1 line provided excellent security because it physically connected the two offices and could be accessed only from the two endpoints.

VPN provides the same point-to-point connection as a private leased line, but does it over the Internet instead of through expensive dedicated lines. To create the tunnel that guarantees privacy of the data as it travels from one end of the VPN to the other, the data is encrypted using special security protocols.

The most important of the VPN security protocols is *Internet Protocol Security* (IPSec), which is a collection of standards for encrypting and authenticating packets that travel on the Internet. In other words, it provides a way to encrypt the contents of a data packet so that only a person who knows the secret encryption keys can decode the data. And it provides a way to reliably identify the source of a packet so that the parties at either end of the VPN tunnel can trust that the packets are authentic.

Another commonly used VPN protocol is Layer 2 Tunneling Protocol (L2TP). This protocol doesn't provide data encryption. Instead, it's designed to create end-to-end connections — *tunnels* — through which data can travel. L2TP is actually a combination of two older protocols: Layer 2 Forwarding Protocol (L2FP, from Cisco), and Point-to-Point Tunneling Protocol (PPTP, from Microsoft).

Many VPNs today use a combination of L2TP and IPSec: L2TP over IPSec. This type of VPN combines the best features of L2TP and IPSec to provide a high degree of security and reliability.

Understanding VPN servers and clients

A VPN connection requires a VPN *server* — the gatekeeper at one end of the tunnel — and a VPN client at the other end. The main difference between the server and the client is that the client initiates the connection with the server, and a VPN client can establish a connection with just one server at a time. However, a server can accept connections from many clients.

Typically, the VPN server is a separate hardware device, most often a security appliance such as a Cisco ASA security appliance. VPN servers can also be implemented in software. For example, Windows Server includes built-in VPN capabilities even though they're not easy to configure. And a VPN server can be implemented in Linux and macOS as well.

Figure 24-3 shows one of the many VPN configuration screens for a Cisco ASA appliance. This screen provides the configuration details for an IPSec VPN connection. The most important item of information on this screen is the Pre-Shared Key, which is used to encrypt the data sent over the VPN. The client will need to provide the identical key in order to participate in the VPN.

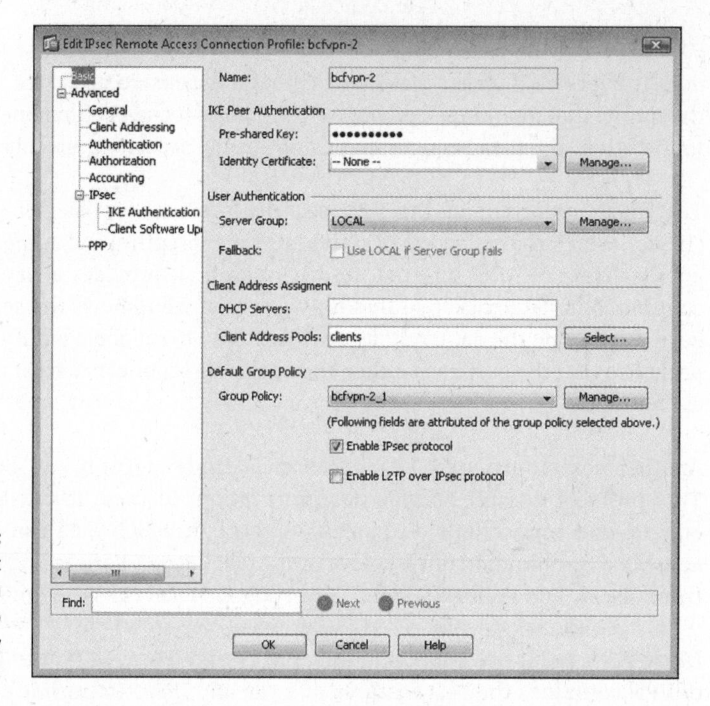

FIGURE 24-3: An IPSec configuration page on a Cisco ASA security appliance.

A VPN client is usually software that runs on a client computer that wants to connect to the remote network. The VPN client software must be configured with the IP address of the VPN server as well as authentication information such as a username and the Pre-Shared Key that will be used to encrypt the data. If the key used by the client doesn't match the key used by the server, the VPN server will reject the connection request from the client.

Figure 24-4 shows a typical VPN software client. When the client is configured with the correct connection information (which you can do by clicking the New button), you just click Connect. After a few moments, the VPN client will announce that the connection has been established and the VPN is connected.

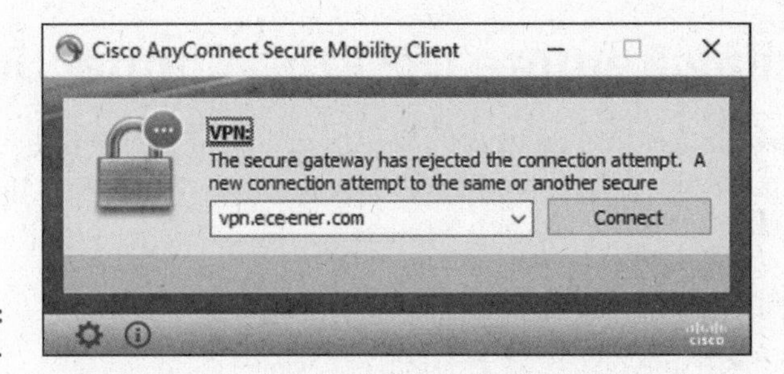

FIGURE 24-4: A VPN client.

TIP

A VPN client can also be a hardware device, like another security appliance. This is most common when the VPN is used to connect two networks at separate locations. For example, suppose your company has an office in Pixley and a second office in Hooterville. Each office has its own network with servers and client computers. The easiest way to connect these offices with a VPN would be to put an identical security appliance at each location. Then you could configure the security appliances to communicate with each other over a VPN.

Connecting with Remote Desktop Connection

Remote Desktop Connection (RDC) is designed to let you log into a Windows computer from a remote location. This is useful for accessing your work computer from home (or vice versa), as well as for managing virtual servers that have no physical console.

Note that in order to remotely connect to a computer on a domain, your computer must have access to the domain. The easiest way to accomplish that is to connect to the domain with a VPN. When you're connected with the VPN, the Remote Desktop client will be able to find the computer you're trying to connect to.

TECHNICAL STUFF

Remote Desktop Connection utilizes a protocol called *Remote Desktop Protocol* (RDP). Strictly speaking, RDP refers to the protocol and not the connection itself. However, the term *RDP* is often used as a substitute for RDC. For example, here's how you would use it in a sentence: "I don't really feel like driving to work today, so I think I'll just RDP in." If your boss approves of your plan, you can work this way for days without showering.

In the following sections, you'll learn how to enable and use Remote Desktop Connection to connect to work remotely.

Enabling Remote Desktop Connection

Before you can use RDP to access a server, you must enable remote access on the server. To do that, follow these steps (on the server computer, not your desktop computer):

1. **Open the Control Panel and click System.**

   This step brings up the System settings page.

2. **Click the Remote Settings link.**

   The remote access options appear, as shown in Figure 24-5.

3. **Select the Allow Remote Connections to This Computer radio button.**
4. **Click OK.**

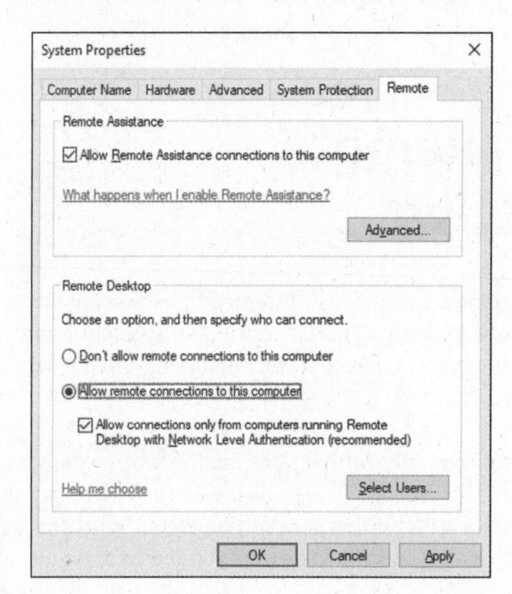

FIGURE 24-5: Configuring remote access.

You're done!

You can click the Select Users button to create a list of users who are authorized to access the computer remotely. Note that the account you're logged in with when you enable RDP is automatically granted access, so you don't have to add yourself.

WARNING

There's no question that RDP is convenient and useful. It's also inherently dangerous, however. Don't enable it unless you've taken precautions to secure your account by using a strong password. Also, you should already have a firewall installed to keep unwanted visitors out of your network. For more information on account security, see Chapter 19.

Connecting remotely

After you've enabled remote access on a server, you can connect to the server by using the remote desktop client that's automatically installed with Windows. Here's the procedure:

1. **Click the Start button, type the word** Remote**, and then click the Remote Desktop Connection icon.**

   The Remote Desktop Connection client comes to life, as shown in Figure 24-6.

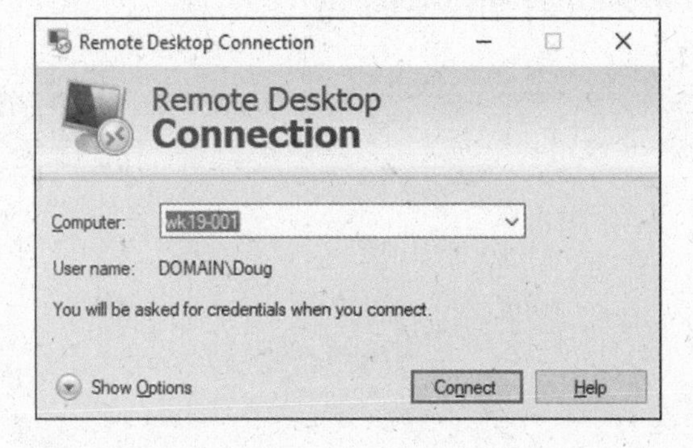

FIGURE 24-6: Connecting with Remote Desktop Connection.

2. **Enter the name of the computer you want to connect to.**

   Alternatively, you can use the drop-down list to select the computer from the list of available computers.

   You can also enter the IP address of the computer you want to connect to.

3. **Click the Connect button.**

   You're connected to the computer you selected, and then prompted for login credentials, as shown in Figure 24-7.

TIP

If you don't see this screen, but instead you're greeted by a message that says the computer couldn't be contacted, check to make sure that you entered the correct computer name or IP address. If the name or address is correct, perhaps you failed to first connect with the VPN. The Remote Desktop Client won't be able to find the computer unless it can access the domain via the VPN.

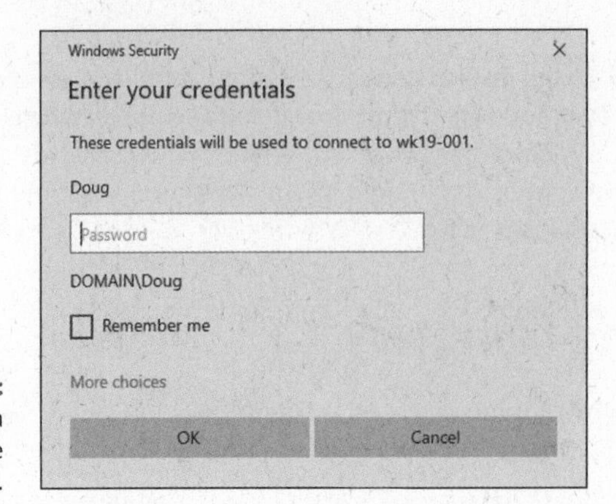

FIGURE 24-7: Logging in to a remote computer.

4. **Enter your username and password, and then click OK.**

   Assuming you enter valid credentials, the desktop of the remote computer is displayed, as shown in Figure 24-8.

5. **Now you can use the remote computer!**

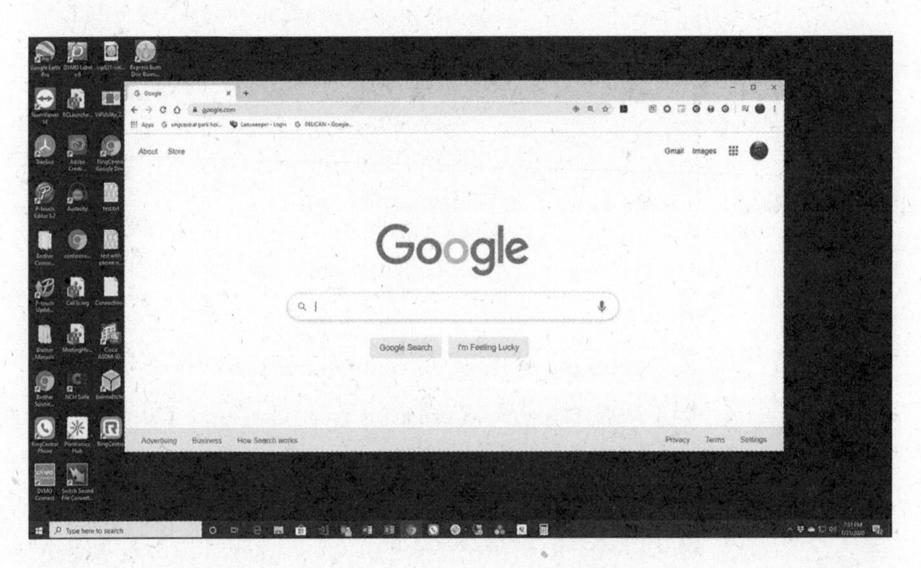

FIGURE 24-8: You're connected!

You may notice in Figure 24-8 that the Remote Desktop window is not large enough to display the entire desktop of the remote computer. As a result, scroll bars appear to allow you to scroll the desktop horizontally or vertically. You can always maximize the Remote Desktop window to see the entire desktop of the remote computer.

TIP

Here are a few other tips for working with Remote Desktop Connection:

- **Remote Desktop allows only one user at a time to log in to the remote computer.** If another user is remotely logged in when you try to connect, you'll get a notice informing you that another user is already logged on. You can either cancel or attempt to barge in on the other user's remote session. If you choose the latter option, the other user will see a message stating that you want to connect. If the other user accepts your request, the other user is logged off and you're logged in. If the other user denies your request, your attempt to log in is cancelled. If the user does not respond, eventually Windows will kick the other user off and let you in.
- **When you're using the Remote Desktop Connection client, you can't just press Alt+Tab to get another program running on the client computer.** Instead, you must first minimize the RDC client's window by clicking its minimize button. Then you can access other programs running on your computer.
- **If you minimize the RDC client window, you have to provide your logon credentials again when you return.** This security feature is there in case you forget that you have an RDC session open.

TIP

If you use RDC a lot on a particular computer (such as your own desktop computer), I suggest that you create a shortcut to RDC and place it on the Desktop, at the top of the Start menu, or in the Quick Launch portion of the taskbar.

Using keyboard shortcuts for Remote Desktop

When you're working in a Remote Desktop session, some of the standard Windows keyboard shortcuts don't work exactly as you expect them to. Table 24-1 lists the special keyboard shortcuts you can use in a Remote Desktop session.

TABLE 24-1 Keyboard Shortcuts for Remote Desktop

Shortcut	What It Does
Ctrl+Alt+Break	Toggles between full screen and windowed views.
Ctrl+Alt+Pause	Similar to Ctrl+Alt+Break, but instead of maximizing the remote window to full screen, it displays the remote window against a black background.
Alt+Insert	Cycles between applications running on the remote desktop, the same as Alt+Tab on your local machine.
Alt+PageUp	Same as Alt+Insert.
Alt+PageDown	Similar to Alt+Insert, but reverses the order in which applications are cycled. This is the same as Alt+Shift+Tab on your local machine.
Ctrl+Alt+End	Sends a Ctrl+Alt+Del to the remote desktop.
Alt+Home	Brings up the Start menu on the remote system.
Alt+Delete	Opens the Windows menu on a window in the remote desktop. (The Windows menu is the one at the top left of every window, with options to move, resize, minimize, maximize, and close the window.)
Ctrl+Alt+Plus Sign (+)	Captures a screen image of the entire remote desktop and saves it to the Clipboard. This is the same as pressing Print Screen on your local machine.
Ctrl+Alt+Minus Sign (–)	Captures an image of the current window and saves it to the Clipboard. This is the same as pressing Alt+Tab on your local machine.

IN THIS CHAPTER

- Examining the basics of cloud computing
- Looking at three kinds of cloud computing services
- Understanding the pros and cons of cloud computing
- Perusing a few major cloud computing service providers

Chapter 25
Life in Cloud City

The world's two most popular science-fiction franchises — *Star Wars* and *Star Trek* — both feature cities that are suspended in the clouds. In *Star Wars Episode V: The Empire Strikes Back*, Han takes the *Millennium Falcon* to Cloud City, hoping that his friend Lando Calrissian can help repair their damaged hyperdrive. And in the original *Star Trek* series episode "The Cloud Minders," the crew of the *Enterprise* visits a city named Stratos, which is suspended in the clouds.

Coincidence? Perhaps. Or maybe Gene Roddenberry and George Lucas both knew that the future would be in the clouds. At any rate, the future of computer networking is rapidly heading for the clouds. Cloud computing, to be specific. This chapter is a brief introduction to cloud computing. You discover what it is, the pros and cons of adopting it, and what services are provided by the major cloud computer providers.

Introducing Cloud Computing

The basic idea behind cloud computing is to outsource one or more of your networked computing resources to the Internet. "The cloud" represents a new way of handling common computer tasks. Following are just a few examples of how the cloud way differs from the traditional way:

- **Email services**
  - *Traditional:* Email services are provided with a local server running Microsoft Exchange. Then your clients can connect use Microsoft Outlook to connect to the Exchange server to send and receive email.
  - *Cloud:* Contract with an Internet-based email provider, such as Google Mail (Gmail) or Microsoft's Exchange Online. Cloud-based email services typically charge a low monthly per-user fee, so the amount you pay for your email service depends solely on the number of email users you have.
- **Disk storage**
  - *Traditional:* Set up a local file server computer with a large amount of shared disk space.
  - *Cloud:* Sign up for an Internet file storage service and then store your data on the Internet. Cloud-based file storage typically charges a small monthly per-gigabyte fee, so you pay only for the storage you use. The disk capacity of cloud-based storage is essentially unlimited.
- **Accounting services**
  - *Traditional:* Purchase expensive accounting software and install it on a local server computer.
  - *Cloud:* Sign up for a web-based accounting service. Then all your accounting data is saved and managed on the provider's servers, not on yours.
- **Virtual servers**
  - *Traditional:* Purchase expensive computers, install a virtualization platform such as VMware's ESXi or Microsoft's Hyper-V, and create virtual machines.
  - *Cloud:* Sign up for virtual machines hosted on a cloud provider's servers. Skip the part about buying expensive computers and installing ESXi or Hyper-V.

Looking at the Benefits of Cloud Computing

Cloud computing is a different — and, in many ways, better — approach to networking. Here are a few of the main benefits of moving to cloud-based networking:

- **Cost-effective:** Cloud-based computing typically is less expensive than traditional computing. Consider a typical file server application: To implement a file server, first you must purchase a file server computer with enough disk space to accommodate your users' needs, which amounts to 1TB of disk storage. You want the most reliable data storage possible, so you purchase a server-quality computer and fully redundant disk drives. For the sake of this discussion, figure that the total price of the server — including its disk drive, the operating system license, and the labor cost of setting it up — is about $10,000. Assuming that the server will last for four years, that totals about $2,500 per year.

  If you instead acquire your disk storage from a cloud-based file sharing service, you can expect to pay about one fourth of that amount for an equivalent amount of storage.

  The same economies apply to most other cloud-based solutions. Cloud-based email solutions, for example, typically cost around $5 per month per user — well less than the cost of setting up and maintaining an on-premises Microsoft Exchange Server.

- **Scalable:** So what happens if you guess wrong about the storage requirements of your file server, and your users end up needing 20TB instead of just 10TB? With a traditional file server, you must purchase additional disk drives to accommodate the extra space. Sooner than you want, you'll run out of capacity in the server's cabinet. Then you'll have to purchase an external storage cabinet. Eventually, you'll fill that up, too.

  Now suppose that after you expand your server capacity to 20TB, your users' needs contract back to just 10TB. Unfortunately, you can't return disk drives for a refund.

  REMEMBER

  With cloud computing, you pay only for the capacity you're actually using, and you can add capacity whenever you need it. In the file server example, you can write as much data as you need to the cloud storage. Each month, you're billed according to your actual usage. Thus, you don't have to purchase and install additional disk drives to add storage capacity.

- **Reliable:** Especially for smaller businesses, cloud services are much more reliable than in-house services. Just a week before I wrote this chapter, the tape drive that a friend uses to back up his company's data failed. As a result, he was unable to back up data for three days while the tape drive was repaired. Had he been using cloud-based backup, he could have restored his data immediately and wouldn't have been without backups for those three days.

The reason for the increased reliability of cloud services is simply a matter of scale. Most small businesses can't afford the redundancies needed to make their computer operations as reliable as possible. My friend's company can't afford to buy two tape drives so that an extra is available in case the main one fails.

By contrast, cloud services are usually provided by large companies such as Amazon, Google, Microsoft, and IBM. These companies have state-of-the-art data centers with multiple redundancies for their cloud services. Cloud storage may be kept on multiple servers so that if one server fails, others can take over the load. In some cases, these servers are in different data centers in different parts of the country. Thus, your data will still be available even in the event of a disaster that shuts down an entire data system.

- **Hassle-free:** Face it, IT can be a hassle. With cloud-based services, you basically outsource the job of complex system maintenance chores, such as software upgrade, patches, hardware maintenance, backup, and so on. You get to consume the services while someone else takes care of making sure that the services run properly.
- **Globally accessible:** One of the best things about cloud services is that they're available anywhere you have an Internet connection. Suppose that you have offices in five cities. Using traditional computing, each office would require its own servers, and you'd have to carefully design systems that allowed users in each of the offices to access shared data.

  With cloud computing, each office simply connects to the Internet to access the cloud applications. Cloud-based applications are also great if your users are mobile because they can access the applications anywhere they can find an Internet connection.

Detailing the Drawbacks of Cloud Computing

Although cloud computing has many advantages over traditional techniques, it isn't without its drawbacks. Here are some of the most significant roadblocks to adopting cloud computing:

- **Entrenched applications:** Your organization may depend on entrenched applications that don't lend themselves especially well to cloud computing — or that at least require significant conversion efforts to migrate to the cloud. For example, you might have use an accounting system that relies on local file storage.

Fortunately, many cloud providers offer assistance with this migration. And in many cases, the same application that you run locally can be run in the cloud, so no conversion is necessary.

- **Internet connection speed:** Cloud computing shifts much of the burden of your network to your Internet connection. Your users used to access their data on local file servers over gigabit-speed connections; now they must access data over slower bandwidth Internet connections.

REMEMBER

Although you can upgrade your connection to higher speeds, doing so will cost money — money that may well offset the money you otherwise save from migrating to the cloud.

- **Internet connection reliability:** The cloud resources you access may feature all the redundancy in the world, but if your users access the cloud through a single Internet connection, that connection becomes a key point of vulnerability. Should it fail, any applications that depend on the cloud will be unavailable. If those applications are mission-critical, business will come to a halt until the connection is restored.

  Here are two ways to mitigate this risk:

  - *Make sure that you're using an enterprise-class Internet connection.* Enterprise-class connections are more expensive but provide much better fault tolerance and repair service than consumer-class connections do.
  - *Provide redundant connections if you can.* That way, if one connection fails, traffic can be rerouted through alternative connections.

- **Security threats:** You can bet your life that hackers throughout the world are continually probing for ways to break through the security perimeter of all the major cloud providers. When they do, your data may be exposed.

  The best way to mitigate this threat is to ensure that strong password policies are enforced and make sure that your data is backed up.

Examining Three Basic Kinds of Cloud Services

Three distinct kinds of services can be provided via the cloud: applications, platforms, and services (infrastructure). The following paragraphs describe these three types of cloud services in greater detail.

Applications

Most often referred to as *Software as a Service* (SaaS), fully functional applications can be delivered via the cloud. One of the best-known examples is *Google Apps*, which is a suite of cloud-based office applications designed to compete directly with Microsoft's traditional office applications, including Word, Excel, PowerPoint, Access, and Outlook. Google Apps can also replace the back-end software often used to support Microsoft Office, including Exchange and SharePoint.

When you use a cloud-based application, you don't have to worry about any of the details that are commonly associated with running an application on your network, such as deploying the application and applying product upgrades and software patches. Cloud-based applications usually charge a small monthly fee based on the number of users running the software, so costs are low.

Also, as a cloud-based application user, you don't have to worry about providing the hardware or operating system platform on which the application will run. The application provider takes care of that detail for you, so you can focus simply on developing the application to best serve your users' needs. Your users can access the application using any operating system that has a standard web browser.

Platforms

Also referred to as *Platform as a Service* (PaaS), this class of service refers to providers that give you access to a remote virtual operating platform on which you can build your own applications.

At the simplest level, a PaaS provider gives you a complete, functional remote virtual machine that's fully configured and ready for you to deploy your applications to. If you use a web provider to host your company's website, you're already using PaaS: Most web host providers give you a functioning Linux system, fully configured with all the necessary servers, such as Apache or MySQL. All you have to do is build and deploy your web application on the provider's server.

More-complex PaaS solutions include specialized software that your custom applications can tap to provide services such as data storage, online order processing, and credit card payments. One of the best-known examples of this type of PaaS provider is Amazon.

REMEMBER

When you use PaaS, you take on the responsibility of developing your own custom applications to run on the remote platform. The PaaS provider takes care of the details of maintaining the platform itself, including the base operating system and the hardware on which the platform runs.

Infrastructure

If you don't want to delegate the responsibility of maintaining operating systems and other elements of the platform, you can use *Infrastructure as a Service* (IaaS). When you use IaaS, you're purchasing raw computing power that's accessible via the cloud. Typically, IaaS provides you access to a remote virtual machine. It's up to you to manage and configure the remote machine however you want.

Public Clouds versus Private Clouds

The most common form of cloud computing uses what is known as a *public cloud* — that is, cloud services that are available to anyone in the world via the Internet. Google Apps is an excellent example of a public cloud service. Anyone with access to the Internet can access the public cloud services of Google Apps: Just point your browser to `http://gsuite.google.com`.

A public cloud is like a public utility, in that anyone can subscribe to it on a pay-as-you-go basis. One of the drawbacks of public cloud services is that they're inherently insecure. When you use a public cloud service, you're entrusting your valuable data to a third party that you cannot control. Sure, you can protect your access to your public cloud services by using strong passwords, but if your account names and passwords are compromised, your public cloud services can be hacked into, and your data can be stolen. Every so often, we all hear news stories about how this company's or that company's back-door security has been compromised.

Besides security, another drawback of public cloud computing is that it's dependent on high-speed, reliable Internet connections. Your cloud service provider may have all the redundancy in the world, but if your connection to the Internet goes down, you won't be able to access your cloud services. And if your connection is slow, your cloud services will be slow.

A *private cloud* mimics many of the features of cloud computing but is implemented on a private hardware within a local network, so it isn't accessible to the general public. Private clouds are inherently more secure because the general public can't access them. Also, they're dependent only on private network connections, so they aren't subject to the limits of a public Internet connection.

TIP

As a rule, private clouds are implemented by large organizations that have the resources available to create and maintain their own cloud servers.

A relative newcomer to the cloud computing scene is the *hybrid cloud*, which combines the features of public and private clouds. Typically, a hybrid cloud system uses a small private cloud that provides local access to the some of the applications and the public cloud for others. You might maintain your most frequently used data on a private cloud for fast access via the local network and use the public cloud to store archives and other less frequently used data, for which performance isn't as much of an issue.

Introducing Some of the Major Cloud Providers

Hundreds, if not thousands, of companies provide cloud services. Most of the cloud computing done today, however, is provided by just a few providers, which are described in the following sections.

Amazon

By far the largest provider of cloud services in the world is Amazon. Amazon launched its cloud platform — Amazon Web Services (AWS) — in 2006. Since then, hundreds of thousands of customers have signed up. Some of the most notable users of AWS include Netflix, Pinterest, and Instagram.

AWS includes the following features:

- **Amazon CloudFront:** A PaaS content-delivery system designed to deliver web content to large numbers of users.
- **Amazon Elastic Compute Cloud:** Also called Amazon EC2. An IaaS system that provides access to raw computing power.
- **Amazon Simple Storage Service:** Also called Amazon S3. Provides web-based data storage for unlimited amounts of data.
- **Amazon Simple Queue Service:** Also called Amazon SQS. Provides a data transfer system that lets applications send messages to other applications. SQS enables you to build applications that work together.
- **Amazon Virtual Private Cloud:** Also called Amazon VPC. Uses virtual private network (VPN) connections to connect your local network to Amazon's cloud services.

Google

Google is also one of the largest providers of cloud services. Its offerings include the following:

- **Google Apps:** A replacement for Microsoft Office that provides basic email, word processing, spreadsheet, and database functions via the cloud. Google Apps is free to the general public and can even be used free by small business (up to 50 users). For larger businesses, Google offers an advanced version, Google Apps for Business. For $5 per month per user, you get extra features, such as 25GB of email data per user, archiving, and advanced options for customizing your account policies.
- **Google Cloud Connect:** A cloud-based solution that lets you work with Google cloud data directly from within Microsoft Office applications.
- **Google App Engine:** A PaaS interface that lets you develop your own applications that work with Google's cloud services.
- **Google Cloud Print:** Allows you to connect your printers to the cloud so that they can be accessed from anywhere.
- **Google Maps:** A Global Information System (GIS).

Microsoft

Microsoft has its own cloud strategy, designed in part to protect its core business of operating systems and Office applications against competition from other cloud providers, such as Google Apps.

The following paragraphs summarize several of Microsoft's cloud offerings:

- **Microsoft Office 365:** A cloud-based version of Microsoft Office. According to Microsoft's website, Office 365 provides "anywhere access to cloud-based email, web conferencing, file sharing, and Office Web Apps at a low predictable monthly cost." For more information, check out `www.office365.com`.
- **Windows Azure:** A PaaS offering that lets you build websites, deploy virtual machines that run Windows Server or Linux, or access cloud versions of server applications such as SQL Server.
- **Microsoft Business Productivity Suite:** A SaaS product that provides cloud-based access to two of Microsoft's most popular productivity servers: Microsoft Exchange and Microsoft SharePoint. The suite lets you deploy these servers without having to create and maintain your own local servers.

Getting into the Cloud

After you wrap your head around just how cool cloud computing can be, what should you do to take your network toward the cloud? Allow me to make a few recommendations:

- » **Don't depend on a poor Internet connection.** First and foremost, before you take any of your network operations to the cloud, make sure that you're *not* dependent on a consumer-grade Internet connection if you decide to adopt cloud computing. Consumer-grade Internet connections can be fast, but when an outage occurs, there's no telling how long you'll wait for the connection to be repaired. You definitely don't want to wait for hours or days while the cable company thinks about sending someone out to your site. Instead, spend the money for a high-speed enterprise-class connection that can scale as your dependence on it increases.
- » **Assess what applications you may already have running on the cloud.** If you use Gmail rather than Exchange for your email, congratulations! You've already embraced the cloud. Other examples of cloud services that you may already be using include a remote web or FTP host, Dropbox or another file sharing service, Carbonite or another online backup service, a payroll service, and so on.
- » **Don't move to the cloud all at once.** Start by identifying a single application that lends itself to the cloud. If your engineering firm archives projects when they close and wants to get them off your primary file server but keep them readily available, look to the cloud for a file storage service.
- » **Go with a reputable company.** Google, Amazon, and Microsoft are all huge companies with proven track records in cloud computing. Many other large and established companies also offer cloud services. Don't stake your company's future on a company that didn't exist six months ago.
- » **Research, research, research.** Pour yourself into the web, and buy a few books. *Hybrid Cloud For Dummies,* by Judith Hurwitz, Marcia Kaufman, Dr. Fern Halper, and Daniel Kirsch (Wiley), is a good place to start.

» **Looking at the benefits of going hybrid**

» **Dealing with identity management**

» **Creating hybrid cloud infrastructure**

Chapter 26 Going Hybrid

When most of us hear the word *hybrid*, we think of the Toyota Prius and other similar hybrid automobiles. These cars combine a traditional gasoline-powered internal combustion engine with an efficient electric motor powered by rechargeable batteries to produce a vehicle that is better than pure gasoline-powered cars and pure electric-powered cars: A hybrid vehicle has much better fuel efficiency than a gasoline-only car, and much better range than a battery-only car.

Hybrid cars hit the market in 1999. A few decades later, the term *hybrid* became commonplace in computing as well. These days, many companies — large and small — are touting hybrid as one of their major IT initiatives. In this chapter, I explain just what that means and why it's a good idea. I also tell you how you can dip your toe into the hybrid waters.

What Is a Hybrid Cloud?

Most organizations today have two distinct IT engines. The traditional IT engine is based on server computers, storage devices, and network components neatly mounted on racks and locked away behind closed doors, operated and maintained by highly paid professionals.

The other engine is the cloud. Chapter 25 explains how the cloud has revolutionized the world of IT by moving critical workloads away from on-premises computers, eliminating the need for large investments in hardware and staff to support those applications.

However, a key problem with the cloud engine is that it works very differently from the traditional IT engine. Instead of physical computer equipment that you have sole control over, your cloud engine consists of a multitude of cloud providers whose services you subscribe to. You don't have direct control of any of the details you're used to managing.

For example, cloud providers take care of traditional IT details, such as purchasing and maintaining hardware, keeping systems patched and tuned, backing up data, and so on. You don't have to worry about what brand of servers they use, what operating systems they install, what virtualization platform they rely on, or what backup technology they trust. The cloud provider takes care of all that for you.

All you have to do is tailor the cloud services to your needs. You typically do so by using web-based management consoles to set up user accounts, configure services, and so on. Unfortunately, no two cloud providers use the same management consoles. Each has a completely different interface and requires a different set of skills to manage.

Imagine if your hybrid automobile had both a gas and an electric engine, but you had to have a different driver's license to operate each engine. Or if the two engines had separate operating controls — different accelerator pedals, steering wheels, and brake pedals depending on which engine was on. Or if you had to take the vehicle to different mechanics depending on which engine was making a funny noise. Or if the rules of the road were different depending on which engine was in use — the speed limit was 45 miles per hour with the electric engine but 60 miles per hour with the gas engine, or you drove on the right side of the street with the gas engine but drove on the left side with the electric.

That's exactly the situation most organizations put up with for their IT engines. And that's precisely the problem hybrid cloud intends to solve. Hybrid cloud seeks to unify the on-premises IT engine with the cloud engine so that they drive the same.

Figure 26-1 shows what I'm talking about. Here, you can see that all three types of cloud services described in Chapter 25 — applications (Software as a Service, or SaaS), infrastructure (Infrastructure as a Service, or IaaS), and platform (Platform as a Service, or PaaS) — are integrated with the on-premises IT infrastructure to create a cohesive ecosystem called the hybrid cloud.

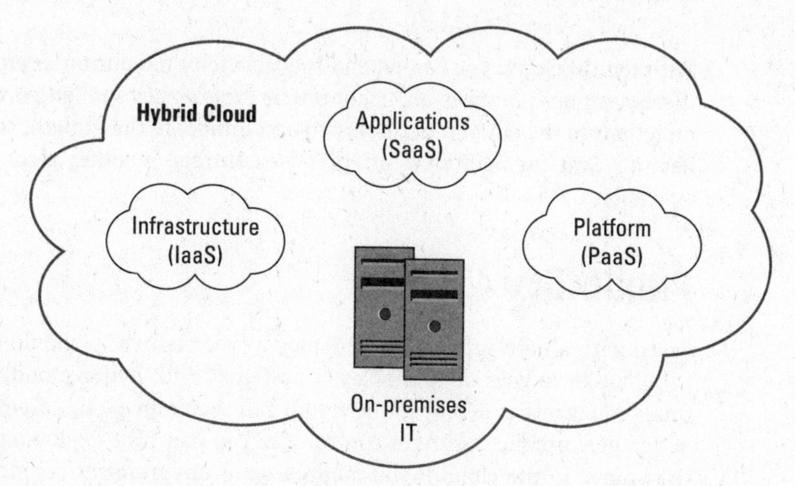

FIGURE 26-1: Hybrid cloud seeks to unify on-premises IT with Cloud services.

This is more challenging that it seems like it should be. But recent trends have brought us much closer to the reality of being able to seamlessly move between on-premises IT and cloud-based IT.

What Are the Benefits of Hybrid Cloud?

Moving the traditional IT engine and the cloud engine closer together provides numerous benefits for your IT organization. I cover a few of the most important ones in the following sections.

Elasticity

As I get older, I find it very helpful to buy pants with elastic waistbands. *Elasticity* is a valuable characteristic in IT systems as well. Elasticity simply means that your IT infrastructure can stretch to accommodate growth and that it can shrink down when your IT loses weight. For example, if you roll out a new line of business or open a new branch office, your IT should be able to expand to accommodate the new work. But if you sell off that line of business or close an office, you should be able to contract.

Cloud services are elastic by nature. For example, a cloud-based email service typically charges a certain amount per user per month. If you add 50 users, you pay more. If you lose 50 users, you pay less.

With hybrid cloud, you can extend this elasticity to your on-premises IT as well. Ideally, an on-premises application that experiences sudden growth can expand directly into the cloud if necessary to accommodate the growth, so that you don't have to wait for additional on-premises storage or other resources to become available.

Flexibility

Faced with a new application to deploy to your network, should you deploy it to the cloud or to your on-premises IT systems? With hybrid cloud, the choice isn't final: You can first deploy to the cloud, and then move it on-premises if you need better performance or more control. Or, you can first deploy on-premises, and then move to the cloud if you outgrow your on-premises resources or find that managing the underlying hardware is too much of a burden.

Agility

Agility refers to the ability to get new applications deployed quickly so that your company benefits from the new applications immediately, instead of having to wait for traditional IT to go through its months- or years-long processes. Hybrid cloud enables you to deploy new applications quickly to the cloud while simultaneously planning for their eventual move to an on-premises platform.

Innovation

Many of the newest technologies are expensive to implement for smaller companies. For example, artificial intelligence and machine learning are growing trends that can significantly improve your business processes and give your company a competitive advantage. But most companies can't afford to spend years to develop these technologies. Cloud providers, however, can. And hybrid cloud can make it easier to integrate these technologies into your systems.

Operational efficiency

If the on-premises IT engine and the cloud engine are managed using completely different tools, possibly by completely different teams, you're paying to do the same thing in two different ways. Your organization will be much leaner if you consolidate the management of on-premises and cloud IT as much as possible.

Integrating Identity

By *identity*, I mean the various systems that are used to authenticate and authorize users on your systems. When you start using cloud applications, managing identity on each of your applications will become a problem. In short, just about every cloud service you subscribe to will have its own identity management system in which you set up users and assign permissions. And each service will have its own password management systems, with different requirements for length and complexity.

All of these identity management systems are in addition to the one you already have: Active Directory.

Wouldn't it be nice if you could unify them all, so that all your cloud services were integrated with your existing Active Directory accounts, and you could manage accounts and authorize access to cloud services through a single management console?

The good news is that you *can* — or at least you can come close. Taking the steps to make this happen will go a long way toward simplifying the relationship between your on-premises IT and your cloud IT.

Azure Active Directory

An obvious place to start with unifying your identity management is *Azure Active Directory* (AAD). AAD is Microsoft's cloud-based version of Active Directory. On its own, AAD won't solve all your cloud identity management problems. But it will enable you to integrate Office 365 identities with Active Directory, and it will extend your on-premises Active Directory into Azure, if you choose to use Azure as a cloud platform.

AAD is sold as a subscription service with four editions:

- **Free:** Includes basic Active Directory features with a limited number of objects in the directory (the current limit is half a million, so the free edition can accommodate a fairly large directory). Unfortunately, this edition is sparse on features — many of the features you've come to know and love in standard on-premises Active Directory require premium subscriptions.
- **Office 365:** This version is included with an Office 365 subscription and provides unlimited directory size, all the basic features of the free edition, plus self-service password management, company branding for login pages, and a few other bells and whistles.

- **Premium P1:** For $6 per user per month, you get additional features such as custom password restrictions, the ability to limit access to certain computers and/or certain hours, and more.
- **Premium P2:** Even more features for $9 per user per month.

Here are some additional details to ponder about AAD:

- Just as cloud applications are often called SaaS, AAD is sometimes referred to as *Identity Management as a Service* (IDaaS). I think some people just love to tack things in front of "as a Service."
- AAD can run in hybrid mode, in which case it cooperates with your existing on-premises Active Directory servers. Or, you can run it in Standalone mode and eliminate your on-premises Active Directory servers.
- Setting up AAD is not for the faint-of-heart or the uninitiated. You'll need the help of a qualified systems engineer to get the job done right.

Single sign-on

Another way to improve the continuity between your on-premises Active Directory and your cloud identities is to deploy a single sign-on (SSO) solution. As its name suggests, *single sign-on* refers to the ability to log in once to an identity provider and have that identity provider handle your logins to other services. When SSO is properly implemented, you can log in once to Windows using your Active Directory credentials. Then, your SSO platform can automatically log you in to any cloud services you use throughout the day.

Some cloud services provide SSO integration with Active Directory. For example, the popular RingCentral phone system can integrate with Active Directory so that a separate logon to RingCentral is not necessary.

But most cloud services don't provide that level of integration. To achieve real SSO, you'll need to use a third-party tool designed specifically for this purpose. One of the best known is Okta (`www.okta.com`). For a small monthly per-user fee, Okta can manage sign-ons for thousands of cloud applications.

Figure 26-2 shows a typical Okta dashboard, which provides single-click access to a variety of cloud services. These services have been configured by an account administrator to allow the user to access the services that he or she needs. So, Okta allows IT to control access to cloud services and shields the user from the task of managing his or her service account.

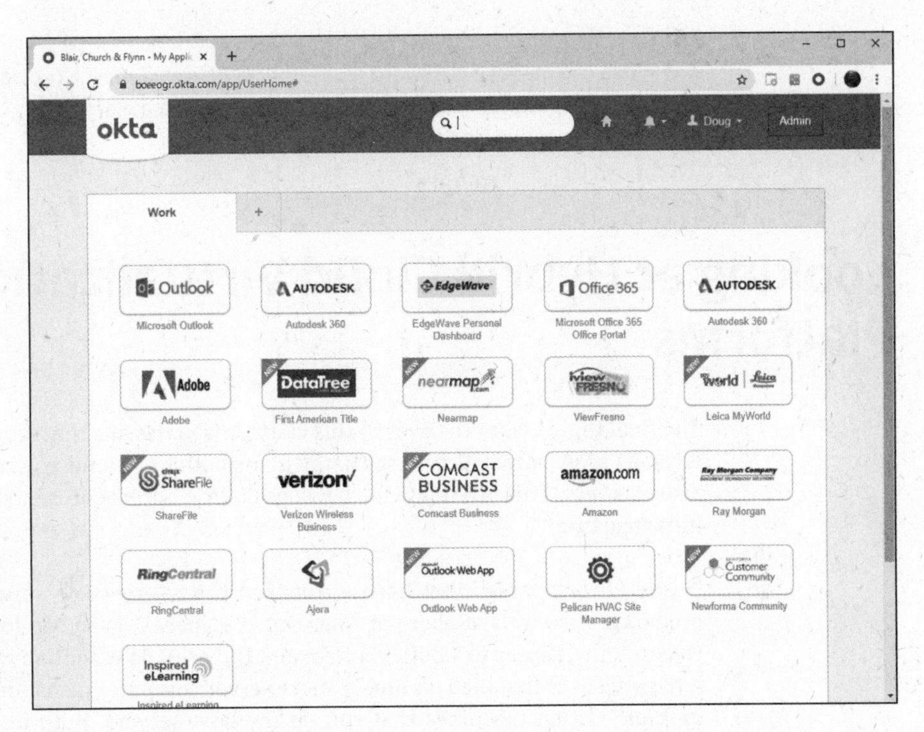

FIGURE 26-2: Hybrid cloud seeks to unify on-premises IT with cloud services.

Here are some additional points to ponder about the abilities of identity management platforms such as Okta:

- Okta does much more than simply remember your usernames and passwords — all modern web browsers have the ability to do that. Okta actually participates with the cloud service provider to negotiate the user authentication process, which often involves more than just supplying the username and password. And unlike built-in password management found in browsers, Okta provides a single point of management for your entire organization.
- An SSO platform such as Okta dramatically improves your password security for one simple reason: Because the SSO platform remembers the password for all of a user's accounts, the user doesn't have to. And because the user doesn't have to remember all those passwords, he or she won't resort to writing them down on a sticky note stuck to the monitor.
- Okta can even be set up so that users never actually *know* their passwords. This feature is incredibly useful because it allows your organization to maintain control of its user accounts. For example, when a user leaves your organization, simply disabling the user in Okta will also prevent the now-former-employee from using cloud subscriptions that your company is paying for. Without an identity manager like Okta, you'll have to manually disable the user's account for every cloud service the user had access to.

» Okta can manage password resets for cloud providers. If a provider requires periodic password resets, Okta can do that automatically. Again, the user doesn't need to know the password.

Looking at Hybrid Cloud Virtualization Platforms

The final topic I want to cover in this chapter is platforms that let you extend your virtualization infrastructure so that it spans both on-premises servers and public cloud servers. This is a relatively new possibility, but it represents the future of IT infrastructure.

Nearly all companies that have virtualized their servers do so with one of two products: VMware's vSphere or Microsoft's Hyper-V. (You can learn more about Hyper-V in Chapter 10.) Both vSphere and Hyper-V do essentially the same thing: A hypervisor is installed on one or more server computers, enabling you to create multiple virtual machines that run on the host servers. Both platforms provide similar capabilities:

» Multiple host servers can be combined to create a single cluster, which can be managed as a cohesive unit.

» Individual virtual machines can be easily moved from one host to another host in the same cluster.

» The virtualization platform can manage storage pools that are connected to the host servers (storage area network, or SAN) or to the network (network attached storage, or NAS). Individual virtual machines are allocated storage from these storage pools.

» The virtualization platform can also manage virtual networks that facilitate communications among the virtual machines, the host servers, and the outside world.

Traditionally, all the hardware utilized to run the virtualization platform is located within your data center. Or, if your organization is large, the hardware may be spread over several data centers connected to one another via dedicated network circuits.

Now consider the advent of IaaS providers who have created similar virtualization platforms that you can access via the cloud. The two best-known providers are Amazon Web Services (AWS) and Microsoft Azure. AWS uses its own homegrown virtualization platform that lets you easily provision new servers, storage, and networks from a variety of preconfigured templates. Azure does essentially the same thing, but its virtualization platform is based on Hyper-V.

Both Microsoft and VMware have recently extended their virtualization platforms to seamlessly embrace cloud infrastructure. Setting this up can be complicated, but after it's done, you can include not just physical servers and storage located on your premises, but also cloud servers and storage within a single, unified cluster. Then, you can easily move virtual machines and data back and forth between your on-premises infrastructure and your cloud infrastructure.

Because cloud infrastructure is much easier to expand than on-premises infrastructure, this can dramatically the simplify growth challenges that most IT organizations face. If you outgrow your local storage, you can move some workloads up to the cloud infrastructure. If you think the growth is relatively permanent, you can then order additional local storage, and when it's installed and configured, move the workloads back to your local infrastructure. But if you think the growth is temporary, you can leave the workload in the cloud until it contracts back to its previous size, and then move it back down.

VMware's approach to this integration is called VMware Cloud. Because VMware is not itself a major cloud provider, it works in concert with other cloud providers — most notably, AWS. In short, you can use VMware Cloud to configure VMware hosts, storage, and networks on AWS and integrate these resources with your on-premises vSphere clusters. The whole thing is managed from the familiar vCenter console, so you don't need a separate team of specialists to manage the cloud side.

Microsoft's approach is pretty much the opposite of VMware's: Instead of extending the virtualization platform up into the cloud, Microsoft extends its Azure IaaS platform down to your on-premises Hyper-V environment. Its product is called Azure Stack. The end result is the same: A single virtualization platform that integrates on-premises and cloud-based servers, storage, and networking into a seamless environment.

Truthfully, building a true hybrid cloud is far more complicated than I've let on here, because there are many details that have to be accounted for when planning and implementing a true hybrid-cloud environment. Security and privacy are obviously huge concerns, and in many cases regulatory compliance is as well (especially if you're in the medical or financial industries). Backup and disaster recovery is a major issue that needs to be addressed.

Cost is also a huge concern. The whole idea of hybrid cloud is to streamline operations so that systems actually end up costing less while providing more. But cloud contracts can be very complicated. You'll usually be charged not just for the amount of CPU clock cycles and data storage you consume but also for the amount of data you transfer, which can quickly become a major expense. A basic difference between on-premises IT and cloud IT is that you pay for on-premises IT upfront, and then hope you didn't overspend. With cloud IT, you pay as you go based on actual consumption. The result can be a nice reduction in cost. But you can also get hit by a huge bill you weren't expecting, so you need to make sure you thoroughly understand the pricing model so that you can predict your costs.

7

The Part of Tens

IN THIS PART . . .

Learn the unwritten rules of networking.

Find out the common networking mistakes you should avoid.

Review a list of ten handy networking things you should keep in your closet.

Chapter 27
Ten Networking Commandments

"Blessed is the network manager who walks not in the council of the ignorant, nor stands in the way of the oblivious, nor sits in the seat of the greenhorn, but delights in the Law of the Network and meditates on this Law day and night."

— NETWORKS 1:1

And so it came to pass that these Ten Networking Commandments were handed down from generation to generation, to be worn as frontlets between the computer geeks' eyes (taped on the bridges of their broken glasses) and written upon their doorposts with Sharpie markers. Obey these commandments, and it shall go well with you, with your users, and with your users' users.

I. Thou Shalt Back Up Thy Data Religiously

Prayer is a good thing, and I heartily recommend it. But when it comes to protecting the data on your network, nothing beats a well-thought-out schedule of backups followed religiously.

Make sure that you follow the 3-2-1 rule: Have at least *three* backup copies of your data, *two* of which are kept at a different location (off-premises), and *one* of which is not accessible from your network (offline).

For more information about crafting a bulletproof backup plan, refer to Chapter 22.

II. Thou Shalt Protect Thy Network from Infidels

In the classic movie *Independence Day*, aliens from another planet seek to destroy all life on Earth. Unfortunately, their IT Director failed to secure their networks, so Jeff Goldblum was able to hack in and save the world.

In the movie, Jeff Goldblum used his hacking powers for good. In the real world, not everyone is so altruistic. The world is full of bad actors who would love nothing more than to steal your data, hold it for ransom, or destroy it altogether.

To counter these threats, be sure to take precautions such as using a secure firewall to keep intruders out of your network, using antivirus software, and using anti-spam software to keep malicious email out of your users' mailboxes.

For more information about how to secure your network, refer to Chapters 19 and 20.

III. Thou Shalt Train Up Thy Users in the Ways of Safe Computing

All the security protection in the world won't stop users from clicking bad links or opening infected files. In addition to firewalls and security software, it's vital that you train your users in the ways of cybersecurity. Your users are the final and best line of defense against cyberthreats.

Fortunately, many security training opportunities are available on the interwebs. Just use your favorite search engine to look for "cyber security user training" and you'll find tons of possibilities, some free, others paid.

An important part of cybersecurity training is running periodic phishing drills to help your users identify and ignore phishing email. You can find services that help you do that as well. Just search the web for "phish testing" and you'll find loads of options.

Refer to Chapter 21 for more information.

IV. Thou Shalt Keepeth Thy Network Drive Pure and Cleanse It of Old Files

Don't wait until your 10TB network drive is down to just 1GB of free space before you think about cleaning it up. Set up a routine schedule for disk housekeeping, where you wade through the files and directories on the network disk to remove old junk.

V. Thou Shalt Not Tinker with Thine Network Configuration unless Thou Knowest What Thou Art Doing

Networks are finicky things. After yours is up and running, don't mess with it unless you know what you're doing. You may be tempted to log on to your firewall router to see whether you can tweak some of its settings to squeeze another ounce of performance out of it. But unless you know what you're doing, be careful! (Be especially careful if you *think* you know what you're doing. It's the people who think they know what they're doing who get themselves into trouble!)

VI. Thou Shalt Not Covet Thy Neighbor's Network

Network envy is a common malady among network managers. If your network servers are chugging along fine with Windows Server 2016, don't feel compelled to upgrade to Windows Server 2019 right away. Resist the urge to upgrade unless you have a really good reason.

You're especially susceptible to network envy if you're a gadget freak. There's always a better switch to be had or some fancy network-protocol gizmo to lust after. Don't give in to these base urges! Resist the devil, and he will flee!

VII. Thou Shalt Not Take Down Thy Network without Proper Notification

As a courtesy, try to give your users plenty of advance notice before you take down the network to work on it. Obviously, you can't predict when random problems strike. But if you know you're going to patch the server on Thursday morning, you earn points if you tell everyone about the inconvenience two days before rather than two minutes before. (You'll earn even more points if you patch the server Saturday morning.)

VIII. Thou Shalt Keep an Adequate Supply of Spare Parts

There's no reason that your network should be down for two days just because a cable breaks. Always make sure that you have at least a minimal supply of network spare parts on hand. (As luck would have it, Chapter 29 suggests ten things you should keep in your closet.)

IX. Thou Shalt Not Steal Thy Neighbor's Program without a License

How would you like it if Inspector Clouseau (from the classic *Pink Panther* movies) barged into your office, looked over your shoulder as you ran Excel from a network server, and asked, "Do you have a *liesaunce*?"

"A *liesaunce*?" you reply, puzzled.

"Yes. of course, a *liesaunce* — that is what I said! The law specifically prohibits the playing of a computer program on a network without a proper *liesaunce*."

You don't want to get in trouble with Inspector Clouseau, do you? Then make sure you have the correct licenses for the applications you run on your network.

If you haven't seen Peter Sellars as Inspector Clouseau, stream *The Return of the Pink Panther* right away.

X. Thou Shalt Write Down Thy Network Configuration upon Tablets of Stone

Network documentation should be written down. If you cross the River Jordan, who else will know diddly-squat about the network if you don't write it down somewhere? Write down everything, put it in an official binder labeled *Network Bible,* and protect the binder as if it were sacred.

Your hope should be that 2,000 years from now, when archeologists are exploring caves in your area, they find your network documentation hidden in a jar and marvel at how meticulously the people of our time recorded their network configurations.

They'll probably draw ridiculous conclusions, such as we offered sacrifices of burnt data packets to a deity named TCP/IP and confessed our transgressions in a ritual known as "logging," but that makes it all the more fun.

Chapter 28
Ten Big Network Mistakes

Just about the time you figure out how to avoid the most embarrassing computer mistakes (such as using your CD drive's tray as a cup holder), the network lands on your computer. Now you have a whole new list of dumb things you can do, mistakes that can give your average computer geek a belly laugh because they seem so basic to him. Well, that's because he's a computer geek. Nobody had to tell *him* not to fold the floppy disk — he was born with an extra gene that gave him an instinctive knowledge of such things.

Here's a list of some of the most common mistakes made by network novices. Avoid these mistakes and you deprive your local computer geek of the pleasure of a good laugh at your expense.

Skimping on Hardware

Good computer equipment is not cheap. You can walk into your local electronics store and buy everything you need to set up a cheap network: cables, switches, and computers to use as servers. But you get what you pay for. Professional-grade equipment costs much more, and in a business environment, it's worth it.

Why? Because professional-grade equipment is designed with performance, reliability, and centralized management in mind.

Professional server computers typically include redundancy in all the key systems — duplicate power supplies, duplicate network ports, duplicate disk controllers, and often even duplicate CPUs and motherboards. So, if one component fails, the server can continue operating.

Professional switches typically include management features that let you pinpoint problems on your network, segment your network for better performance, and monitor your employees' usage of the network.

You may also be tempted to cut costs by stringing inexpensive cable directly from the switches to each computer on the network. In the long run, though, the Scrooge approach may actually prove to be more expensive than investing in a good cable installation in the first place. A professionally installed cable infrastructure will last much longer than the computers it services, and will be considerably more reliable.

Turning Off or Restarting a Server Computer While Users Are Logged On

The fastest way to blow your network users' accounts to kingdom come is to turn off a server computer while users are logged on. Restarting it by pressing its reset button can have the same disastrous effect.

If your network is set up with a dedicated file server, you probably won't be tempted to turn it off or restart it. But if your network is set up as a true peer-to-peer network, where each of the workstation computers — including your own — also doubles as a server computer, be careful about the impulsive urge to turn off or restart your computer. Someone may be accessing a file or printer on your computer at that very moment.

So, before you turn off or restart a server computer, find out whether anyone is logged on. If so, politely ask her to log off.

REMEMBER

Many server problems don't require a server reboot. Instead, you can often correct the problem just by restarting the particular service that's affected.

Deleting Important Files on the Server

Without a network, you can do anything you want to your computer, and the only person you can hurt is yourself. (Kind of like the old "victimless crime" debate.) Put your computer on a network, though, and you take on a certain amount of responsibility. You must find out how to live like a responsible member of the network society.

Therefore, you can't capriciously delete files from a network server just because you don't need them. They may not be yours. You wouldn't want someone deleting your files, would you?

Be especially careful about files that are required to keep the network running. For example, some versions of Windows use a folder named `wgpo0000` to hold email. If you delete this folder, your email is history. Look before you delete.

WARNING

The first time you accidentally delete an important file from a network share, you may be unpleasantly surprised to discover that the Recycle Bin does not work for network files. The Recycle Bin saves copies of files you've deleted from your computer's local hard disk, but it does *not* save copies of files you delete from network shares. As a result, you can't undelete a file you've accidentally deleted from the network.

Copying a File from the Server, Changing It, and Then Copying It Back

Sometimes working on a network file is easier if you first copy the file to your local hard drive. Then you can access it from your application program more efficiently because you don't have to use the network. This is especially true for large database files that have to be sorted to print reports.

You're asking for trouble, though, if you copy the file to your PC's local hard drive, make changes to the file, and then copy the updated version of the file back to the server. Why? Because somebody else may be trying the same thing at the same time. If that happens, the updates made by one of you — whoever copies the file back to the server first — are lost.

Copying a file to a local drive is rarely a good idea.

Sending Something to the Printer Again Just Because It Didn't Print the First Time

What do you do if you send something to the printer and nothing happens?

- **Right answer:** Find out why nothing happened and *fix it.*
- **Wrong answer:** Send it again and see whether it works this time.

Some users keep sending it over and over again, hoping that one of these days, it'll take. The result is rather embarrassing when someone finally clears the paper jam and then watches 30 copies of the same letter print. Or when 30 copies of your document print on a different printer because you had the wrong printer selected.

Assuming That the Server Is Safely Backed Up

Some users make the unfortunate assumption that the network somehow represents an efficient and organized bureaucracy worthy of their trust. This is far from the truth. Never assume that the network jocks are doing their jobs backing up the network data every day, even if they are. Check up on them. Conduct a surprise inspection one day: Burst into the computer room wearing white gloves and demand to see the backup tapes. Check the tape rotation to make sure that more than one day's worth of backups is available.

If you're not impressed with your network's backup procedures, take it upon yourself to make sure that you never lose any of your data. Back up your most valued files to a flash drive. Or purchase an inexpensive 2TB or 4TB portable hard drive and back up your critical data to it.

Connecting to the Internet without Considering Security Issues

If you connect a non-networked computer to the Internet and then pick up a virus or get yourself hacked into, only that one computer is affected. But if you connect a networked computer to the Internet, the entire network becomes vulnerable.

Beware: Never connect a networked computer to the Internet without first considering the security issues:

- How will you protect yourself and the network from viruses?
- How will you ensure that the sensitive files located on your file server don't suddenly become accessible to the entire world?
- How can you prevent evil hackers from sneaking into your network, stealing your customer file, and selling your customer's credit card data on the black market?

For answers to these and other Internet-security questions, see the chapters in Part 5.

Plugging in a Wireless Access Point without Asking

For that matter, plugging any device into your network without first getting permission from the network administrator is a big no-no. But wireless access points (WAPs) are particularly insidious. Many users fall for the marketing line that wireless networking is as easy as plugging in one of these devices to the network. Then, your wireless notebook PC or handheld device can instantly join the network.

The trouble is, so can anyone else within about one-quarter mile of the WAP. Therefore, you must employ extra security measures to make sure hackers can't get into your network via a wireless computer located in the parking lot or across the street.

If you think that's unlikely, think again. Several underground websites on the Internet actually display maps of unsecured wireless networks in major cities. For more information about securing a wireless network, see Chapter 8.

Thinking You Can't Work Just Because the Network Is Down

A few years back, I realized that I can't do my job without electricity. Should a power failure occur and I find myself without electricity, I can't even light a candle and work with pencil and paper because the only pencil sharpener I have is electric.

Some people have the same attitude about the network: They figure that if the network goes down, they may as well go home. That's not always the case. Just because your computer is attached to a network doesn't mean that it won't work when the network is down. True — if the wind flies out of the network sails, you can't access any network devices. You can't get files from network drives, and you can't print on network printers. But you can still use your computer for local work — accessing files and programs on your local hard drive and printing on your local printer (if you're lucky enough to have one).

Running Out of Space on a Server

One of the most disastrous mistakes to make on a network server is to let it run out of disk space. When you buy a new server with hundreds of gigabytes of disk space, you might think you'll never run out of space. It's amazing how quickly an entire network full of users can run through a few hundred gigabytes of disk space, though.

Unfortunately, bad things begin to happen when you get down to a few gigabytes of free space on a server. Windows begins to perform poorly and may even slow to a crawl. Errors start popping up. And, when you finally run out of space completely, users line up at your door demanding an immediate fix:

- The best way to avoid this unhappy situation is to monitor the free disk space on your servers on a daily basis. It's also a good idea to keep track of free disk space on a weekly basis so you can look for project trends. For example, if your file server has 500GB of free space and your users chew up about 25GB of space per week, you know you'll most likely run out of disk space in 20 weeks. With that knowledge in hand, you can formulate a plan.
- Adding additional disk storage to your servers isn't always the best solution to the problem of running out of disk space. Before you buy more disks, you should
  - Look for old and unnecessary files that can be removed.
  - Consider using disk quotas to limit the amount of network disk space your users can consume.

Always Blaming the Network

Some people treat the network kind of like the village idiot who can be blamed whenever anything goes wrong. Networks cause problems of their own, but they aren't the root of all evil:

- » If your monitor displays only capital letters, it's probably because you pressed the Caps Lock key.

  Don't blame the network.

- » If you spill coffee on the keyboard, well, that's your fault.

  Don't blame the network.

- » If your toddler sticks Play-Doh in the USB ports, kids will be kids.

  Don't blame the network.

Get the point?

Chapter 29
Ten Things You Should Keep in Your Closet

When you first network your office computers, you need to find a closet where you can stash some network goodies. If you can't find a whole closet, shoot for a shelf, a drawer, or at least a sturdy cardboard box.

Here's a list of what stuff to keep on hand.

Duct Tape

Duct tape helped get the crew of *Apollo* 13 back from their near-disastrous moon voyage.

You don't actually need duct tape, but it serves the symbolic purpose of demonstrating that you realize things sometimes go wrong and you're willing to improvise to get your network up and running.

So get yourself a roll of duct tape and display it prominently.

Tools

Make sure that you have at least a basic computer toolkit, the kind you can pick up for $15 from just about any office supply store. At the minimum, you'll need a good set of screwdrivers, plus wire cutters, wire strippers, and cable crimpers for assembling RJ-45 connectors.

Patch Cables

Keep a good supply of patch cables on hand. You'll use them often: when you move users around from one office to another, when you add computers to your network, or when you need to rearrange things at the patch panels (assuming you wired your network using patch panels).

When you buy patch cables, buy them in a variety of lengths and colors. One good way to quickly make a mess of your patch panels is to use 15-foot cables when 3-foot cables will do the job. And having a variety of colors can help you sort out a mass of cables.

TIP

The last place you should buy patch cables is from one of those big-box office supply or consumer electronics stores. Instead, get them online. Cables that sell for $15 or $20 each at chain stores can be purchased online for $3 or $4 each.

Cable Ties and Velcro

Cable ties — those little plastic zip things that you wrap around a group of cables and pull to tighten — can go a long way toward helping keep your network cables neat and organized. You can buy them in bags of 1,000 at big-box home-improvement stores.

Even better than cable ties are rolls of Velcro, which you can cut to the appropriate length to neatly bundle up your cables. I much prefer Velcro over cable ties because Velcro is reusable. To remove a cable tie, you have to cut it.

Twinkies

If left sealed in their little individually wrapped packages, Twinkies keep for years. In fact, they'll probably outlast the network itself. You can bequeath them to future network geeks, ensuring continued network support for generations to come.

In November of 2012, computer geeks throughout the world faced a crisis far more menacing than the end of the Mayan calendar: the possible end of Hostess and Twinkies. Fortunately, the gods intervened, and Twinkies were saved, thus ensuring the continued operation of computer networks throughout the globe.

Replacement Parts

Keep a supply of the parts that most often break on your users' computers so you don't have to order replacement parts when the need arises. I usually keep the following on hand:

- Power supplies
- Monitors
- Keyboards
- Mice
- Spare disk drives
- Battery backup units, as well as replacement batteries
- RAM
- Video cables
- Sound cards
- Network interface cards
- Case fans

If you have enough users to justify it, I recommend you also keep one or more spare computers on hand so that if one of the computers on your network dies, you can quickly swap it out for a spare.

Cheap Network Switches

Keep a couple of inexpensive (about $20) four- or eight-port network switches on hand. You don't want to use them for your main network infrastructure, but they sure come in handy when you need to add a computer or printer somewhere, and you don't have an available network jack. For example, suppose one of your users has a short-term need for a second computer, but there's only one network jack in the user's office. Rather than pulling a new cable to the user's office, just plug a cheap switch into the existing jack and then plug both of the computers into the switch.

The Complete Documentation of the Network on Tablets of Stone

I mention several times in this book the importance of documenting your network. Don't spend hours documenting your network and then hide the documentation under a pile of old magazines behind your desk. Put the binder in the closet with the other network supplies so that you and everyone else always know where to find it. And keep backup copies of the Word, Excel, Visio, or other documents that make up the network binder in a fireproof safe or at another site.

WARNING

Don't you dare chisel passwords into the network documentation, though. Shame on you for even thinking about it!

TIP

If you decide to chisel the network documentation onto actual stone tablets, consider using *sandstone.* It's attractive, inexpensive, and easy to update (just rub out the old info and chisel in the new). Keep in mind, however, that sandstone is subject to erosion from spilled Diet Coke. Oh, and make sure that you store it on a reinforced shelf.

The Network Manuals and Disks

In the Land of Oz, a common lament of the Network Scarecrow is, "If I only had the manual." True, the manual probably isn't a Pulitzer Prize candidate, but that doesn't mean you should toss it in a landfill, either.

REMEMBER

Put the manuals and disks for all the software you use on your network where they belong — in the closet with all the other network tools and artifacts.

Ten Copies of This Book

Obviously, you want to keep an adequate supply of this book on hand to distribute to all your network users. The more they know, the more they stay off your back.

Sheesh, 10 copies may not be enough — 20 may be closer to what you need.

Index

C

D

E

F

G

H

I

J

K

L

M

N

R

S

T

U

V

W

Z

About the Author

Doug Lowe has written enough computer books to line all the birdcages in California. His other books include *Networking All-in-One For Dummies*, 7th Edition; *Microsoft PowerPoint 2019 For Dummies; Java All-in-One For Dummies*, 5th Edition; and *Electronics All-in-One For Dummies*, 2nd Edition (all published by Wiley).

Although Doug has yet to win a Pulitzer Prize, he remains cautiously optimistic. He is hopeful that Claude-Michel Schönberg and Alain Boublil will turn this book into a musical, titled *Les Réseau Miserables.* (Hopefully, the role of the vengeful network administrator will be played by someone who can sing.)

Doug lives in sunny Fresno, California, where the nearby Sierra Nevada mountains are visible through the smog at least three or four glorious days every year.

Dedication

This one is for Emmi.

Author's Acknowledgments

I'd like to thank my project editor, Elizabeth Kuball, who did a great job of managing all the editorial work that was required to put this book together, and my acquisitions editor, Ashley Coffey, who made the whole project possible. I'd also like to thank Dan DiNicolo and Dwight Spivey, who gave the entire manuscript a thorough technical review and offered many excellent suggestions. As always, thanks to all the behind-the-scenes people who chipped in with help I'm not even aware of.

Publisher's Acknowledgments

Acquisitions Editor: Steve Hayes

Project Editor: Elizabeth Kuball

Copy Editor: Elizabeth Kuball

Technical Editors: Dan DiNicolo, Dwight Spivey

Proofreader: Debbye Butler

Production Editor: Mohammed Zafar Ali

Cover Image: © bluebay2014/Getty Images

Leverage the power

Dummies is the global leader in the reference category and one of the most trusted and highly regarded brands in the world. No longer just focused on books, customers now have access to the dummies content they need in the format they want. Together we'll craft a solution that engages your customers, stands out from the competition, and helps you meet your goals.

Advertising & Sponsorships

Connect with an engaged audience on a powerful multimedia site, and position your message alongside expert how-to content. Dummies.com is a one-stop shop for free, online information and know-how curated by a team of experts.

- Targeted ads
- Video
- Email Marketing
- Microsites
- Sweepstakes sponsorship

VISITORS PER MONTH

43%
OF ALL VISITORS ACCESS THE SITE
VIA THEIR MOBILE DEVICES

700,000 NEWSLETTER SUBSCRIPTIONS
TO THE INBOXES OF
300,000 UNIQUE INDIVIDUALS EVERY WEEK

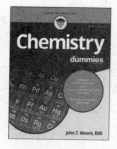